**CHEMICAL
PLANT DESIGN
WITH
REINFORCED PLASTICS**

CHEMICAL PLANT DESIGN WITH REINFORCED PLASTICS

JOHN H. MALLINSON

Plant Engineer, FMC Corporation
American Viscose Division, Front Royal, Virginia

McGraw-Hill Book Company

New York   St. Louis   San Francisco   London   Sydney
Toronto   Mexico   Panama

**CHEMICAL
PLANT DESIGN
WITH
REINFORCED PLASTICS**

39793

1234567890 MAMM 754321069

Foreword

The development of nearly any book in the technical field stems predominantly from the work of others. This is certainly true in this book on Reinforced Plastics. The author trusts that all of the material sources which have been consulted have been acknowledged and identified. But knowledge and sources extend far beyond the written word—the spoken word, in visits and interviews with many colleagues in the RP field have been the source of much additional material. Helpful criticisms and suggestions have been received from a variety of sources not only from colleagues but from other people in the RP industry.

Each chapter has been reviewed by a knowledgeable individual with experience in the particular area involved. This has covered not only the field of Reinforced Plastics per se but areas of related fields which overlap such as purchasing, safety, medicine, and toxicology, as they relate to Reinforced Plastics. The writer felt that some of the RP fields could be better developed by specialists. We have been most fortunate in having sections of these areas contributed by men with knowledge on the frontiers of the technology. These contributions have been suitably identified but may be summarized here:

Name	Company	Type of Material Contributed
Boggs, H. D.	Amercoat Corp. Brea, California	Proposed standards on filament wound piping and filament wound tanks.
Hanszen, E. W.	Hanszen Plastics Dallas, Texas	Ductwork calculations.
Kraus, Norbert	Justin Enterprises Fairfield, Ohio	Wound in place RP tanks.
Munger, C. G.	Amercoat Corp. Brea, California	Epoxy piping.
Steelman, C.	Heil Process Equipment Co. Cleveland, Ohio	(a) Vacuum collapse pressures on polyester piping. (b) Wt per ft polyester piping.

Name	*Company*	*Type of Material Contributed*
Stone, D. R.	Metalcladding, Inc. Tonawanda, New York	Kabe-o-rap[R] Tanks.
Szymanski, W. A.	Durez Division Hooker Chemical Corp. North Tonawanda, New York	(a) Attack mechanisms on chlorinated polyesters. (b) Chemical service recommendations on chlorinated polyesters.
	The Fibercast Co. Sand Springs, Oklahoma	Service recommendations on Fibercast epoxy pipe.

Certainly to other organizations appreciation is expressed for their generous help and guidance:

American Cyanamid (Ed Kerle)
Atlas Chemical Industries (Joe Costello)
Beetle Plastics (Paul Silva)
Carolina Fiberglass (Bob Whiteman)
Ceilcote (Bill Smith, Ron McMahon, Tony Fonda, Jack Galloway)
DowSmith (Mark Kelly) (Note: DowSmith is now Smith Plastics)
DuVerre Division, Amercoat (Raul Bernhoft)
Heil Process (Fred Arndt, Clarence Steelman)
Owens-Corning Fiberglass (Chet Stafford, Bob Pistole)
Society of the Plastics Industry (RP Division)

I am also indebted to the corporate officials of FMC Corp., American Viscose Division, for permitting me to draw on company experience gained in this new and developing field.

Mr. J. S. Wetzel is responsible for the technical photographs of RP equipment. Miss Vera Miles performed the large task of typing and working out the seemingly endless details of manuscript preparation. Mrs. Hazel Longerbeam typed many of the standards. Mr. S. W. Foglesong prepared the preliminary drawings on which the finished drawings were based. And to my patient wife, a debt of gratitude for her assistance in all phases of manuscript preparation.

Although every effort has been made to avoid errors and misstatements of fact, it is not reasonable to believe that some have not crept in. The author will be most grateful if any of these are brought to his attention.

Finally, the opinions expressed in this book are the sole responsibility of the author. Identification of product names are intended for illustration only—other products currently marketed may be equally acceptable.

John H. Mallinson

Preface

Many of the great discoveries of the past have come about by what could be termed a learned accident. That is, a knowledgeable person seeking an end target but keenly aware that along the way he may discover things of greater interest and value than his original quest. So it was some twenty-five years ago that a young engineer inadvertently spilled a thermosetting resin onto some glass cloth. Left to its own devices it set overnight and when re-examined in the morning light the field of reinforced plastics was born. Today this exploding field is tailor-making a variety of materials to meet a wide area of conditions in shapes, and in properties, which are directable and controllable. The engineer thus has at his disposal a wide area of formulations that span a breadth envisioned by only the most imaginative. Off we go into the field of missiles, aircraft, archery, automobile bodies, boating, caskets, golf carts, piping, railroad cars, tanks, swimming pools, etc. Even the field of competitive sports has succumbed to this new material. New pole vault records are being created as the fiberglass pole literally hurls the vaulter like the missile in the sling shot to new world records.

The engineer of today is indeed fortunate to have at his disposal a new generation of materials which are capable of solving materials problems that ten to fifteen years ago could only be solved by difficult, cumbersome means at a relatively high cost, or sometimes not at all.

The entire field of reinforced plastics is so broad that it could only be dealt with by an encyclopedic series of volumes. Here we are concerned with one volume of this encyclopedia so as to provide a reference text for the student, engineer, designer, and user of chemical resistant reinforced plastic materials, as applied to the chemical process industries. We are interested primarily in chemical plant design with reinforced plastics. There will be no attempt here to cover the phases related to small multiple reinforced plastic molding or the use of these materials in building panels, boats, truck or auto bodies, baptistries, or their use solely as a structural material. In general, it will be found that the widest use today in the field of chemical resistant reinforced plastics exists in polyester, epoxy and phenolic resins. In the spectrum of chemical resistance these resins are found to generally cover the field of corrosion resistance, with some few exceptions. Reinforced with glass mat, glass cloth, asbestos, dynel, polypropylene, etc., they are able to develop exceptionally high strength/weight ratios.

Today entire new techniques must be learned by the engineer, designer, and installation crews. A new vocabulary speaks in strange terms not heard a few short years ago. New skills and concepts must be learned which join the know-how of many trades.

A considerable amount of material exists today in the chemical industry covering the applications of reinforced plastics to its problems but it is widely scattered over a large number of sources. This book would make every attempt to provide a one source reference for the engineer.

The application of these materials by the engineer needs to be thoroughly understood so that he can become a part of the design and not have to be an onlooker and rely on the good faith of the fabricator or guess work on his part. The designer needs to be put in the position of being able to design and apply, with confidence, these reinforced plastic materials that his wide list of corrosion problems present. He needs to be able to do this with the same degree of sureness that he now does for steel structures. He needs to learn that he is working with a new generation of materials and while similar methods of application may apply, the physical characteristics of the material are sufficiently different that it is almost a whole new science. Typical design problems are presented and solved to permit the engineer to use these as a reference area in his own design work and give him a basis for specification and quotation evaluation.

Fabricators and manufacturers in the RP industries (like everyone else) may ballyhoo their successes but their failures are quietly ignored. The writer's experience has been that in nearly every case, failures in RP systems can be traced to a "sin of omission" or something which was done wrong. The fabricator's literature is filled with things to do. There is almost a total lack of things not to do. "Accentuating the positive" is fine and ninety per cent of this book will be "accentuating the positive," but ten per cent will be aimed at "eliminating the negative." We are dealing with the ingenuity of man, and man is an ingenious creature. Give him a set of directions to follow and experience proves that before the ink is dry he will be thinking up new ways, which in his mind, are better than the methods he has been shown. *Training, techniques, tools, and adherence to specifications are the keys to successful application.* It is surprising how competent craftsmen can develop poor methods through improper training or infrequent training. Training must be given and then continually reinforced.

The dream of the corrosion engineer for a tough material, lighter than aluminum, and with superb corrosion resistance is one step nearer reality for a whole area of troublesome corrosive solutions. Such wide spanning industries as caustic and chlorine, canning, chemicals, electroplating and pickling, fertilizers, food products, petrochemicals, pharmaceuticals, pulp and paper, steel, and textiles, have all felt the impact of this advancing chemical technology.

Chemical piping systems in many areas are well suited to the use of these materials. Limited experience in reinforced plastics (RP) piping systems goes back over a decade with experience in depth somewhat less. There is a saying in the trade that if the installation lasts twelve months it is good forever. Stretching it a bit perhaps, but experience has shown there is more than a grain of truth in it. They have also invaded the utility fields in soft water, water treatment, and condensate piping. Leakproof chemical sewer systems in small (2″) and big (72″) piping are a natural for these materials.

The storage of corrosive solutions has long been a worrisome, costly business. Now storage tanks built from glass reinforced polyesters and epoxies are commonly constructed from 100 gallon midgets to 250,000 gallon behemoths. It is unfortunately true that in metals chemical attack often occurs from the outside in. The engineer may spend an infinite amount of care to design and specify a top flight lining system on a storage tank only to be completely defeated by the Trojan Horse of external corrosion—the lined metal tank has failed from the outside in. Not so with RP tanks. It makes little difference if they are overflowed daily, the outside is nearly as corrosion proof as the inside.

The field of process vessels is being penetrated more each day as the Chemical Engineer and Process Designer learn more about the applications of these materials. Crystallizers, filters, injection heaters, barometric condensers, trays, sluices, extraction columns, scrubbing towers, rotary blenders, flash chambers, are all ready candidates for economical applications.

The field of exhaust systems from chemical solutions can and has been a rough area. Here corrosive gases moving at high speeds have combined the problem of blistering corrosion with the need for fire retardancy. A new era of Chlorinated Polyesters with excellent fire retardant properties have created enviable records in handling corrosive gases. To these have been added fans and exhausters of the same materials plus great stacks and scrubbers for fume removal and dispersal.

The use of these materials to solve knotty problems is limited mainly by the imagination of the user. But like all successful programs, imagination must be accompanied by good engineering application. The safe use and fabrication of the materials are something to be concerned with and this is gone into in detail. Good specifications are one of the prime necessities. Start with them and your problems will be greatly reduced.

Procurement methods and stocking programs for the user are

covered.   The method in deciding which system to use is evaluated. Corrosion resistance charts are given.   Design criteria to use in the installation of piping systems are explained and typical illustrative examples given.

This work is intended to be a distillation of theory and practical experience combined and designed to permit the chemical plant users of reinforced plastic material to proceed with a greater degree of firmness and assurance in the solution of his process problems.   All of us learn a great deal more from adversity than we ever do from success.   Unfortunately, there are many ways in which failure can occur so that the list of "Don'ts" is equally as important as an assembly of "Do's."

The safe use and proper working environment in the assembly of these materials are necessary for a successful program.   They have been studied in depth and the benefits of these studies are detailed.   Only by adherence to safe working procedures can personnel and property be adequately protected.   Such safety instructions are neither restrictive nor rigorous but they are essential.

In terms of cost reduction over many normal systems the application of these materials can result in mouth watering savings which may range from 20%–80% over what used to be considered a standard design system.   The stainless steels, the high nickel alloys, lead, and expensively lined equipment, may all require a second look as the engineer and designer become more familiar with the use of these new materials.   In today's increasingly competitive markets there is but little choice to find better and more economical ways to accomplish the objectives.   This is the continuing mandate of industrial management.

". . . he who does not advance falls back; he who stops is overwhelmed . . . ; he who ceases to grow becomes smaller . . . the stationary condition is the beginning of the end."

—Henry Amiel, 1821–1881

John H. Mallinson

Contents

CHEMICAL
PLANT DESIGN
WITH
REINFORCED PLASTICS

1

The Why of
Reinforced Plastics

1.1 INTRODUCTION AND A SHORT HISTORY

The why of reinforced polymers is, basically, an extension of man's curiosity and his continuing search for the performance of materials with better physical and chemical properties. Being constructively dissatisfied is the first step toward improvement. The battle against corrosion and decay did not begin in the last decade or even in the last century— the evolution of vegetation and animal matter on earth is a living testimonial to reinforcing materials in a covering of organic resins. Bone, hair, and fingernails are all examples of this same process. Bone is a composite of the mineral apatite and the protein collagen. Wood is a composite of cellulose and lignin.[1]*

One of the great steps in man's development was the discovery that he could change the nature of materials. Early man must have felt the Jehovah complex as he turned clay to stone by simply heating it.[2] Thus constructive action against deterioration by our ancestors began thousands of years ago. The ancient Egyptians practiced the art of preservation when they wrapped human bodies in cloth saturated with natural resins. Today, as our civilization has become more highly industrialized and complex, the problems of preservation are changed, and emphasis with it. Economic competition forces the engineer to strive constructively for the lowest unit cost over the service life of the equipment.

Nearly all engineering materials are composites of some type. Steel, for example, is painted to resist the ravages of corrosion. Early cannon barrels were once made of wood bound with brass to endure better under internal pressure. The resourceful corrosion engineer can generally conceive composite materials which are lighter, stiffer, stronger, or more corrosion-resistant than those of twenty years ago.

* Superscript numbers refer to sources listed at the end of each chapter.

Probably our basic process problems began with the Industrial Revolution in the early 1800s, when our general civilization changed from a predominantly agricultural civilization to one of urban growth. Beginning in the early 1800s chemical plants, admittedly small, began to appear in many of the civilized countries. As the Industrial Revolution broadened and steadily drew within its scope of influence much of the civilized world, human beings became more and more dependent upon industry for many of the necessities and nearly all the luxuries of civilization.

Man's entire approach to the discoveries of fundamental facts had undergone a profound change. Only within the last 200 years has experimental work changed from empirical to scientific fact. No one knew why things happened. Prior to that time experimentation was done without understanding, and results by accident produced knowledge at an infinitesimal rate. Experiment is the foundation of theory.[3] Following its development, theory must be tested repeatedly to prove its validity. Early man speculated and formed theories, but then failed to test them. Most of the early theories could not be confirmed experimentally. For thousands of years religion and science were at loggerheads and religious authorities were powerful. Many scientists were exiled, burned at the stake, had their laboratories wrecked and burned by mobs, or were executed for their convictions.[4]

As late as the 1600s one of the first truly modern scientists was born— Robert Boyle. His theories stood up to experimental confirmation, and from them Boyle's law was formulated. No scientist had done this before.[5] In the latter 1700s three outstanding scientists, Cavendish, Priestley, and Lavoisier, established the basis of modern science by performing a series of experiments which culminated in basic discoveries on the composition of air and water.

The infant chemical industry saw the manufacture of sulfuric acid in 1740, when the acid was made in glass globes. Chamber acid was concentrated in a series of silica vessels or retorts. Gay-Lussac towers were packed with coke.[6] Acetic acid was made in wood towers packed with beechwood shavings. Hydrochloric acid in 1836 was made in equipment constructed of slabs of sandstone boiled in tar, and the joints sealed with tar and pitch. The packing of such acid towers was also coke. Nitric acid was made in towers 60 to 80 ft high and packed with broken quartz. In 1791 Leblanc founded the modern alkali industry. In 1823 Muspratt erected a plant at Liverpool for the manufacture of sodium carbonate.[7] One hundred years ago the embryonic chemical engineer was even then dissatisfied with the rapid oxidation of iron and had to turn to other materials, but as the chemical industry grew quickly, the products of its mills and factories vastly expanded its corrosion problems. Each one of man's advances seems to bring with it new material problems. It was not long before it was discovered that rubber linings or coverings represented good protection against some of the more moderate oxidizing acids and most acids, if in a diluted form. Later, the discovery that chrome and nickel, particularly together, happily changed the corrosion characteristics of steel was an event of great importance, and an entire industry began with the manufacture of various grades of stainless steel in many forms, such as sheet, piping, and cast material. The discovery that molybdenum further enhanced corrosion properties of the stainless steels and the chrome-nickel alloys was still another step forward. It is quite apparent that each step in the evolution of materials technology was built on the shoulders of its predecessors.

Man, of course, for centuries has worked with iron and bronze. Perhaps 20,000 years ago he learned to make containers from the various materials available to him, which at that time were mainly ceramic.[8] These glassy residues from burning fires were one of his earliest discoveries. From these residues early pottery was made. Certainly centuries must have passed between man's first observance of this glassy residue and the time he was able to work it into a useful form. With imperceptible progress man came down through the Bronze Age, which began about 3000 B.C., and into the Iron Age about 1500 B.C.[9] For two hundred centuries man's progress crawled at a snail's pace, until trial and error was replaced with thought and understanding some two hundred years ago. Since then learning has galloped forward in an exponential fashion.

Each engineering decision that is made in balance is an economic choice. Economics, however, is generally the independent variable. Design is for expected life. Performance demands of a vessel or process system are usually met by different parts. Each part in the engineer's choice is the optimum type for the local function.[10] Rarely if ever does a single material become a blanket prescription for success.

Considered in broad terms, alkali materials at moderate temperatures have never represented great corrosion problems. The really great corrosion problems are generally found in severely oxidizing or reducing atmospheres or combinations of both. Under these conditions the life of steel pipe may be measured in a few short days. Type 304 stainless steel fares little better; Type 316 stainless steel might last for several months; and only the highest grade Cr-Ni-Mo alloys could be said to be satisfactory. For many years the engineer attempted to solve the problem in this area through the use of high Cr-Ni-Mo alloys or by providing some type of lining on steel equipment. Twenty years ago such linings took many forms, such as lead, rubber, a few of the elastomers, and some baked phenolic linings. Since a solution had to be found at any price, the solution which was found was a high-priced one.   Man continually sought a material which would be as strong as steel, as light as aluminum, and with a chemical resistance equal to the expensive Cr-Ni-Mo alloys. As he attempted to tailor a new-found material to fit his corrosion problems, he found that many answers were possible.

1.2   THE THERMOPLASTS

It is difficult to say exactly when the field of synthetic polymers was born.   Certainly, nature had the natural polymers in abundance, but man's level of technical learning had to be sufficiently developed to recognize that here lay the great materials key to the future.   In the latter 1800s some synthetic polymers were created in the laboratories, such as cellulose and Bakelite.   But only since the beginning of the twentieth century, when advanced new instruments for investigation came into being, such as the electron microscope, x-ray diffraction apparatus, and the ultracentrifuge, has polymer investigation really advanced.[11] Some of the molecules involved, being composed of building blocks based on monomers, have extremely large molecular weights, running into the millions of units.

An entire series of thermoplastic resins were uncovered over many years:

Chlorinated polyethers	Polystyrene
Fluorinated hydrocarbons	Polyvinyl chlorides
Polyamides	Vinylidenes
Polyethylene	Vinyls

Many of the early thermoplastics had relatively low tensile strength and a limited resistance to heat.   The search went on.

1.3   THE THERMOSETS

Another group of substances was of interest, that is, the thermoset compounds.   These are, initially, a liquid at room temperature; then, by means of a catalyst or accelerator, they are changed into a rigid product which sets, or cures, into its final shape.   Typical of the polymer thermoset resins are:

Diallyl phthalates	Phenolics
Epoxies	Polyesters
Furanes	Polyurethanes

The field of reinforced plastics as we know it today really began after World War II, at which time radomes were made. The boating industry, however, was the first large-scale user and producer of reinforced plastic material. Without the wide acceptance of this material by the boating industry, reinforced plastics would not be where it is today. Originally, in the young industry, delamination was a severe problem, and the first boats frequently came apart. No published information existed simply because there was no information to be had, but finally, many of the design data on which the base for the industry was formed came from the boat builders.

Now, the cost-conscious engineer is faced with the formidable task of reducing materials expenditures caused by the ravages of corrosion. Today it is estimated that corrosion costs American industry some 6 to 10 billion dollars a year.[12] It is also estimated that 60 percent of the output of America's gigantic steel industry is going into replacement products. Certainly, it is not reasonable to expect any single material of construction to eliminate this problem, but in the last decade the field of high polymers, coupled with the use of a suitable reinforcing material such as fibrous glass or other reinforcements, provides a useful solution in various other areas of process corrosion. In this field of high polymers the engineer has at his disposal such a wide range of modifications that he is able, generally, to tailor-make the solution to his specific problem.

If, for example, we devise a simple table such as Table 1-1, comparing the properties of several of these high polymers reinforced with glass with four other common materials of construction in the process industries, the potential use of such a material becomes more evident. Because of the strength considerations involved, most of the resins used today in the reinforced plastics industry are thermosetting by nature. In terms of usage in the construction of reinforced plastics equipment for the chemical industry, the polyesters, epoxies, and the furanes have been most widely adopted. The engineer will, however, wish to consider combinations of thermosetting thermoplast materials, which is covered in some detail in Chap. 5. Certainly, what has been said here does not preclude or prejudge the use of phenolics and DAP (diallyl phthalate) compounds for literally thousands of molding operations widely used in chemical plants.

Reinforced plastic pipe is commonly made of epoxy or polyester resins, the polyester resins predominating in the larger sizes. Generally, custom-made tanks, ductwork, and structures are made from high-chemical-grade reinforced polyesters. The reason for this is that the polyesters are much easier to work, easier to repair, and less expensive to manufacture. They are also resistant in a wide variety of chemical conditions.

1.4   REINFORCING MATERIAL

Of the materials used for reinforcing the thermosetting materials, such as the polyesters, epoxies, and furanes, fibrous glass (in the E, C, R, and S grades*) is most widely preferred, followed by asbestos. This, however, does not preclude the use of other materials for reinforcing, such as:

Boron nitride	Modacrylic fiber
Ceramic fibers	Polyester fiber
Graphite	Polypropylene fiber
Jute	Quartz
Metallic wire or sheet	Sapphire whiskers

Boron, graphite, quartz, and sapphire whiskers are used principally in aerospace research, and while they are of great interest, they are of limited usefulness in the chemical-industry corrosion program because of their relatively high price. Research in England has produced a wire sheet reinforcement from which piping has been made.[13] Organic-fiber veils are used quite often in glass-resin laminates to provide better abrasion and chemical resistance in certain applications. Modacrylic, polyester, and polypropylene fiber veils are suitable for alkalis.[14] A more detailed review of organic-fiber veiling is given in Chaps. 2 and 16.

Polypropylene fiber also has been used as a wall-reinforcing agent in reinforced plastic structures. If the resin and reinforcing agent are resistant to the service conditions, the laminate is then completely resistant throughout. While the polypropylene fiber does not develop the strength of glass, successful applications have been made in RP tanks and structures.[15]

The use of jute fabrics as a reinforcing material is of considerable interest, especially in the Asiatic countries. Normally, the price of jute laminating cloth is about two-thirds that of glass-fiber chopped-strand mat. In addition, a jute polyester laminate may weigh up to 25 percent less than a glass polyester laminate. Generally, however, jute fabrics are used as a core reinforcement of sandwich construction in conjunction

---

* Although glass was used by the Phoenicians 3,000 years ago, fiber glass itself is a recent innovation. Some work was done in Germany and in this country in the 1920s on such a product, but, actually, fiber glass was not commercialized until the late 1930s, at which time it was shown at the New York World's Fair. Fiber glass is made by a number of different processes, such as melt spinning process or drawn from a marble. E and C glass presently predominate, but work is being done on other glasses which, hopefully, will combine the high strength of E glass and the chemical resistance of C glass at a price approaching E glass. To improve the adhesion of these glasses to resins, various so-called binders have been developed, the most common of which are the silanes.

with glass-fiber chopped-strand-mat reinforcement. This provides the necessary rigidity. Glass-fiber laminate in the surface lay-up must be used to protect the jute core. At present jute is not used extensively in this country for RP reinforcement in the chemical industry. It does have potential application for machine guards, trays, vehicle panels, etc.[16]

1.5 ADVANTAGES, SCOPE, AND USE OF THE THERMOSETS

By changing the amount of reinforcing in a particular structure the designer has at his command the ability to modify physical and chemical characteristics. High-glass-content structures provide maximum physical strengths, while high-resin-content structures will provide maximum chemical resistance. The designer can obviously combine the two elements in the form of a modified composite structure which will result in an optimum design. Thus the designer has taken a giant step forward in mastering his environment with a new generation of materials which a few short years ago his predecessors only dreamed about. Here he has a family of materials:

- Costing less than stainless steel
- As light as aluminum
- Possessing exceptional strength
- Purchasable with a short lead time
- Easily altered or repaired
- Possessing exceptional chemical resistance
- Providing in some areas the performance of Hastelloy for less than the price of stainless steel
- Not requiring painting
- With very high strength/weight ratios, exceeding steel
- Easily formed into many different shapes
- Of low maintenance cost
- Producing, with the addition of chlorine into the molecule, an excellent fire-retardant material from which entire duct systems may be designed to handle corrosive gases

If the designer permits his imagination to wander, the use of this unique family of materials can be extended to cover a great variety of applications. Instant landing pads for tactical helicopters made of glass-reinforced plastics have been demonstrated successfully in Vietnam. Both the Polaris submarine and the Minuteman missile are improved through the use of reinforced plastic materials. RP fairwaters for modern submarines have been in service for over eleven years.[17] The Corvette and Avanti utilized RP bodies. Many trucks are equipped with RP cabs and fenders. The field of sporting goods feels its impact everywhere: swimming pools, fishing rods, and pleasure boats use RP material, widely attesting to its high strength/weight ratio. The U.S. Navy has currently under construction PGM gunboats, 165-ft vessels fabricated of reinforced polyester for use in shallow coastal waters where light draft is essential.

As we move into the field of chemical-process-plant design, applications cover piping, tanks, pumps, ductwork, agitators, crystallizers, scrubbers, hoods, fans, stacks, thermocompressors, troughs, distribution trays, filters, evaporator parts, condensers, flumes, heaters, tank cars, water softeners, rain spouting, tank lids, and pressure bottles, etc.

Reinforced plastic structures possess a very high strength/weight ratio, which is stronger than most metals on a pound-for-pound basis. Reference is made to Table 1-1 for a comparison of some common materials. Here it will be noted that a glass-mat laminate has a strength/weight ratio of 300,000 compared with 230,000 for steel. On the same basis some filament-wound structures can be pushed up to strength/weight ratios as high as 1,500,000. Lest the reader become overenthused by this comparison, let us remember that the optimum combination of corrosion resistance and strength lies in the medium ratios of strength/weight, lying in the 500,000 area. (Glass-mat–filament-wound composite laminates will provide excellent chemical resistance coupled with high-strength performance.) Composite structures made of glass mat and woven roving will exhibit a strength/weight ratio of 300,000 or higher.

One item of particular importance to the designer is an understanding of how reinforced plastics break; their break is completely different from metals. They do not bend or deform as a metal commonly does, but rupture instantly at the ultimate tensile value. This is why the yield strength and the tensile strength are the same for reinforced plastics, while for metals the yield strength is only a fraction of the tensile strength.[18] To illustrate the toughness of reinforced plastics, there are stories of towed boats being in automobile wrecks, skidding down the highway, and being repaired for a few dollars.

The designer should also take particular note of the difference in modulus of elasticity. The metallic materials are considerably stiffer than reinforced plastics. Steel, for example, is eight to twelve times stiffer than reinforced plastics. This is both a plus and a minus, which must be coped with in each individual design problem. The low modulus of elasticity virtually eliminates the use of expansion joints, which are commonly required with metals. On the other hand, this low modulus of elasticity must be kept in mind at all times since it limits the upside pressure limit in large structures. Large tanks, however, may be pro-

COMPARATIVE PHYSICAL PROPERTIES OF METALS AND REINFORCED PLASTICS[18] (Room Temperature)

Table 1-1

	Carbon steel 1020	Stainless steel 316	Hastelloy C	Aluminum	Glass-mat laminate	Composite-structure glass-mat woven roving	Glass-reinforced epoxy, filament-wound*
Density, lb/in.³	0.283	0.286	0.324	0.098	0.050	0.065	0.065
Coefficient of thermal expansion, in./(in.)(°F)(10⁻⁶)	6.5	9.2	6.3	13.2	17	13	9–12
Modulus of elasticity, psi × 10⁶, in tension (Young's modulus)	30.0	28.0	26.0	10.0	0.7–1.0†	0.8–1.5	4.0–4.5
Tensile strength, psi × 10³	66	85	80	12	9–15†	12–20	100
Yield strength, psi × 10³	33	35	50	4	9–15†	12–20	100
Thermal conductivity, Btu/(hr)(ft²)(°F/ft)	28.0	9.4	6.5	135	1.5	1.5	1.5–2.0
Strength/weight ratio, 10³	230	300	250	122	300	308	1,500

NOTE: The physical-strength figures used here for the glass-reinforced plastic laminates are most conservative. For example, some filament-wound epoxy tensiles will run to 300,000 lb, giving them phenomenal strength/weight ratios of $4,500 \times 10^3$.

* Data on glass-reinforced filament-wound epoxy have been drawn from a variety of sources. Filament winding, in general, be it polyester or epoxy, will result in much higher physical strengths.

† These data are from the Recommended Product Standard for Custom Contact Molded Reinforced Polyester Chemical Resistant Process Equipment, TS-122C, Sept. 18, 1968.

9

vided with additional reinforcing by winding the tank exterior with wire, as will be described in Chap. 12.   The reader must always bear in mind that reinforced plastic material follows Hooke's law, which states that stress is proportional to strain.

The purpose of glass reinforcing is to provide strength and dimensional stability, which is not possible with the resin alone.   Then one of the big pluses is the savings in insulation, since, generally, it requires none. There appears to be no tendency to become brittle at low temperatures as with some of the thermoplasts, and the laminate is actually stronger at 0°F than at room temperature.[18]

There is no doubt that in the future polymer physicals will be improved as man continues to learn to tailor-make the molecule to suit his desired end purposes.   Tensiles will increase, elasticity where desired will be improved, and melting or softening points will be raised.   Chemical resistance will be enhanced, particularly at elevated temperatures. The crystallization of polymers containing polar groups generally provides a stronger and more heat-resistant material.   Chemical cross-linking of the long chains is another method used for strengthening and toughening organic polymers.   A third method being continually investigated is the production of long stiff chains which provide a high degree of inflexibility.   Phenylenes are under investigation in this regard. It is quite conceivable that combinations of crystallization, cross-linking, and stiff chains may provide additive results.[11]

Most composite materials now being manufactured are in the form of fiber glass.   The fact that the fibers remain largely intact is one of the chief advantages of glass-reinforced plastics.   The resin in liquid form can be made to flow around the fibers at room temperature and pressure. Large pieces of reinforced polyester can be built up layer by layer to any desired shape or size.   The glass itself does not remain strong at temperatures much in excess of 400°C.   Above that, another type of reinforcing would be necessary, such as boron, carbon, or silicon carbide. Carbon fibers in epoxy resin provide compressor blades in lightweight jet engines, and boron in epoxy resin is used for helicopter rotor blades that turn at high speed.   The advantage of reinforcement may also be extended to the metals themselves, so that tungsten fibers may be used in cobalt and nickel.[19]

1.6   RELATIVE COST TRENDS AND USES OF SOME COMMON CHEMICAL–PROCESS MATERIALS OF CONSTRUCTION

One of the interesting areas for study by the cost-conscious engineer is the relative cost trends of metallic pipe versus reinforced plastic pipe.

Replacement cost ratios are available for metallic pipe extending over almost any period of time that the user chooses. For this purpose, we have gone back to 1955 and prepared a table of replacement cost ratios. Since reinforced plastic pipe has been available only for the past seven years *in competitive quantity*, it is interesting to compare the relative cost trends of the metallic pipe with the reinforced epoxy and polyester pipe. For metallic pipe we have arbitrarily chosen a mix of stainless steel, black iron, lead, lead-lined steel, rubber-lined steel, and copper, which make up the general run-of-the-mill metallic material used by many chemical plants for a wide diversity of applications. The general trend of this metallic aggregate in piping systems is shown in Table 1-2. In the same table we have also shown the replacement cost ratios of reinforced plastic pipe extending over a period of 7 years, beginning with the year 1961, the first year that such pipe could be said to be available in large competitive supply. We are immediately confronted with several obvious conclusions. For example, if a piping system of metallic pipe was installed for $1,000 in 1961, the same system cost $1,340 in 1967; whereas a reinforced plastic system costing $1,000 in 1961 could be

PIPE AND FITTINGS[20]

Table 1-2

Year of investment	Replacement cost ratio	
	Metals	Plastic
1955	1.76	
1956	1.55	
1957	1.37	
1958	1.29	
1959	1.22	
1960	1.23	
1961	1.34	0.77
1962	1.20	0.75
1963	1.25	0.86
1964	1.23	0.85
1965	1.16	0.88
1966	1.03	0.99
1967	1.00	1.00

EXAMPLE: In 1967 it cost $120 to purchase and install a small piping system in metallic material which cost $100 in 1962. In 1967 a reinforced plastic system which cost $100 in 1962 could be installed for $75.

bought in 1967 for $770. This table is not intended to measure the cost trend of any single vendor in the industry. Rather, it is intended to measure a multiplicity of factors, such as cost trends, changes in competitive positions, and, most important, the "buyer awareness" of the cost-conscious engineer. Today, in reinforced plastics, it is quite possible to buy material for at least 23 percent less than the price in 1961 and receive a better product, closely allied to superior quality control. The general trend of reinforced plastic material is "down." The general trend of metallic material is "up." As a matter of information, other accessory materials that go into piping systems, such as fasteners, show a slight decrease in price of perhaps 10 percent over the last 10 years, while gasketing materials show a slight price increase, amounting, roughly, to 10 percent over the last 10 years.

One more important factor: in the early 1960s reinforced plastics in one large chemical plant confronted with severe corrosion problems amounted to an insignificant expenditure in the area of corrosion-resistant pipe and fittings. Today, 27 percent of the material costs in pipe and fittings is spent in the reinforced plastics area, and corrosion-resistant RP ductwork has gone from almost nothing in 1960 to over 80 percent in 1967. Tanks and storage vessels in reinforced plastics have gone from virtually zero in 1960 to over 50 percent in 1967.

Metallic pipe and fittings represent one of the areas which has escalated most rapidly in the last 10 to 12 years, even exceeding machine parts in the rate of cost rise. In attacking high-cost areas in large plants, the engineer has little choice in singling out pipe and fittings as one of the potential areas in which large cost reductions may be achieved.

It is difficult to generalize in many of these installations because each installation is a separate cost study in itself, but one installation was studied extensively which was typical of that found in many process-plant installations. This particular installation of 6-in. pipe studied showed the following relative cost index on an installed-cost basis:[20]

```
Steel.....................................  1.0
Reinforced polyester.....................  1.43
Reinforced epoxy.........................  1.94
304 stainless steel, Schedule 5..........  2.23
Lead-lined steel.........................  2.67
316 stainless steel, Schedule 5..........  3.08
```

One of the ramifications in studying these various materials of construction showed that it lies within the power of the designer to materially affect the cost of any system simply by the mechanics of putting it together. An exploration of the factors involved is covered in some

detail in Chaps. 3 and 4.   The appropriateness of including 304 stainless
steel in the above compilation is certainly questionable since it is satisfactory for only the mildest of applications and it will even rust.

Ease of handling   If you have ever watched a section of 10-in. rubber-
lined or lead-lined steel pipe being installed 10 or 12 ft in the air, you
will have noticed that it takes chain falls, rigging, and a great deal of
perspiration to get it there.   But take a piece of 10-in. reinforced plastic
pipe flanged at both ends, put up two ladders, let a man pick up each

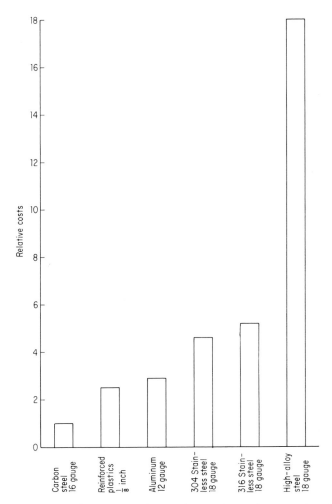

FIG. 1-1   Corrosion-resistant ductwork.[18]   Comparative fabrication prices of
12-in. ductwork.

end of it, and the entire section will go into place in minutes. Complete sections of reinforced plastic pipe can be prepared, all ready to go into position, on 2 hr notice or less. Minimizing of down time in expensive processes is tremendously important. The pipe can be bought many times over when equated to the value of the cost of product lost. The long-range view must be taken: not only the cost of the piping or equipment at hand and the labor of installation, but the value of product lost through machine shutdown. One of the keys to good profits is the highest equipment utilization possible. Equipment outage is costly and expensive. If a piece of equipment returns $500/hr profit, then high reliability, continuous performance, ease of repair, and reasonable first cost are certainly barometers by which equipment performance should be judged.

Today, we are most fortunate to have available at the designer's command material which a few short years ago existed only in a test tube or a struggling engineer's imagination. It is a case of "working smarter, not harder." Some of the breakthroughs involving reinforced plastics equipment, coupled with other new material, have permitted the construction of equipment with twice the productivity at half the cost, and permitted the equipment to be put on stream in half the time taken 10 years ago. Then, as a bonus, we find lower operating costs, lower maintenance costs, and a much higher performance reliability.

Throughout the book cost comparisons are made to provide the engineer with accurate guidelines on relative prices of the various materials of construction. The limiting parameters of reinforced plastics are also carefully explained in the various types of applications. Guidelines are provided on good installation practices and, most important, things to avoid.

Figure 1-1 shows competitive fabrication prices for 12-in. ductwork.[18] Here again it is illustrated that reinforced plastics offer a minimum-cost approach in corrosive applications.

REFERENCES

[1] Whitehouse, A. A. K.: Glass Fibre Reinforced Polyester Resins, paper 10, at Symposium on Plastics and the Mechanical Engineer, London, Oct. 7–8, 1964.

[2] Smith, C. S.: Materials, *Scientific American*, September, 1967, p. 69.

[3] Coles, Leonard A.: "The Book of Chemical Discovery," p. 33, George G. Harrap & Co., Ltd., London, 1933.

[4] *Ibid.*, p. 34.

[5] *Ibid.*, p. 59.

[6] *Ibid.*, pp. 208–211.

[7] *Ibid.*, p. 222.

[8] *Ibid.*, pp. 30–31.

[9] *Ibid.*, pp. 29–31.

[10] Ref. 2, p. 79.

[11] Mark, Herman F.: The Nature of Polymeric Materials, *Scientific American*, September, 1967, pp. 149–154.

[12] Estimated from a number of sources.

[13] Jaray, F. F.: Behavior of Wire Reinforced Plastic Using Wire Sheet as Reinforcing Material, paper presented before the Society of the Plastics Industry, Washington, D.C., February, 1967.

[14] Hoehn, R., and W. Heling: "Organic Fiber Veils for Glass-resin Laminates," TN 7-9110, Pellon Corp., New York.

[15] Haveg Industries, Inc., *Bulletin* CED-65-12, Wilmington, Del., December, 1965.

[16] Stout, H. P.: Why Plastics Designers Are Turning to Jute, *Textile Mercury International*, June 24, 1966.

[17] Fried, N., and W. R. Graner: Durability of Reinforced Plastic Structural Materials in Marine Service, *Marine Technology*, vol. 3, no. 3, pp. 321–327.

[18] Controlling Corrosion with Reinforced Atlac 382, Atlas Chemical Co., *Bulletin* LP 29, March, 1966, p. 14, Wilmington, Del.

[19] Kelly, Anthony: Nature of Composite Materials, *Scientific American*, September, 1967, p. 174.

[20] Loftin, E. E., FMC Corporation, Front Royal, Va., personal communication, August, 1967.

2

Basic Application Principles

2.1 INTRODUCTION

The successful application of any corrosion-resistant material begins with a complete and accurate definition of the problem to be solved. In order to consider any material as a potential solution to the problem, the engineer must be intimately familiar with the material to be applied. He needs to know its strong points and its weaknesses and, above all, the limits on its successful application. We must therefore determine, first, with the general classes of reinforced plastic material available, where each can be used successfully and where it should be avoided.

Many different types of polyester and epoxy chemical-resistant resins (and furanes and vinyl esters) are being manufactured today, both in the United States and abroad. Although, for the purposes of illustration, a specific resin may be referred to occasionally, this mention is for the purpose of illustration. It is not meant to imply that other resins, manufactured by other vendors, are not equally good or equally acceptable.

Generally, we are dealing with families of resins which have the following characteristics:

* An upper wet-temperature limit of about the boiling point of water or perhaps a little higher (250°F)
* A dry limit of perhaps 350°F
* Remarkable resistance to many of the oxidizing acids, up to fairly strong concentrations
* Potentially, superior alkaline resistance
* Good solvent resistance in some areas but limited resistance in others
* In variegated shapes, most useful in the areas of low pressure or vacuum

• Can provide complete piping systems at relatively high operating pressures and in most of the common sizes up to 150 psig
• Can be tailor-made for ablative conditions where service under extended temperatures for short periods of time is desired
• Low thermal and electrical conductivity so that use can be made of these properties

The reinforcement, which is generally glass, asbestos, or a synthetic fiber, may be furnished by other suppliers.

These ingredients are processed into the end product needed by still another processor, generally referred to as the fabricator. Fabricators vary from large corporations to those of intermediate sizes down to the very small, a shop of 5,000 ft^2 or less being commonplace.

There is some feeling that this diversity of source material has been a handicap to the orderly advancement of the industry. While the resin and glass manufacturers have behind them the resources of relatively large corporations, the fabricators, except for a few, have relatively limited capital. This limitation of capital precluded the use of large applied research budgets which could be used to solve application problems on a broad scale. Fortunately, there are signs that knowledge of this problem exists and that the necessary applied research in specific areas is forthcoming. This is particularly true in the field of tankage. As proven manufacturing techniques are applied, good progress is being made in reducing tank costs. It is a reasonable conjecture that these same management techniques will ultimately be applied to many of the other problems in the industry, which will certainly result in better-product availability at a reduced price.

There are strong indications that previously published figures on the usage of reinforced plastic material in industry are considerably short of the amount actually being used. RP consumption figures for the immediate past are being revised strongly upward. This will have a like effect on an upward revision of future projections.

2.2 COMMON FABRICATION METHODS

Contact-molding—hand lay-up[1] The most widely used manufacturing method is the contact-molded hand lay-up. For purposes of illustration, we shall take the building of a tank. Normally, a single steel mold is prepared. After mold preparation, it is polished. When a tank is being built, either a release agent is applied to the mandrel or the mandrel is wrapped with a Mylar or cellophane film. This is essential to provide a good surface on the finished part. First, a 10-mil gel coat of unreinforced resin is applied. Into the gel coat is generally worked some light

C grade surfacing mat.   After it has gelled, successive layers of 1½-oz mat saturated with resin are applied.   It is common practice to have at least two layers of 1½-oz mat backing up the surfacing system.   Following this, layers of woven roving and chopped-strand mat are commonly

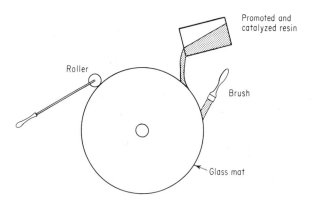

FIG. 2-1   Contact molding—hand lay-up on a rotating-tank mandrel.

used, depending upon the thickness required.   Each layer of 1½-oz chopped-strand mat may be considered as having approximately 0.045-in. finished thickness, and each layer of woven roving, 0.085-in. finished thickness.   Localized areas may be strengthened at will.   Metallic inserts are used where necessary.   Nozzles and manholes are generally strapped in when the shell has been completed.   The exterior finish may again revert to a surfacing mat and hot coat, or simply a hot coat, depending upon the practices being followed.   Figure 2-1 is a diagrammatic illustration of a contact-molding hand-lay-up operation.

Contact molding—spray lay-up[1]   Mandrel preparation is the same as in the case above.   The special spray gun mixes three components some distance beyond the surface of the guns; they are (1) a catalyzed resin, (2) a promoted resin, and (3) chopped roving.   In this manner the glass is saturated by the blending of the catalyzed promoted resin.   The advantage of this technique is that it reduces the labor in the entire operation, although there is no doubt that the necessary labor must be of a higher skill.   Tank caps, some shells, and many items, such as boats, tote boxes, ducts, and hoods, are made by this technique.   Figure 2-2 illustrates contact molding using spray-lay-up operation.

Continuous extrusion process[1]   Considering man's ingenuity, it could have been predicted that production of shapes with regular cross sections should be adaptable to some type of machine.   Shapes such as rods, bars,

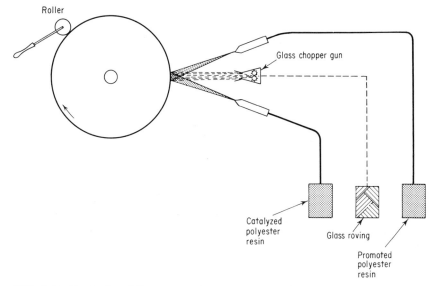

FIG. 2-2   Contact molding—spray lay-up on a rotating-tank mandrel.

and piping can be constructed by saturating roving with the particular resin desired and then forcing it through a simple die.   These shapes illustrate the type of process involved, used also in the manufacture of ductwork, angles, Z bars, I beams, channels, etc.   Final cure is in an oven, a process also sometimes referred to as pultrusion.   Figure 2-3 shows a continuous extrusion process.

Filament winding   Filament winding generally requires an extensive capital investment in the winding machine, which permits adjustment

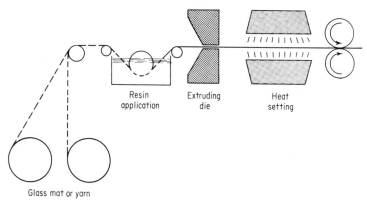

FIG. 2-3   Continuous extrusion process.

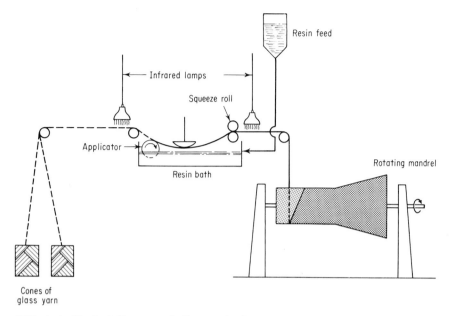

FIG. 2-4   Typical filament-winding methods.

of the helix angle on an automatic basis.   Continuous-filament glass is taken from a creel and passed over a base of infrared lamps, then over an applicator which is spinning constantly in a resin bath.   The warmed yarn provides a better resin pickup.   The squeeze roll leaving the bath squeezes out the bulk of the excess resin.   Another bank of lamps may be used to promote further resin penetration into the fiber, after which it passes onto the winding head, and from there onto the rotating mandrel. The winding head may be manually operated or programmed to perform to the design requirements.   Epoxy resins are a favorite in this type of application and are used in the electrical and space fields.   Slower setup time for the epoxy resins makes their use more applicable in this type of construction, along with high-glass-content and low-resin-content design. Some polyester applications are, however, done in similar manner, slightly modified to take into account the decrease in time available for setting the resin.

The ability to wind in a loose or a tight manner has many advantages for the fabricator.   Most epoxy winds are relatively tight, perhaps 70 to 80 percent glass and 20 to 30 percent resin.   In the chemical industry, however, this does not provide the optimum corrosion resistance.   A relatively loose wind in which the winder is able to achieve 50 to 60 percent glass and 40 to 50 percent resin represents a much more desirable combination of strength and corrosion resistance.   This loose winding is

2.3 SIX GENERIC TYPES OF RESINS

Table 2-1

Resin type	Basic organic formula	Comments	End uses
General-purpose polyester...	Reacted fumaric group	Generally not used for wet chemical service	Tote boxes, baptistries, boats, fishing poles, building materials, car and truck bodies
Isophthalic polyester...		Good chemical resistance in many environments. Less expensive than those below	Gasoline tanks, composite chemical tanks, refinery-tank linings
Bisphenol polyester...		Excellent overall chemical resistance. One of the major work horses of the industry	Chemical piping, tanks, and structures
Hydrogenated bisphenol-A polyester		Excellent chemical resistance in many areas	Chemical piping, tanks, and structures
Chlorinated polyester...	Chlorinated molecular structure	Excellent chemical resistance in many severe services, particularly chlorine	Chemical piping, tanks, and structures; with small quantities of Sb_2O_3 added provides highest fire retardancy, for ductwork
Epoxy...	Bisphenol group	Provides good chemical resistance when heat-cured. More expensive than polyesters	Chemical piping and tanks; superior alkaline resistance; good solvent resistance in some areas

Ⓝ Hydrogen-saturated bisphenol molecule.

used by fabricators in the construction of corrosion-resistant equipment built at least in part by the filament-winding method.

A variation of filament winding is practiced by winding with tape and is commonly referred to as "tape wrapping." Figure 2-4 shows a typical filament-winding method.

The foregoing methods are those commonly in use in manufacturing reinforced plastic material for the chemical-process industries. There are other methods of rather limited use, such as vacuum bag molding, pressure bag molding, and vacuum injection processes.

2.4 THE GRADES OF REINFORCING GLASS

Although at the present time the great bulk of reinforced plastic fiberglass reinforcing is made from an E grade glass, some interest is being shown in producing filament-wound products with an all-C-glass construction. This type of construction would provide a smooth inner surface reinforced with a C glass veil, followed by sufficient C glass filament winding to provide the necessary structural strength and, finally, by an outer C glass veil. The advantage of such a construction is that it would possess increased chemical resistance throughout the laminate, and generally, laminate degradation with time would be considerably less than with E glass. The manufacture of C glass in filament form, however, is relatively small at this time and is made only on demand. Obviously, should the demand ultimately warrant it, filament C glass would become more widely available. Today, filament C glass costs approximately three times as much as filament E glass. Assuming use of the same high-chemical-resistant resin in both cases, a filament-wound C glass vessel containing 75 percent glass and 25 percent resin would command a healthy premium of about 70 percent. If a complete random-mat lay-up were made from C glass so that the final laminate contained 25 percent glass and 75 percent resin, the premium for the use of C glass would be approximately 20 percent.

Work is going on to develop, particularly for use with the high-chemical-resistant resins, a grade of glass which will possess the high-physical-strength characteristics of the E glass and excellent chemical resistance of the C glass. Present indications are that such a hybrid would command a premium over the E glass but cost only about half the price of C glass. This, then, would permit the production of laminates from the hybrid glass at an overall premium price estimated at 7 to 10 percent. Competition, however, is the final test, and experience will have to justify that such a premium results in a lower cost per year of service life.

Although Table 1-1 indicates a comparison of some of the physical properties of the various types of fibrous glass available for reinforcement, the reader will find that only in laminate surfacing and structural systems can the end use be made of the properties of the various glasses available. Through the various options available in laminate design, the engineer has at his disposal wide variations in strength from tensiles of 15,000 psi in an all-mat lay-up to tensiles of 120,000 psi in filament-wound construction. But strength per se is not the whole story. It is a combination of corrosion resistance and adequate strength to perform the necessary task.

2.5 CHEMICAL–RESISTANT LAMINATE SURFACING SYSTEMS

Table 2-2

Type of surface layer	Thickness range	Description	Specific uses
Gel coat......	10–20 mils common, but may go to 60 mils with some vendors	Unreinforced layer of resin	Commonly used in many epoxy piping systems. Little used in hand-laid-up polyester construction. Less resistance to cracking and crazing than with a reinforced laminate.
Type C Surface mat	20–25 mils	10-mil type C mat commonly employed, adding strength and stability to the resin-rich surface	Most commonly used with reinforcement, for chemical-plant applications. Particularly necessary on tanks, ductwork, etc. Commonly used in polyester piping. About 10% glass/90% resin.
Organic veil...	20–30 mils	Dynel, dacron, acrylic, polypropylene	Good weathering characteristics. Improves abrasion resistance and impact strength in many cases. Some provide complete transparency. Standard specification in HF* work or caustic applications. About 10% veil/90% resin.
Asbestos mat	40–60 mils	Crocidolite asbestos mat	Good acid and alkaline resistance. Used as an inner liner in some commercial epoxy pipe. Could equally be used with polyesters. Good solvent resistance.

Table 2-2 *(Continued)*

Type of surface layer	Thickness range	Description	Specific uses
Oriented flake-lined	25–35 mils	Small glass flakes, resin-impregnated, parallel to laminate surface	Penetrant path increased manyfold between corrosive liquid and understructure. Provides excellent resistance to cracking and crazing and improvement in strong oxidizing environments.

*HF stands for hydrofluoric acid or hydrogen fluoride when it occurs in a gaseous form.

2.6 CHEMICAL–RESISTANT LAMINATE STRUCTURAL SYSTEMS

Table 2-3

Type of structural system	Glass content	Description	End uses
All chopped-strand mat	25–35%	1½-oz random-mat cloth, resin-saturated and built up to the desired wall thickness	Widely used in piping, duct, and tank construction.
Chopped-strand mat and woven roving	30–50%	Many varieties of construction can be designed. Hand-laid-up polyester pipe employs woven roving 4-in. in diameter and above.	Used extensively in tank construction for high impact resistance and strength. Woven roving generally 24-oz weight.
Cloth.........	40–50%	Generally, successive layers of cloth, resin-impregnated, are built up to the desired thickness. An effective surfacing system must be used.	Various special weaves are used for piping construction. Sometimes used in building intricate shapes where drape is a necessity. Also used where glass-content ranges of this magnitude are required.

Table 2-3 *(Continued)*

Type of structural system	Glass content	Description	End uses
Filament winding	70–85%	Develops high wall strength through keeping glass in tension. Must be protected with a good surfacing system.	Commonly used in epoxy piping. Some tanks built by this method.
Chopped-strand mat and filament winding	Varies across wall: 25–35% in laminate, 70–85% in filament winding	Combines the best of the construction processes to provide a high-strength chemical-resistant lay-up	Industry is showing a growing trend toward this type of construction for both piping and tanks.

2.7 LAMINATE CONSTRUCTION

A study of laminate design and construction, be it epoxy or polyester, fire-retardant or non-fire-retardant, of high chemical resistance, isophthalic, or orthophthalic, reveals a wide range of opinion as to resin specifications, glass/resin ratio, filament winding versus hand lay-up, type of glass to be employed, and even as to the laminate design itself. To give the reader some background in the extremes in differences in thinking, some views are listed below:

PRO

1. Satisfactory chemical resistance can be achieved with a 10- to 20-mil chemical-resistant inner surface suitably reinforced with a C glass.

2. The optimum combination of physical and chemical characteristics can be achieved by a high glass/resin ratio, such as 25 to 30% glass and 70 to 75% resin. The wall can then be suitably designed

CON

1. Good chemical resistance can be achieved only by a 60- to 120-mil inner liner. One group would vote to reinforce the liner with asbestos or other reinforcing compound. Another group would provide no liner reinforcing whatsoever.

2. The key to good design is to use a suitable inner liner followed by filament winding for maximum strength. Such a design will weep but never suffer a catastrophic failure. Glass/resin

to take care of whatever physical demands are placed on it. Such a design will not weep.

3. The best laminate design is one solely made up of random mat, with the exception of the 10-mil inner veil and 10-mil outer veil. This provides the optimum in chemical resistance.

4. All fabricated equipment should be finished with a hot coat for maximum protection and sealing of the exterior.

5. To provide for degradation with time, design safety factors of 10:1 in tanks and vessels and piping and 5:1 in ductwork should be used.

6. Woven roving should always be encased with random mat to provide the utmost in protection of the continuous-filament roving.

7. Reinforcement of the corrosion-resistant inner barrier is essential.

8. Continuous support of tanks is best provided by means of a sand bed for a firm upward pressure on the tank bottom at all points.

9. In tankage the key to the lowest possible costs lies in the standard-tank approach.

ratios may run 75% glass/25% resin.

3. The best design is one combining the best properties of random mat and woven roving to provide not only good chemical resistance but maximum impact strength.

4. The hot coat is a waste of time and money and adds nothing to the overall chemical resistance of the part in question.

5. Degradations of 5:1 and 10:1 are vastly overrated, and an allowance of one-third is sufficient.

6. Woven roving on the outside of a pipe or vessel does not need to be encased with anything but a good hot coat. In fact, in piping which is to be encased in concrete, the dimpled roughened structure of the roving gives a good adhesive bond with the concrete.

7. Reinforcement of the corrosion-resistant inner barrier is not essential.

8. The use of a sand bed for supporting a tank is a delusion. It exists until the sand is washed out, which will probably occur in a matter of months, leaving a strain on connecting piping. It is much better to use a brick-covered concrete base.

9. Unless you are building a plant in a cornfield or a relatively nonrestricted area, the use of standard tanks is impractical.

10. The bisphenol resins or hydrogenated bisphenol resins are demonstrably better than any other resins on the market.

10. Needless premiums are paid for high-chemical-resistant resins when the isophthalic would perform equally well at a reduced cost. Even admitting the corrosion superiority of the bisphenols, economize with composite construction of 125 mils bisphenol interior and the remainder isophthalic resin.

The pros and cons listed above are typical of the divergences of opinion currently held by segments of the industry, many of which have received firm financial backing in the strong belief that they are absolutely correct. Anyone fresh to the industry and seeking guidance is justifiably confused, wondering where to turn. However, this is a relatively young industry, which is evolving and developing experience and know-how at a fast rate. There is an element of truth and reason in many of the positions taken, both pro and con. Ultimately, as the industry continues to develop, there will probably be a merging of the filament-wound approach and the hand-laid-up approach to provide the best qualities of each. It is quite possible to visualize that such a union will produce a synergistic effect, so that one plus one will no longer equal two, but more likely, in combined effect, three. This union of filament winding, providing extremely high physical strength, and hand-laid-up techniques, providing the maximum chemical resistance, is beginning to be seen now in a small portion of the industry output in tanks, ducts, and piping.

The entire aspect of piping, duct, and tank design, along with resin specification, is discussed in detail in other sections of the book. Fire retardancy and how to achieve it, laminate design, and all the other factors necessary in design, purchasing, and installing reinforced plastic equipment are discussed for the engineer's guidance. If the engineer realizes that most failures in glass-reinforced plastic equipment are physical or mechanical, and not too often chemical, then it can be concluded that they are preventable and that a successful application is an attainable objective. Most failures in the RP field are generally due to "sins of omission," compounded by lack of knowledge, usually on the part of the purchaser.

2.8   SERVICE–APPLICATION GUIDE

For the successful application of the reinforced plastics in the field of corrosion-resistant process equipment, the following summaries have been prepared, covering a variety of common chemical environments. The remarks on the application to environmental conditions are meant to serve as a guide for the engineer only. For details to meet any specific corrosion condition, the designer should consult either the fabricator or the original resin producer. Quite often the exercise of a little ingenuity, especially in the field of composite materials, can be most rewarding. For additional details in the field of composite piping, tanks, and structures, the reader is referred to Chap. 5, where the union of thermosets/thermosets and thermoplasts/thermosets can both expand the field of usefulness and the areas of application and, quite often, save money. Here the use of general-purpose and isophthalic resins is explored in some detail.

Bisphenol and hydrogenated bisphenol polyesters[2]   The bisphenol and hydrogenated bisphenol polyesters have an exceptionally high degree of chemical resistance, superior to both the general-purpose and isophthalic classes of polyester resins. Depending upon the particularly severe chemical environment, the bisphenol, hydrogenated, and chlorinated chemical-resistant-type polyesters represent the current ultimate in resins available for severe chemical service. These three classes of resins cost approximately twice as much as a general-purpose resin and one-third more than an isophthalic resin. Only a fraction of this premium, however, carries over into the final fabricated cost of the equipment. These high-class chemical-resistant resins are very much easier to fabricate than epoxies. They show superior acid resistance to the epoxies. The bisphenol-type resins show good performance with moderate alkaline solutions and excellent resistance to the various categories of bleaching agents. All the polyesters and the epoxies break down under highly concentrated acids or alkalis, such as 93% H_2SO_4 and 73% NaOH. Failure with the concentrated acid is by dehydration.

Recommended areas of application are as follows:

Acids, to 200°F

25% acetic acid	Maleic acid
Benzoic acid	Oleic acid
Boric acid	Oxalic acid
Butyric acid	80% phosphoric acid
25% chloroacetic acid	Stearic acid
5% chromic acid	50% sulfuric acid
Citric acid	Tannic acid
Fatty acids	Tartaric acid
10% hydrochloric acid	50% trichloroacetic acid
Lactic acid	

Salts, solutions to 200°F

All aluminum salts	Low sodium and potassium salts,
Most ammonium salts	except the high-alkaline salts
Calcium salts	Zinc salts
Copper salts	Most plating solutions
Iron salts	

Solvents

All the solvents shown under	Alcohols at ambient temperature
the isophthalic resins	Glycerin
Sour crude oil	Linseed oil

Alkalis, to 160°F

5% ammonium hydroxide	25% potassium hydroxide
25% calcium hydroxide	25% sodium hydroxide

Bleaches, to 200°F

20% calcium hypochlorite	Hydrosulfite
15% chlorine dioxide	Sodium hypochlorite (up to stable
Chlorite	temperature)
Saturated chlorine water	Textone

Gases, to 200°F

Carbon dioxide	Sulfur dioxide (dry)
Carbon monoxide	Sulfur dioxide (wet)
Chlorine, dry	Sulfur trioxide
Chlorine, wet	Rayon waste gases (at 150°F)

This type of resin has been found to be unsatisfactory in:

Solvents. Test any organic solvent other than those listed above. Solvents such as benzine, carbon disulfide, ether, methyl ethyl ketone, toluene, xylene, etc., are not satisfactory.

93% sulfuric acid.

73% NaOH.

30% chromic acid.

Chlorinated chemical-resistant-type polyesters[2]* Although it is a generalization, and generalizations have many exceptions, it can be said that room-temperature-cured chlorinated chemical-resistant-type polyesters, reinforced with fiber glass, possess unique chemical-resistant characteristics. Outlined in the table below are findings based on almost twelve

* This section on Chlorinated Chemical-resistant-type Polyesters was prepared by W. A. Szymanski of Hooker Chemical Corp., North Tonawanda, N.Y.

years of highly successful applications throughout industry and on extensive laboratory and field investigations. Resistance, in many instances, can be improved by elevated-temperature postcures. Additionally, these resins possess the highest heat resistance of any chemical-resistant polyester and are inherently fire-retardant. A noncombustible rating of 20, by the ASTM E-84 tunnel test, can be achieved, making this the safest possible polyester for stacks, hoods, fans, ducts, etc., or wherever a fire hazard might exist.

Environment	*Remarks*
Water: demineralized, distilled, deionized, steam, and condensate	Lowest absorption of any polyester. Resistant to 212°F.
Alkaline solutions:	
pH greater than 10	Not ordinarily recommended for continuous exposure, but resistant to 20% ammonium and 10% sodium hydroxides, to about 150°F.
pH greater than 10, intermittent pH less than 10	Suitable to about 200°F, depending upon exposure. Resistant to, and often higher than, about 200°F.
Amines: aliphatic, primary aromatic	Can cause severe attack, depending upon conditions.
Amides, other alkaline organics	Can cause severe attack, depending upon conditions.
Salts: alkaline, such as sodium sulfide, trisodium phosphate	Can cause severe attack, depending upon conditions.
Salts, neutral	Resistant to, and in some instances higher than, about 250°F.
Salts, acid	Resistant to, and in some instances higher than, about 250°F.
Acids: mineral, nonoxidizing	Resistant to, and in some instances higher than, about 250°F.
Acids, organic	Resistant to, and in some instances higher than, about 250°F, including glacial acetic to about 120°F.
Acids: organic; certain high-molecular-weight acids; sulfonic, amino, and sulfinic acids	Can cause severe attack, depending upon conditions.
Phenol and phenols	Not ordinarily recommended.
Acid halides	Not ordinarily recommended.
Mercaptans	Resistant to about 180°F.
Ketones	Resistant to about 180°F.
Aldehydes	Resistant to about 180°F.
Alcohols	Resistant to about 180°F.
Glycols	Resistant to about 180°F.
Esters, organic	Resistant to about 180°F.
Fats and oils	Resistant to about 200°F.
Solvents: aliphatic, aromatic, and chlorinated	Uniquely resistant to many, such as CCl_4 to 110°F; trichlorethylene to

Environment	*Remarks*
	180°F; monochlorbenzene to 150°F; petroleum ether, benzene, toluene, heptane, CS_2, dimethyl sulfoxide, polyvinylidene chloride at room temperature.
Oxidizing acids and solutions	Uniquely resistant to many, such as 35% HNO_3 to 140°F; 70% HNO_3 at room temperature; 40% chromic at 210°F; chlorine water to 285°F; wet Cl_2 and ClO_2 to 220°F; 15% hypochlorites to 110°F; 6% hypochlorites to 140°F. Concentrated H_2SO_4 is severely destructive.
Gases: wet and dry	Resistant to about 350°F, with intermittent exposure to as high as 425°F. SO_3 can cause severe attack.

General-purpose polyesters[3]   General-purpose polyester resins are normally not recommended for use in chemical-process equipment. Their use in the finished fabrication represents a potential savings of 10 to 20 percent. Only the purchaser can make the ultimate decision as to the premium he wishes to pay for chemical resistance. These resins are generally adequate for use with nonoxidizing mineral acids and corrodents that are relatively mild. This is the resin which predominates in boat building, so that, obviously, its resistance to water of all types, including seawater, is more than adequate. Test work has indicated satisfactory application in the following areas up to 125°F.

Acids

10% acetic acid	Oleic acid
Citric acid	Benzoic acid
Fatty acids	Boric acid
1% lactic acid	

Salts

Aluminum sulfate	Ferrous chloride
Ammonium chloride	Magnesium chloride
10% ammonium sulfate	Magnesium sulfate
Calcium chloride (saturated)	Nickel chloride
Calcium sulfate	Nickel nitrate
Copper sulfate (saturated)	Nickel sulfate
Ferric chloride	Potassium chloride

Ferric nitrate   Potassium sulfate
Ferric sulfate   10% sodium chloride

Solvents

Amyl alcohol   Kerosene
Glycerin   Naphtha

General-purpose resins have been found to be unsatisfactory in:

Oxidizing acids
Alkaline solutions, such as calcium hydroxide, sodium hydroxide, and sodium carbonate
Bleach solutions, such as 5% sodium hypochlorite
Solvents such as carbon disulfide, carbon tetrachloride, gasoline, distilled water

Where the use of general-purpose resin is contemplated, an environmental test program should be inaugurated to determine if the resin will be satisfactory. All contemplated applications above 125°F should receive rigorous testing.

Isophthalic polyesters[3]   The isophthalic polyesters offer better chemical resistance, in certain specific areas, to the general-purpose resins, at a slightly higher cost. They definitely show much better resistance to attack in the solvent areas, and are used extensively in the manufacture of underground gasoline tanks, where a satisfactory service life in the storage of gasoline and under the varied conditions of ground-soil corrosion are met successfully. The following general usage of the isophthalic-type resins is given as a guide in applications up to 125°F.

Acids

10% acetic acid   Oleic acid
Benzoic acid   25% phosphoric acid
Boric acid   Tartaric acid
Citric acid   10% sulfuric acid
Fatty acids   25% sulfuric acid

Salts

Aluminum sulfate   Iron salts
10% ammonium carbonate   5% hydrogen peroxide
Ammonium chloride   Magnesium salts
Ammonium nitrate   Nickel salts
Ammonium sulfate   Sodium and potassium salts which

Barium chloride
Calcium chloride (saturated)
Copper chloride
Copper sulfate

do not have a high-alkaline
   reaction
Dilute bleach solutions

Solvents

Amyl alcohol
Ethylene glycol
Formaldehyde

Gasoline
Kerosene
Naphtha

Isophthalic resins have been found to be unsatisfactory in:

Acetone
Amyl acetate
Benzene
Carbon disulfide
Solutions of alkaline salts of potassium and sodium
Hot distilled water
Higher concentrations of oxidizing acids

Vinyl ester resin[3]   One of the recent additions to the field of corrosion-resistant resins is the new vinyl ester resin, currently available on a limited basis in a piping form only.   These vinyl ester resins are styrene-diluted and free-radical-initiated.   Catalyst and promoter systems similar to those used with polyesters may be used with the vinyl ester resins. Since these resins are so new, limited test data are available.   The resins are reported to be useful in chlorine-caustic and oxidizing acids at elevated temperatures.   A few of the service conditions in which tests have been run indicate useful applications up to 200°F in the following areas:

Acids

Benzene sulfonic acid
Benzoic acid
Boric acid
25% butyric acid
5% chromic acid
Citric acid
Fatty acids
50% hydrobromic acid
Hydrochloric acid (0–30%)
Hypochlorous acid, 10%
   (to 150°F)

Lactic acid
25% maleic acid
10% nitric acid (to 150°F)
Oxalic acid
Phosphoric acid (0–80%)
50% sulfuric acid (to 150°F)
Tannic acid
Tartaric acid
Trichloroacetic acid

Bases

0–50% sodium hydroxide (to 150°F)

Gases

Chlorine, wet or dry

Solvents

Crude oil, sweet and sour (to 150°F)
40% formaldehyde
Kerosene (room temperature)

Miscellaneous

Calcium hypochlorite (0–20%)
15% sodium hypochlorite (to 150°F)
Distilled water

Epoxies[3]   Many different epoxy resins will provide outstanding service in chemical-process equipment under severe conditions. For the purpose of illustration, the specific chemical-resistance characteristics of one resin are described. This resin, Epon 828,* has been widely used in chemical service, and as such it is backed by a large number of successful case histories. Other epoxy resins with equally successful case histories may be considered for similar applications.

Resins of this type generally require elevated-temperature post-cures. This is common practice in the manufacture of piping, tanks, and structures made from epoxy resins. This type of resin is inherently more expensive, and the manufacturing process is considerably slower, so that even simple molds can be turned over only twice weekly. The epoxies are nearly always combined with filament winding to produce an extremely strong product. The absence of room-temperature cures makes the use of the high-chemical-grade epoxy such as this difficult for "do-it-yourself" plant application—an area in which the polyester largely predominates; reinforced epoxies are generally confined to piping systems and tanks.

Piping systems in reinforced epoxy have been well engineered, and their use is prevalent. Flanged and adhesive assembled systems are commonly used.

The following services are applicable to 200°F unless otherwise stated:

Acids	*Salts*
10% acetic acid (to 150°F)	Aluminum
Benzoic acid	Most ammonium salts

* Registered trademark of Shell Chemical Co., New York.

Boric acid
50% butyric acid
25% chloroacetic acid
5% chromic acid
Citric acid
Fatty acids
25% hydrochloric acid
Hypochlorous acid
10% nitric acid (to 150°F)
Oxalic acid
30% perchloric acid (to 75°F)
Phosphoric acid (80%)
25% sulfuric acid

Barium
Calcium
Iron
Magnesium
Potassium
Sodium

Bases

50% calcium hydroxide (to 150°F)
Magnesium hydroxide
50% sodium hydroxide
Trisodium phosphate (150°F)

Solvents

Alcohol: methyl, ethyl, isopropyl (to 150°F)
Kerosene
Turpentine (to 150°F)
Vinyl acetate (to 150°F)
Good in many strong solvents—test specific use

Miscellaneous

Freshwater
Seawater
Jet fuel

This type of resin has been found to be unsatisfactory in:

93% sulfuric acid
73% NaOH
Aqua regia
Wet chlorine gas
Wet SO_2

Furanes[3]   The range of application of furane resins reinforced with fiber glass or used as a composite structure is described in detail in Sec. 5.5.

2.9   METALLIC CORROSION

A brief review of the more common types of metallic corrosion shows deterioration by a number of methods:

General corrosion............ General degradation of the metal, which may be accelerated in localized areas and cause perforations. Rusting is a typical phenomenon.

Galvanic corrosion........... A flow of current between two dissimilar metals which results in the deterioration of the least noble metal.

Aerobic corrosion............. Generally bacterial in nature.

Other types of common metallic corrosion which occur may include pitting, dezincification, and intergranular corrosion. It is interesting to note that reinforced plastics are not subject to any of these symptoms, so prevalent in metals. This does not mean, however, that reinforced plastics are not subject to attack. They certainly are, but in a completely different manner.

2.10   CHEMICAL ATTACK OF REINFORCED POLYESTERS OR EPOXIES—THE COUNTERPART OF METALLIC CORROSION[4]

If and when reinforced polyesters or epoxies are misapplied, chemical attack, which is a relatively complicated phenomenon for these materials, may occur in several ways. These may be broadly classified as follows:

1. Disintegration or degradation of a physical nature due to absorption, permeation, solvent action, etc.
2. Oxidation, where chemical bonds are attacked
3. Hydrolysis, where ester linkages are attacked
4. Dehydration (rather uncommon)
5. Radiation
6. Thermal degradation, involving depolymerization, and possibly repolymerization
7. Combinations of these mechanisms, and possibly others

As a result of such attacks, the material itself may be affected in one or more different ways; for example, it may be embrittled, softened, charred, crazed, delaminated, discolored, dissolved, blistered, swelled, etc. Although all polyesters will be attacked in essentially the same manner, certain chemical-resistant types suffer negligible attack, or exhibit significantly lower rates of attack, under a wide variety of severely corrosive conditions, due primarily to the unique molecular structure of the resins, including built-in steric protection of ester groups, etc. However, because of the complicated way in which attack does occur, knowledge of the chemical structure of the resin does not preclude actual testing in most environments to determine resistance.

Cure, of course, plays an important part in the chemical resistance developed by a polyester or epoxy, as does the construction of the laminate itself and the type of glass or reinforcing used. The degree and nature of the bond between the resin and the glass or other reinforcement also plays an important role.

All these various modes of attack affect the strength of the laminate in different ways, depending on the environment and other service conditions and the mechanism or combinations of mechanisms that are at work. Absorption, for example, in certain instances can weaken the polymer network by relieving internal strains or stresses. Certain environs may weaken primary and/or secondary polymer linkages, with resultant depolymerization. Other environs may cause swelling or microcracking. And of course others may hydrolyze ester groupings or linkage. Extraction can occur in certain environments. In still other environments, repolymerization can occur, with resultant change in structure. Chain scission and a decrease in molecular weight can occur under certain conditions. Simple solvent action may occur. Also, absorption or attack at the interface between the glass and the resin will result in weakening.

All in all, the mechanisms involved are rather complicated and certainly not well understood. In addition to the combinations of various mechanisms that can occur at the same time, it is believed that synergism probably enters the picture in many instances also.

Quite fortunately, chemical attack on the reinforced polyesters and epoxies can be described, in military parlance, as a "go–no go" proposition. Attack on reinforced plastics in many of the ways mentioned above may occur in a relatively short time with improper environment. In reaching the extreme upper limit of its temperature range the phenomenon known as "blistering" may take nearly twelve months to be evident. This, however, is a physical, rather than a chemical phenomenon. Experience has indicated that if an installation has been soundly engineered and has operated satisfactorily for twelve months, the probability is good that operation will continue in a completely satisfactory manner for a substantial period of time. Fortunately, all this can be found out in advance with a well-conceived test program, so that the engineer's liability at any time can be suitably limited.

We are dealing here, of course, with only the high-grade chemical-resistant polyesters and epoxies. In severe chemical service these highly resistant resins will normally be the material of choice as opposed to either a general-purpose polyester or an isophthalic polyester.

2.11   A PRACTICAL TEST PROGRAM FOR PIPING SYSTEMS

The practical test program—coupons to big systems   A well-equipped laboratory can run virtually any test program required, but many companies will find it convenient to make simple tests themselves and let the fabricator supplement this with additional field services.   A practical program is suggested below, from conception to successful achievement and installation.   Any such program must have good guidelines, which must limit the financial involvement so that investment liability is commensurate with risk.   This assures the program of a continually sound footing.

Initial tests.   Start out with:

1. Test panels measuring 4 × 5 × ¼ in.
2. Piping sections measuring 2 in. in diameter by 4 in. long.   Make sure the raw edges are sealed.

The initial observations should be

1. Visual.
2. Strength tests (by vendor).
3. Ten hardness tests, using a Shore Durometer.   Discard the four outside ones.   Average the six remaining.
4. The panel or pipe fitting.

Expose the test samples under field conditions.   This is much better than submergence in a bottle at room temperature.   *Take special care to protect the field samples from loss.*

Expose the samples for a period of:

1. Thirty days
2. Sixty days
3. Ninety days
4. Six months
5. One year

Tests 1 to 3 can be visually checked and hardness-checked.   If the sample still looks good and hardness is holding up well, continue on. At 6 months and 1 year, add to the visual and hardness checks a weighing and a strength check by the vendor furnishing the sample.   (He will cut out rings for testing and furnish data on "before and after" strength comparisons.)

In short term, that is, 72 hr, boiling tests of the samples, a weight gain of more than 1 percent spells a poor product. High weight gains mean not only that the resin is absorbing moisture, but also that moisture is penetrating along the glass-fiber surface, which ultimately will destroy adhesion and strength. Nonporosity should be demanded in your specifications; it is an absolute necessity for success.

Test program: second stage. If after 90 days the checks look good, consider a "spool" program. This is a short (4 to 6 ft) section of pipe with flanges and a 90° ell on one end. This permits an evaluation of flange and fitting design. Run this test section for 6 to 9 months. Return the section to the vendor for physical evaluation after making your own visual and hardness checks.

One sure way to set up a "guaranteed to fail" program is to jam a spool of reinforced plastic pipe in the center of a long run of a lined steel pipe. The difference in expansion modulus and strength is so great that the spool of reinforced plastic pipe will attempt to operate as an expansion joint. Only ultimate failure can result. Considering the great difference in modulus of elasticity, it is not hard to see why. All the stresses in the metal system are transported into the short reinforced plastic spool. Failure under compression is the result.

Minor installation. Choose a small installation consisting of 50 to 100 ft of pipe with assorted fittings. Make sure this permits you to evaluate joining systems, adhesives, flanges, fittings, hanging, expansion, and anchoring requirements. Obtain joining cost data if possible. See how the field group handles it, and react to it. Evaluate the safety features. Permit it to run for a year. A good liner will not blister in heated service (above 180°F). Blistering is rarely seen in short-term tests, but tests of a year or more will pick it up. Naturally, blistering will destroy the resin-rich inner liner and permit destruction of the remaining structure. This is why heavy-duty liners of 40 to 60 mils, well compacted and air-free, have much to recommend them for severe service. Perhaps add another minor installation during this period for further evaluation.

Often, by this time, sufficient information has been accumulated so that we are ready to proceed to the last stage of the program.

Major installation and stocking program. Management may decide to limit the size of the first major installation by either a dollar value or by system productive capacity. This is sometimes done irrespective of the excellent information obtained in the foregoing testing program. It generally is done to limit financial liability in case difficulty occurs. The major installation needs to be worked out in conjunction with the vendor so that the best engineering knowledge is available. This type of approach has been used in the past successfully. It is conservative, but solid, and sometimes takes too long for the vendor selling the material. It might be shortened if historical data on a similar process are available for field inspection. Quite often inspection on a nonrelated or noncompetitive process can be arranged through the vendor. This has the advantage of permitting firsthand information to be obtained. A plant visit to observe a successful installation and for interviews with the engineers and field personnel who installed the equipment is invaluable. Here the good and bad can be determined. This approach is highly recommended.

The following are actual field examples of a simplified test program of this sort, conducted jointly by the purchaser and several vendors. Both of these test programs were carried on to long-term successful conclusions and resulted in major satisfactory installations.

EPOXY (FMC, American Viscose Division). Rings were tested for the purchaser. Rings were cut from 5-in.-long sections of 2-in. Chemline pipe which had been immersed in a spin-bath solution. Exposure duration was 240 days.

Test	Sample 1	Sample 2	Sample 3	Average
Stress, psi.....................	28,500	26,800	27,900	27,733
Stress retention, %*...........	97.6	91.8	95.5	94.9
Strength retention, %.........	104.7	97.4	100.2	100.8

* Retentions based on control rings with ultimate stress 29,220 psi, strength 2,845 lb.

POLYESTER. The following table shows resistance of chlorinated polyesters to rayon spin bath, after about 496 days (Nov. 10, 1964, to Mar. 21, 1966), giving average results of two samples.

	% weight change	% volumetric change	Barcol hardness	Flex strength	Flex modulus
Hetron 72........	+0.8	+1.2	100	96	100
Hetron 92........	+0.5	+1.2	91	87	97

Rayon spin bath may be said to consist of any solution of chemicals, lying within the following range:

5–10%.......... H_2SO_4
1– 8%.......... $ZnSO_4$
10–20%.......... Na_2SO_4
Saturated with H_2S and CS_2
Temperatures of 20–75°C

RESULT: Both laminates were very slightly discolored, 72 to a yellow, 92 to a green. There was slight wicking along edges. Resin-rich surfaces were excellent.

Nondestructive testing of reinforced plastics[5]   It is obvious that one of the greatest needs of the industry is to develop a nondestructive testing procedure for reinforced plastics. Research has been done along this line by ultrasonic means. The device used was a Sonoray ultrasonic thickness tester, Model S-30A, made by Branson Instruments, Stamford, Conn. This unit was equipped with a 2.5-megacycle transducer which gave reliable readings up to $\frac{3}{4}$-in. laminate. As the operator became experienced with the unit, it was even possible to identify different types of laminate construction. The instrument, with an experienced operator, can be used to show air voids in the laminate as well as to test for laminate thickness. Minute air bubbles do not affect the test.

Other testing techniques

1. On all tests the test specimens should be taken from the center of the sample to avoid end wicking.
2. When the sample is removed from the test, it should be washed in clean water and a Barcol hardness obtained. The sample should then be permitted to dry for 24 hr and the Barcol hardness repeated.
3. The examination of samples in a glow box affords a better look at the resin.

To be finite in our resistance specifications, strength tests on field immersed samples should show at least the following strength retention:[6]*

1 month............ 80%
4 months........... 70%
16 months.......... 60%

In addition to meeting the strength test, the sample should not suffer an appreciable hardness loss, become "punky," blister, or show evidences of resin erosion, checking, or "blushing."

This type of practical test program is equally applicable to ductwork, tanks, and other process vessels constructed from RP material.†

---

* For further elaboration on testing programs, see Appendix A.
† Mr. R. M. Jackman, Manager, Research and Development, A. O. Smith Corporation, Smith Plastics Division, in a personal communication dated May 1, 1968, suggests the following criteria:

1 month................. 90% retention (no recommendation made at this short time interval)
3 months................ 75% retention

2.12   SELECTING THE PIPE AND THE VENDOR

In plant maintenance, reinforced plastic products are used extensively in plant sewers, process circulating systems, condensate lines, softwater piping where iron contamination would endanger product quality, bleach systems, and many other areas.

Criteria for selecting piping[7]   The important elements to be considered in a piping installation where corrosive liquids are being handled are:

- Strength and corrosion resistance
- Cost of a given type of pipe
- Cost of fittings
- Insulation
- General ease of installation
- Joining costs
- Ability of the piping installation to provide uninterrupted service
- Expected service life of the piping system
- Ease of modification and repair
- Delivery and field inventory
- Relative safety

Strength and corrosion resistance   A pipe of sufficient strength is required to withstand operating pressures and physical abuse while providing maximum resistance to corrosion. These vital qualities cannot be separated in glass-reinforced plastic pipe. The component materials used in the formulation of the pipe and the method of fabrication determine both the strength and degree of corrosion resistance that the finished pipe will have.

A number of good chemical-resistant polyester and epoxy resins lend themselves well to the fabrication of chemical equipment. The choice of the particular resin depends primarily upon the conditions to be encountered. A bisphenol-A fumarate resin or chlorinated polyester represents a good choice for oxidizing acids. Each type of resin selection

---

6 months.................. 65% retention
12 months................. 57% retention

An SPI task group is currently reviewing the correlation between ID-exposed test samples only and those fully exposed. Longer test samples are being used, and the trend is going from 3 to 5 in. Each manufacturer should establish his own acceptable criterion, based on the leveling off of the strength-loss curve. In lieu of this, a 50 percent retention after 1 year is a proposed tentative minimum.

may be optimized to handle the particular chemicals in question. This does not mean, however, that other resins in the field should be neglected. Testing programs should be continually developed with other resins from reputable manufacturers.

There has been much discussion on the glass/resin ratio used in pipe construction. Some vendors push this ratio up to 75 to 85 percent glass with 15 to 25 percent resin. With this formulation they are able to build a tremendously strong pipe, reduce the wall thickness, and sell their product at a cost below that of some other types of RP pipe.

Evaluate all grades and types of glass-reinforced epoxy and polyester pipe, from those with extremely high glass/resin ratios to those with the more moderate ratios. It is presently believed that the maximum corrosion and strength characteristics are obtained with glass/resin ratios of 25 to 40 percent glass to 60 to 75 percent resin. *The corrosion resistance of any reinforced plastic pipe is created by the resin, and the strength by the glass.* Pipe made according to these specifications still has a safety factor of 10:1.

Some tests made by responsible investigators have shown that, in extremely corrosive conditions, structures made with a ratio of 75 percent glass and 25 percent resin lose a great deal of their strength in a relatively short time, so that they rapidly become weaker than the 25 to 40 percent glass and 60 to 75 percent resin laminates. Recent manufacturing advances, however, in winding techniques, resin impregnation, and construction methods seem to have improved this picture. Tests by the author show excellent strength retention in high-glass laminates with field tests for periods up to 3 years covered by the tests and for corrosive conditions. This is accomplished by a composite, or two-part, laminate, consisting of a surfacing system followed by a structural system. Nearly all vendors agree with this in principle, but their methods of attaining it may vary widely. For details of the various methods used by vendors, the reader is referred to Chaps. 3 and 4. In field experience, if the E grade glass is exposed in any appreciable degree to a corrosive solution, a high-glass-content pipe will rapidly disintegrate.[8]

Tests have shown that high-glass-content pipe fail in several months when the glass fibers became exposed. Glass reinforcement in pipe must always be liberally covered with resin to protect it from reagents. Industry could well use a high-strength chemical-grade glass fiber as the reinforcing material for RP pipe, and some investigative work is now being done in this area by persons who recognize that the problem exists.

Where glass-reinforced piping is completely submerged in corrosive solutions, particular care should be given to laminate construction. High-glass–resin pipe (75:25 percent) made from continuous filaments will not perform well unless such a pipe is specifically engineered with heavy resin protection inside and out.   Some products are so engineered. A low-glass–resin pipe (25:75 percent) will perform exceptionally well under these conditions because sufficient protective barriers have been built into the pipe through the medium of the chopped-strand mat so that total submergence does little harm.

The use of other reinforcements, of course, shows a different set of criteria.   For example, a polypropylene-reinforced polyester will largely eliminate the type of failures experienced with the E glass even when there is exposure to a corrosive solution, provided the polypropylene is completely resistant to the corrosive solution.[9]   C glass will also generally provide improved corrosion resistance.   Pipe of poly-propylene-reinforced plastic does not have as great a physical strength as pipe of E glass–reinforced plastic.

But, as in so many other fields, there are gray areas.   The degree of attack is often related to the solution temperature.   In acidic solutions where high temperatures (above 180°F) exist, a completely different type of problem is encountered.   In high-temperature solutions (because both the epoxies and polyesters are excellent insulators), the inside of the pipe wall is trying desperately to expand while the outside of the pipe wall, being relatively cool, is doing its utmost to limit the expansion. In a pipe of moderate glass/resin ratio of, let us say, 25% glass/75% resin, the physical strains set up within the pipe wall may be sufficient to produce piping failure in the small pipe sizes, such as 2, 3, and 4 in.

Under a stereomicroscope, an examination of the resin-rich inner surface reveals that failure has occurred through intensive internal stress, even though no chemical attack on the resin has taken place.   As the pipe sizes are increased to 6 in. and above, this phenomenon (even in glass/resin ratios of 25:75) no longer occurs.   Nor is it observed in the smaller pipe sizes of 2 to 4 in. when the glass/resin ratio is increased to 40% glass/60% resin and higher.   This indicates that optimum materials of piping construction in high-temperature systems above 180°F should be a pipe of relatively high glass/resin ratio or a composite system, especially in the small sizes.   This can be extended to tank nozzles where the contents of the tank are extremely hot.   It is feasible for a fabricator to adjust the glass/resin ratio in this type of construction to obtain the desired results.

Buying RP pipe   But, as in so many other fields of endeavor, experience and learning may moderate conclusions.   In the field of RP piping most problems occur in the joining systems or in the fittings, and rarely in the pipe.   Emphasis on dependability of assembly is therefore most important.   Each major type of flanging and joining system has its enthusiastic supporters and predominant areas of application. Safety also enters the picture with hazardous solutions and areas. The engineer should specify the most dependable assembly in critical areas as there can be no compromise with the safety of the individual. For service of a less rigorous nature, such as gravity conditions, and with a cost evaluation, conclusions may be different. Flanges and fittings made-on in the factory are less expensive than those in the field and provide high reliability.

2.13 POLYESTER PUTTY—A FILLING MATERIAL[10]

Quite often in the fabrication and installation of ductwork, piping, tanks, and polyester structures, it is necessary to fill voids in fittings, cracks, or crevices, to form a smooth structure. This is commonly done with polyester putty, which is generally referred to as "talc" by workers in the fabricating plants. It is made by adding CAB-O-SIL to an uncatalyzed resin until the desired consistency is reached. The mixture is then catalyzed with methyl ethyl ketone peroxide (MEKPO) on the same basis as an equivalent amount of resin would be. The putty is excellent for filling voids and forming shapes. It should be used with care, however, because it has little intrinsic strength. If used in any mass, it should be covered later with fiber glass. Since it is quite brittle, it will develop stress cracks when curing if used alone.

The addition of a very short fiber glass to the polyester putty or resin tends to reinforce it and make it considerably stronger. A putty such as this may be purchased from a fabricator or prepared on a "do-it-yourself" basis.

2.14 BASIC RECOMMENDED PRODUCT STANDARD, TS-122C, WITH COMMENTS[11]

The Society of the Plastics Industry (SPI), through its membership, has worked aggressively to standardize reinforced polyester chemical-resistant process equipment. At the present time they have issued a proposed Product Standard for Custom Contact Molded Reinforced Polyester Chemical Resistant Process Equipment, known as TS-122C. It is proposed to make this the Commercial Standard in the near future, to be issued by the U.S. Department of Commerce. This Standard will be quoted from time to time throughout the book. It is essential that the engineer thoroughly understand its meaning and know the areas it covers.

The Standard covers the following basic areas:

1. Laminate lay-up
2. The design requirements which may be expected of reinforced polyester laminates
3. Round and rectangular ducts, with standardization of ducts and allowable vacuum and pressure ratings
4. Piping, including sizes, fitting standardization, and wall thickness versus pressure
5. Overlays for butt joints
6. Flange specifications
7. Recommended piping-installation practice
8. Fastening and gaskets
9. Reinforced polyester tanks
    a. Cylindrical, both vertical and horizontal, with suggested wall thicknesses for various-sized tanks
    b. Rectangular
10. Test methods

For the purpose of immediate reference and clarification the sections of the Standard will be quoted and used in the applicable chapter. For example, that which refers to piping will be found in Chap. 3; that which refers to duct systems will be found under Chap. 11; etc. In this manner the reader will have an immediate reference, with additional suggestions by the author on how the Standard may be applied. The reader should be further aware that, for the most part the Standard has been prepared by the fabricators themselves, and in general not by the end user of the equipment. The Standard, of course, cannot meet every situation; so the author, through additional written material, has supplemented the Standard to provide guidelines based on chemical-plant experience, to widen the horizon of application. The Standard does not guide the engineer in the things to avoid—this the author will attempt to do.

Except by a most loosely applied interpretation, the Standard does not cover filament-wound piping, ductwork, or equipment. No doubt this will be rectified in time since it is fairly apparent that future progress will combine the best features of custom-contact-molded and the filament-wound equipment to produce a composite structure, amplifying the most desirable qualities of each method.

To guide the reader on filament-wound equipment, suggested standards have been prepared by an authority, and by permission are reproduced in Appendix D.

The Recommended Product Standard, TS-122C, however, is of utmost importance, since this is the standard by which most of the reinforced polyester chemical-resistant process equipment is built. At least, the engineer should meet this standard. There will be occasions when he will wish to exceed it in order to meet a particular condition.

TS 122C

September 18, 1968

(Replaces TS 122B)

RECOMMENDED PRODUCT STANDARD FOR

CUSTOM CONTACT-MOLDED REINFORCED POLYESTER

CHEMICAL-RESISTANT PROCESS EQUIPMENT

(Based on an original proposal submitted to the National Bureau of Standards, U.S. Department of Commerce by the Society of the Plastics Industry, Inc., for development under the Product Standards Procedures and adjusted in accordance with comment from the trade.)

1. PURPOSE

1.1   The purpose of this Product Standard is to establish on a national basis the standard sizes and dimensions and significant quality requirements for commercially available glass-fiber-reinforced chemical-resistant process equipment for chemical service. The information contained in this Product Standard will be helpful to producers, distributors, and users, and will promote understanding between buyers and sellers.

2. SCOPE

2.1   This Standard covers materials, construction and workmanship, physical properties and methods of testing reinforced polyester materials for process equipment and auxiliaries intended for use in aggressive chemical environments, including but not limited to pipe, ducts and tanks. An identifying hallmark and recommended statement of compliance are included.

2.2   This Standard is based on the technology of fabrication by hand lay-up or contact pressure molding.

2.3   Work is in progress by the industry to prepare standards covering other resins, reinforcing materials, laminate constructions, fabrication methods, and design considerations that are not covered by this Standard.

3. REQUIREMENTS

3.1   *General*

3.1.1   *Terminology*—Unless otherwise indicated, the plastics terminology used in this Standard shall be in accordance with the definitions given in ASTM Designation D883-64bT, Tentative Nomenclature Relating to Plastics.*

3.1.2   *General description*—This Standard describes glass-fiber-reinforced process equipment for chemical service. Other materials may be used for reinforcement of the surface exposed to the chemical environment.

* Later issues of the ASTM publication may be used providing the requirements are applicable and consistent with the issue designated. Copies of ASTM publications are obtainable from the American Society for Testing and Materials, 1916 Race Street, Philadelphia, Pa. 19103.

This standard is not intended to cover selection of the exact resin or reinforcement combination for use in specific chemical and structural conditions. For recommended chemical resistance test procedures, see Appendix A.

3.2   *Materials*

3.2.1   *Resin*—The resin used shall be of a commercial grade and shall be evaluated as a laminate by test (see Appendix A for a recommended test) or previous service to be acceptable for the environment.

3.2.2   *Fillers and pigments*—The resins used shall not contain fillers except as required for viscosity control or fire retardance. Up to 5 percent by weight of thixotropic agent which will not interfere with visual inspection may be added to the resin for viscosity control. Resins may contain pigments and dyes by agreement between fabricator and purchaser, recognizing that such additions may interfere with visual inspection of laminate quality. Antimony compounds or other fire retardant agents may be added as required for improved fire resistance.

3.2.3   *Reinforcing material*—The reinforcing material shall be a commercial grade of glass fiber having a coupling agent which will provide a suitable bond between the glass reinforcement and the resin.

3.2.4   *Surfacing materials*—Unless otherwise agreed upon between fabricator and purchaser, material used as reinforcing on the surface exposed to chemical attack shall be a commercial grade chemical-resistant glass having a coupling agent.

NOTE: The use of other fibrous materials such as acrylic and polyester fibers and asbestos may affect the values obtained for the Barcol hardness of the surface.

3.3   *Laminate*—The laminate shall consist of an inner surface, an interior layer, and an exterior layer or laminate body. The compositions specified for the inner surface and interior layer are intended to achieve optimum chemical resistance.

3.3.1   *Inner surface*—The inner surface shall be free of cracks and crazing with a smooth finish and with an average of not over 2 pits per square foot, providing the pits are less than $\frac{1}{8}$ inch diameter and not over $\frac{1}{32}$ inch deep and are covered with sufficient resin to avoid exposure of inner surface fabric. Some waviness is permissible as long as the surface is smooth and free of pits. Between 0.010 and 0.020 inch of reinforced resin-rich surface shall be provided.* This surface may be reinforced with glass surfacing mat, synthetic fibers, asbestos or other material as usage requires.

3.3.2   *Interior layer*—A minimum of 0.100 inch of the laminate next to the inner surface shall be reinforced with not less than 20 percent nor more than 30 percent by weight of non-continuous glass strands (see 4.3.1), e.g., having fiber lengths from 0.5 to 2.0 inches.

---

* This resin-rich surface layer will usually contain less than 20 percent of reinforcing material. A specific limit is not included because of the impracticability of determining this value in the finished product.

3.3.3   *Exterior layer*—The exterior layer or body of the laminate shall be of chemically resistant construction suitable for the service and providing the additional strength necessary to meet the tensile and flexural requirements. Where separate layers such as mat, cloth or woven roving are used, all layers shall be lapped a minimum of one inch. Laps shall be staggered as much as possible. If woven roving or cloth is used, a layer of chopped strand glass shall be placed as alternate layers. The exterior surface shall be relatively smooth with no exposed fibers or sharp projections. Hand work finish is acceptable, but enough resin shall be present to prevent fiber show.

     3.3.3.1   When the outer surface is subject to a corrosive environment the exterior surface shall consist of a chopped strand glass over which shall be applied a resin-rich coating as described in 3.3.1. Other methods of surface protection may be used as agreed upon between buyer and seller.

     3.3.4   *Cut edges*—All cut edges shall be coated with resin so that no glass fibers are exposed and all voids filled. Structural elements having edges exposed to the chemical environment shall be made with chopped strand glass reinforcement only.

     3.3.5   *Joints*—Finished joints shall be built up in successive layers and be as strong as the pieces being joined and as crevice free as is commercially practicable. The width of the first layer shall be 2 inches minimum. Successive layers shall increase uniformly to provide the specified minimum total width of overlay which shall be centered on the joint (see 3.3.1, 3.4.6.1, 3.5.6, and 3.6.5). Crevices between jointed pieces shall be filled with resin or thixotropic resin paste, leaving a smooth inner surface (see 3.3.1). The interior of joints may also be sealed by covering with not less than 0.100 inch of reinforced resin-rich surface as described in 3.3.1 and 3.3.2.

     3.3.6   *Wall thickness*—The minimum wall thickness shall be as specified in the tables under the appropriate sections, but in no case less than $\frac{1}{8}$ inch in the case of duct and $\frac{3}{16}$ inch in pipes and tanks regardless of operating conditions. Isolated small spots may be as thin as 80 percent of the minimum wall thickness, but in no case more than $\frac{1}{8}$ inch below the specified wall thickness.

     3.3.7   *Mechanical properties*—In order to establish proper wall thickness and other design characteristics, the minimum physical properties for any laminate shall be as shown in Table 1 and 3.3.7.1. Laminates which do not meet the minimum values of Table 1 are considered acceptable provided they are made to afford the same overall strength that would be obtained with a laminate meeting the specified thickness. For example, if the specified thickness for a laminate is $\frac{1}{4}$ inch, reading from Table 1 a minimum tensile strength of 12,000 psi is required. By multiplying thickness times minimum tensile strength a value of 3,000 lbs. breaking load for a 1 inch wide specimen is obtained. A laminate having a tensile strength of 10,000 psi will therefore be acceptable for the $\frac{1}{4}$ inch requirement if it has an actual thickness of at least 0.3 inch.

REQUIREMENTS FOR PROPERTIES OF REINFORCED–POLYESTER LAMINATES

Table 1

Property at 23°C (73.4°F)	Thickness, inches			
	⅛ to 3⁄16, psi	¼, psi	5⁄16, psi	⅜ and up, psi
Ultimate tensile strength, minimum[1]	9,000	12,000	13,500	15,000
Flexural strength, minimum[2]	16,000	19,000	20,000	22,000
Flexural modulus of elasticity (tangent), minimum[3]	700,000	800,000	900,000	1,000,000

[1] See 4.3.2.
[2] See 4.3.3.
[3] See 4.3.4.

3.3.7.1 *Surface hardness*—The laminate shall have a Barcol hardness of at least 90 percent of the resin manufacturer's minimum specified hardness for the cured resin when tested in accordance with 4.3.5. This applies to both interior and exterior surfaces.

3.3.8   *Appearance*—The finished laminate shall be as free as commercially practicable from visual defects such as foreign inclusions, dry spots, air bubbles, pinholes, pimples, and delamination.

3.3.9   By agreement between buyer and seller, a representative laminate sample may be used for determination of acceptable surface finish and visual defects (see 3.3.1, 3.3.3, and 3.3.8).

The need for commercial standards cannot be overemphasized. Such standards are basically necessary for communication, design, education, general sales, and for using specifications with confidence. In addition to the current Recommended Product Standard, other ASTM and SPI committees individually or jointly are working on additional systems for pipe, fittings, fire retardancy, flanges, etc. The present Recommended Product Standard can be considered only a beginning. The grouping of all the necessary standards which will ultimately have to be produced to cover the many diverse applications of hand-laid-up reinforced plastics indicates how inadequate one small publication must be. These new standards are a cooperative effort of manufacturers, fabricators, consumers, and consultants, to solve common problems. Out of sheer necessity similar standards must follow, to cover filament winding, the epoxies, vinyl esters, and any of the other thermosets which are presently looming on the horizon and awaiting aggressive development.

REFERENCES

[1] Whitehouse, A. A. K.: Glass Fibre Reinforced Polyester Resins, paper 10, Symposium on Plastics and the Mechanical Engineer, London, Oct. 7–8, 1964.

[2] Szymanski, W. A., Hooker Chemical Corp., Durez Div., North Tonawanda, N.Y., personal communication, Apr. 24, 1967.

[3] The results of corrosion testing of the various resins were gathered from a number of sources and represent a wide compilation of material. Source material came from:

Amercoat Corp., Brea, Calif.
American Cyanamid, Wallingford, Conn.
Atlas Chemical Industries, Wilmington, Del.
Ceilcote Corp., Berea, Ohio
Dow Chemical Co., Midland, Mich.
DowSmith Co. (now Smith Plastics), Little Rock, Ark.
Fibercast Corp., Sand Springs, Okla.
Heil Corp., Cleveland, Ohio
Hooker Chemical Corp., Durez Div., North Tonawanda, N.Y.
Interchemical Corp., Clifton, N.J.
Reichhold Chemicals, Inc., White Plains, N.Y.

[4] Szymanski, W. A., Hooker Chemical Corp., Durez Div., North Tonawanda, N.Y., personal communication, Jan. 17, 1967.

[5] Talbot, R.: Paper presented at National Association of Corrosion Engineers, Niagara Falls, N.Y., Aug. 8–9, 1967.

[6] Kelly, M. E., Jr., A. O. Smith Corp., Smith Plastics Div., Little Rock, Ark., personal communication, Oct. 7, 1966.

[7] Mallinson, J. H.: Reinforced Plastic Pipe: A User's Experience, *Chemical Engineering*, Dec. 20, 1965.

[8] Atkinson, H. E.: Reinforced Plastics for Chemical Process Equipment, paper presented at annual meeting of the ASME, Nov. 26–Dec. 1, 1961, *Mechanical Engineering*, vol. 84, no. 6, pp. 62–65, June, 1962.

[9] *Haveg Product Information Bulletin*, November, 1965, Haveg Industries, Inc., Wilmington, Del.

[10] Whiteman, R. L., Carolina Fiberglass Products Co., Wilson, N.C., personal communication, Dec. 12, 1966.

[11] Recommended Product Standard for Custom Contact Molded Reinforced Polyester Chemical Resistant Process Equipment, TS-122C, Sept. 18, 1968.

3

Polyester Piping Systems

3.1 LAMINATE CONSTRUCTION[1]

In field applications it will be found that sound laminate construction is vital to pipe strength and corrosion resistance. In glass-reinforced polyester pipe, the chemical-resistant laminate construction illustrated in Fig. 3-1 starts with an all-important 10- to 20-mil resin-rich interior surface that creates a corrosionproof barrier, reinforced with chemical-grade glass mat or organic veiling. Successive layers of resin-saturated rein-

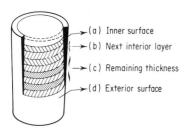

FIG. 3-1 Typical hand-laid-up polyester pipe construction. (*a*) Inner surface: 10- to 20- mil resin-rich inner liner reinforced with C grade surfacing mat or organic veiling. (*b*) Next interior layer: Generally about 100 mils thick consisting of at least two layers of 1½-oz mat reinforcement, 25 to 30 percent glass by weight. (*c*) Remaining thickness: Varies with the laminate strength required; may be additional 1½-oz mat reinforcement, cloth, or woven roving. In heavy construction requiring more than one layer of woven roving, 1½-oz mat is used between layers of roving. Last layer of roving is always covered with 1½-oz mat. (*d*) Exterior surface: Resin-rich surface reinforced with C grade surfacing mat plus an ultraviolet inhibitor.

forcement are then applied to produce the desired thickness and strength. Where pipe-design specifications call for glass cloth, woven roving, or similar materials, a minimum of two layers of $1\frac{1}{2}$-oz chopped-strand mat with a high-solubility binder is used between the inside shell and the first layer of major reinforcement. The outer layer is covered with a resin-saturated glass mat that resists fumes, spillage, and weathering.

Before we leave the subject of corrosion resistance, a point or two should be made concerning the exterior of RP pipe. One of the constant hazards in chemical-plant operations is the alarming number of failures that occur from the outside in. Thus, to be truly corrosionproof, pipe must be resistant both inside and out—and reinforced plastics are.

To provide for the proper cure-out of polyester piping (bisphenol-A fumarate resin), many vendors prevent air inhibition by adding a paraffin wax to the final coat. This produces a beautiful glossy appearance, but may cause subsequent trouble in the development of high wrapped-joint reliability. The removal of this wax is difficult except by using solvents, which all too often are dangerous in themselves. An attempt to sand to provide the proper surface for a wrapped joint only smears the wax and drives it deeper into the pipe. One of the best answers is to eliminate the wax in the final coat and provide a finish coat with one of the glycol–phthalic acid–chlorine additive corrosion-resistant resins. Pipe made up exclusively from the chlorinated resins generally do not suffer from air-inhibition problems. Air inhibition is related to the level of the exotherm and the thickness of the piping wall. High-exotherm resins in general are not troubled with air-inhibition problems. Usually, air inhibition is found in thin laminates such as $\frac{1}{8}$- to $\frac{3}{16}$-in., and less frequently, in heavier laminates, simply because a higher exotherm will develop with the thicker laminates.

In the fabricating shop there are at least four methods by which piping construction may be accomplished:

By one method a resin-rich reinforced interior surface is first laid on a rotating Mylar-covered mandrel. This is followed by successive layers of glass mat or cloth wound at an angle on the mandrel over the slick interior shell to provide a uniform, dependable pipe. Glass fabric or woven roving is wrapped on the mandrel and is immediately saturated with the appropriate chemical-resistant resin to furnish the proper design specifications. Such a pipe is commonly designed and built to handle operating temperatures up to 250°F and pressures to 150 psig. The vendor may fabricate it in lengths of 12, 20, or as long as 40 ft and in diameters from 2 to 60 in. Glass/resin ratios in this type of hand-laid-up pipe normally run about 25% glass/75% resin. A good stock of hand-laid-up polyester pipe is shown in Fig. 3-2.

A second method[2] commonly applied by many vendors also uses a

FIG. 3-2 A well-stocked selection of reinforced polyester pipe stacked on racks in open-field storage.   Well-manufactured pipe of this type is immune to ultra-violet degradation.

Mylar-covered steel mandrel, but by a different method of application, in which the mandrel is intermittently rotating.   A layer of C glass lapped at the joint is applied first, soaked with resin, and rolled out. Before this layer hardens, chopped-strand mat is fed from a roll onto the mandrel simultaneously with resin.   After two revolutions, woven roving is fed to the mandrel, together with the mat and resin (if woven roving is to be used).   This is continued until the required number of layers have been applied, the glass being cut to provide a lap over its starting end.   The wet laminate is thoroughly rolled out to remove air bubbles and then left on the still rotating mandrel until it has hardened.   Radiant heat is applied if necessary.   The mandrel is then removed from the machine, and pipe pulled longitudinally from it.   The pipe ends are cut off to leave a plane square end, the Mylar removed from the interior, and the pipe stored until shipment.   Pipe is not shipped until the Barcol hardness (35 to 40) reaches a value indicating that the laminate is cured. Laminates are made in varying strengths to suit the pressure application desired.   This pipe will also run, approximately, 25% glass/75% resin.

A third method involves seamless reinforced plastic tubing centrifugally cast by an automated process but generally manufactured in short lengths of 7 to 10 ft.   This process uses the same corrosion-resistant resins, but does not employ woven roving in its construction.   In general, roving adds increased impact resistance in piping construction, although piping made without woven roving will still test to satisfactory minimum design strengths.

A fourth process uses continuous glass filaments helically interwoven under tension around a polished mandrel. At the same time each filament is thoroughly coated with a thermosetting polyester resin.   The pipe is generally heat-cured to form a composite pipe of polyester-epoxy or epoxy only.   Good practice insists that such a filament-wound pipe be equipped with an inner corrosion barrier of 20 to 60 mils as a minimum. To this is quite often added one or two layers of $1\frac{1}{2}$-oz random-strand mat.   Continuous filament winding then is employed to provide strength in the wall.   The pipe is finished with an external corrosion barrier.

3.2   PRODUCT STANDARD RECOMMENDED
PURCHASING SPECIFICATIONS

Many of the fabricators of reinforced plastic piping have standardized on the Recommended Product Standard for Custom Contact Molded Reinforced Polyester Chemical Resistant Process Equipment[3] (TS-122C, Sept. 18, 1968) of the Society of the Plastics Industry (SPI) on Piping. This Commercial Standard, as it applies to piping, should be thoroughly understood by the application engineer.

When using and specifying custom-fabricated polyester corrosion-resistant industrial equipment, it is advisable to refer to the Commercial Standard.   Following an initial reference to this Standard, the engineer may then refer to the tables and graphs furnished with this book which serve to amplify the Standard and permit wider usage under conditions which may deviate from those imposed by the Standard.   It will be readily observed that the Standard is limited to hand-laid-up custom-fabricated piping, ductwork, and tanks.   It is not intended to cover filament-wound custom-fabricated equipment, centrifugally cast equipment, or in general, reinforced epoxy equipment made by a multitude of processes.   In other sections of this book will be found guidance to selection of custom piping and equipment made by these other various processes.   Nor does the Commercial Standard intend to cover those pieces of custom-fabricated piping and equipment in which reinforcement other than glass is used, except where such reinforcement may be used as an inner liner or veiling and which contributes solely to the chemical resistance of the part in question and not specifically to the physical strength.

There is, however, one catchall paragraph in the Standard which can be interpreted sufficiently broadly to cover nearly any acceptable specification. See sec. 3.3.7, Mechanical Properties, in the Product Standard, given in Chap. 2.

AUTHOR'S NOTE: The following excerpt is the section dealing with reinforced polyester pipe and is taken from:

RECOMMENDED PRODUCT STANDARD FOR

CUSTOM CONTACT-MOLDED REINFORCED POLYESTER

CHEMICAL-RESISTANT PROCESS EQUIPMENT

TS 122C

September 18, 1968

3.5   *Reinforced-polyester pipe**

3.5.1   *Size*—The standard pipe size shall be the inside diameter in inches. Standard sizes are 2, 3, 4, 6, 8, 10, 12, 14, 16, 18, 20, 24, 30, 36, and 42 inches. The tolerance including out-of-roundness shall be $\pm \frac{1}{16}$ inch for pipe up to and including 6 inch inside diameter, and $\pm \frac{1}{8}$ inch or ± 1 percent, whichever is greater, for pipe exceeding 6 inches in inside diameter. This measurement shall be made at the point of manufacture with the pipe in an unstrained vertical position.

3.5.2   *Length*—The length of each fabricated piece of pipe shall not vary more than $\pm \frac{1}{8}$ inch from the ordered length unless arrangements are made to allow for trim in the field.

3.5.3   *Wall thickness*—The minimum wall thickness of the pipe shall be in accordance with Table 3. See also 3.3.6.

3.5.4   *Squareness of ends*—All unflanged pipe shall be cut square with the axis of the pipe within $\pm \frac{1}{8}$ inch up to and including 24 inches diameter and to within $\pm \frac{3}{16}$ inch for all diameters above 24 inches.

3.5.5   *Fittings*—All fittings such as elbows, laterals, T's, and reducers shall be equal or superior in strength to the adjacent pipe section and shall have the same diameter as the adjacent pipe. The dimensions of fittings shall be as shown in Figure 1. Tolerance on angles of fittings shall be $\pm 1°$ through 24 inches in diameter and $\pm \frac{1}{2}°$ for 30 inches diameter and above. Where necessary, minimum overlay widths may be less than those specified in Table 4, but the joint strength shall be at least equal to the strength of the adjacent pipe.

3.5.5.1   *Elbows*—Standard elbows shall have a centerline radius of one and one-half times the diameter. Standard elbows up to and including 24 inches shall be molded of one piece construction. Elbows of 30 inches diameter and larger may be of mitered construction using pipe for the mitered sections. The width of the overlay on the mitered joint may have to be less than the minimum specified in Table 4 to avoid interference on the inner radius, but the joint strength must be at least equal to the strength of the adjacent pipe. Mitered elbows 45° or less will be one-miter, two

* Rated from full vacuum to 150 psi (see Table 3).

REINFORCED–POLYESTER PIPE WALL THICKNESS

Table 3

Pipe size, inches	Minimum pipe wall thicknesses,[1] at pressure ratings:					
	25 psi, inches	50 psi, inches	75 psi, inches	100 psi, inches	125 psi, inches	150 psi, inches
2	$\frac{3}{16}$	$\frac{3}{16}$	$\frac{3}{16}$	$\frac{3}{16}$	$\frac{3}{16}$	$\frac{3}{16}$
3	$\frac{3}{16}$	$\frac{3}{16}$	$\frac{3}{16}$	$\frac{3}{16}$	$\frac{1}{4}$	$\frac{1}{4}$
4	$\frac{3}{16}$	$\frac{3}{16}$	$\frac{3}{16}$	$\frac{1}{4}$	$\frac{1}{4}$	$\frac{1}{4}$
6	$\frac{3}{16}$	$\frac{3}{16}$	$\frac{1}{4}$	$\frac{1}{4}$	$\frac{5}{16}$	$\frac{3}{8}$
8	$\frac{3}{16}$	$\frac{1}{4}$	$\frac{1}{4}$	$\frac{5}{16}$	$\frac{3}{8}$	$\frac{7}{16}$
10	$\frac{3}{16}$	$\frac{1}{4}$	$\frac{5}{16}$	$\frac{3}{8}$	$\frac{7}{16}$	$\frac{1}{2}$
12	$\frac{3}{16}$	$\frac{1}{4}$	$\frac{3}{8}$	$\frac{7}{16}$	$\frac{1}{2}$	$\frac{5}{8}$
14	$\frac{1}{4}$	$\frac{5}{16}$	$\frac{3}{8}$	$\frac{1}{2}$	$\frac{5}{8}$	$\frac{3}{4}$
16	$\frac{1}{4}$	$\frac{5}{16}$	$\frac{7}{16}$	$\frac{9}{16}$	$\frac{11}{16}$	
18	$\frac{1}{4}$	$\frac{3}{8}$	$\frac{1}{2}$	$\frac{5}{8}$	$\frac{3}{4}$	
20	$\frac{1}{4}$	$\frac{3}{8}$	$\frac{1}{2}$	$\frac{11}{16}$		
24	$\frac{1}{4}$	$\frac{7}{16}$	$\frac{5}{8}$	$\frac{13}{16}$		
30	$\frac{5}{16}$	$\frac{1}{2}$	$\frac{3}{4}$			
36	$\frac{3}{8}$	$\frac{5}{8}$				
42	$\frac{3}{8}$	$\frac{3}{4}$				

[1] The specified wall thicknesses are based upon a 10 to 1 safety factor for the tensile strength listed in Table 1. These ratings are suitable for use up to 180°F (82.2°C). For ratings at higher temperatures consult the manufacturer. For vacuum service see 3.5.9.

section. Elbows above 45° through 90° shall have a minimum of two miters. It will be permissible to incorporate a straight pipe extension on elbows.

3.5.5.2 *Reducers*—Reducers of either concentric or eccentric style will have a length as determined by the diameter of the large end of the reducer as indicated in Figure 1.

3.5.6 *Butt joints*—This type of joint shall be considered the standard

MINIMUM TOTAL WIDTHS OF OVERLAYS FOR REINFORCED–POLYESTER BUTT JOINTS

Table 4

Pipe wall thickness, inches	$\frac{3}{16}$	$\frac{1}{4}$	$\frac{5}{16}$	$\frac{3}{8}$	$\frac{7}{16}$	$\frac{1}{2}$	$\frac{9}{16}$	$\frac{5}{8}$	$\frac{11}{16}$	$\frac{3}{4}$
Minimum total width of overlay, inches	3	4	5	6	7	8	9	10	11	12

MINIMUM FLANGE THICKNESS FOR REINFORCED–POLYESTER PRESSURE PIPE[1,2,3]

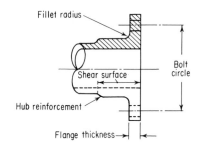

Fillet radius

Bolt circle

Shear surface

Hub reinforcement

Flange thickness

Table 5

Pipe size, inches	Minimum flange thickness at design pressures:					
	25 psi, inches	50 psi, inches	75 psi, inches	100 psi, inches	125 psi, inches	150 psi, inches
2	$\frac{1}{2}$	$\frac{1}{2}$	$\frac{1}{2}$	$\frac{9}{16}$	$\frac{5}{8}$	$\frac{11}{16}$
3	$\frac{1}{2}$	$\frac{1}{2}$	$\frac{5}{8}$	$\frac{11}{16}$	$\frac{3}{4}$	$\frac{13}{16}$
4	$\frac{1}{2}$	$\frac{9}{16}$	$\frac{11}{16}$	$\frac{13}{16}$	$\frac{7}{8}$	$\frac{15}{16}$
6	$\frac{1}{2}$	$\frac{5}{8}$	$\frac{3}{4}$	$\frac{7}{8}$	1	$1\frac{1}{16}$
8	$\frac{9}{16}$	$\frac{3}{4}$	$\frac{7}{8}$	1	$1\frac{1}{8}$	$1\frac{1}{4}$
10	$\frac{11}{16}$	$\frac{7}{8}$	$1\frac{1}{16}$	$1\frac{3}{16}$	$1\frac{5}{16}$	$1\frac{7}{16}$
12	$\frac{3}{4}$	1	$1\frac{1}{4}$	$1\frac{7}{16}$	$1\frac{5}{8}$	$1\frac{3}{4}$
14	$1\frac{3}{16}$	$1\frac{1}{16}$	$1\frac{5}{16}$	$1\frac{1}{2}$	$1\frac{3}{4}$	$1\frac{7}{8}$
16	$\frac{7}{8}$	$1\frac{3}{16}$	$1\frac{7}{16}$	$1\frac{5}{8}$	$1\frac{7}{8}$	
18	$1\frac{5}{16}$	$1\frac{1}{4}$	$1\frac{1}{2}$	$1\frac{3}{4}$	2	
20	1	$1\frac{5}{16}$	$1\frac{5}{8}$	$1\frac{7}{8}$		
24	$1\frac{1}{8}$	$1\frac{1}{2}$	$1\frac{7}{8}$			
30	$1\frac{3}{8}$	$1\frac{7}{8}$				
36	$1\frac{3}{4}$					
42	2					

[1] Based on flat-faced flanges with full-face soft gaskets.

[2] Flange dimensions (except thickness) and bolting correspond to the following standards:

> 2 inch thru 24 inch sizes: ASA Std. B16.5 for 150 lb. steel flanges.
> 30 inch thru 42 inch sizes: ASA Std. B16.1 for 125 lb. C.I. flanges.

[3] The above table is based on a safety factor of 8 to 1 and a flexural strength of 20,000 psi. This latter value is slightly under the minimum flexural strength for laminates of $\frac{3}{8}$ inch and up (see Table 1), due to the manufacturing technique.

means of joining pipe sections and pipe to fittings. The procedure used in making the butt joint will be as outlined in 3.3.5. All pipe 20 inches in diameter and larger shall be overlaid both inside, when accessible, and outside. Pipe less than 20 inches in diameter shall be overlaid outside only, unless the joint is readily accessible. The minimum width of the overlay shall relate to wall thickness and shall be of the dimensions indicated in Table 4. The inside overlay is only for the purpose of sealing the joint and shall not be considered in meeting the strength requirement specified in 3.3.5.

3.5.7 *Flanges*—The use of flanges shall normally be kept to a minimum with the butt joint being used as the standard means of joining pipe sections. All flanges shall be of the minimum thickness given in Table 5 and accompanying illustration. The construction of flanges is the same as that for laminates (see 3.3).

3.5.7.1 *Flange attachment*—The minimum flange shear surface shall be four times the flange thickness indicated in Table 5. The thickness of flange hub reinforcement measured at the top of the fillet radius shall be at least one-half the flange thickness and shall be tapered uniformly the

MAXIMUM SPACING OF PIPE HANGERS FOR REINFORCED-POLYESTER PRESSURE PIPE[1]

Table 6

Pipe I.D., *inches*	*Maximum pipe hanger spacing at pressure ratings:*					
	25 psi, feet	*50 psi, feet*	*75 psi, feet*	*100 psi, feet*	*125 psi, feet*	*150 psi, feet*
2	6.0	6.0	6.0	6.0	6.0	6.0
3	6.5	6.5	6.5	6.5	8.0	8.0
4	7.0	7.0	7.0	8.5	8.5	8.5
6	8.0	8.0	9.0	9.0	10.0	10.5
8	8.5	10.0	10.0	10.5	11.0	11.5
10	9.5	10.5	11.5	12.0	12.5	13.0
12	10.0	11.5	12.5	13.0	13.5	14.0
14	11.5	12.5	13.0	14.0	15.0	15.5
16	12.0	13.0	14.0	15.5	16.5	17.0
18	12.5	14.5	15.0	16.0	16.5	17.5
20	12.5	15.0	15.5	17.0	18.0	18.5
24	8.5	15.0	17.0	18.5	19.0	
30	9.5	17.5	19.5	21.0		
36	10.5	19.5	21.0			
42	8.0	21.0	22.5			

[1] The above table is based on uninsulated pipe containing liquids having a specific gravity of 1.3 and at a maximum temperature of 180°F. For services at temperatures above 180°F (82.2°C) consult the manufacturer relative to hanger spacing.

length of the hub reinforcement.   The fillet radius, where the back of the flange meets the hub, shall be ⅜ inch minimum.

3.5.7.2   *Flange face*—The flange face shall be perpendicular to the axis of the pipe within ½° and shall be flat to ± ⅟₃₂ inch up to and including 18 inch diameter and ± ⅟₁₆ inch for larger diameters.   The face of the flange shall have a chemically resistant surface as described in 3.2.4 and 3.3.1.

3.5.7.3   *Other flange designs*—Other flanges agreed upon between the fabricator and the user are acceptable provided that they produce a tight joint at twice the operating pressure.

3.5.8   *Mechanical properties of pipe*—The minimum mechanical properties of pipe shall be in accordance with Table 1.

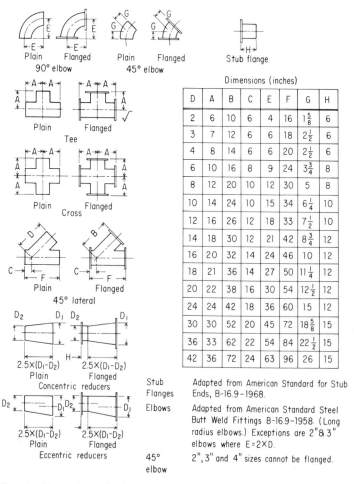

D	A	B	C	E	F	G	H
2	6	10	6	4	16	$1\frac{5}{8}$	6
3	7	12	6	6	18	$2\frac{1}{2}$	6
4	8	14	6	6	20	$2\frac{1}{2}$	6
6	10	16	8	9	24	$3\frac{3}{4}$	8
8	12	20	10	12	30	5	8
10	14	24	10	15	34	$6\frac{1}{4}$	10
12	16	26	12	18	33	$7\frac{1}{2}$	10
14	18	30	12	21	42	$8\frac{3}{4}$	12
16	20	32	14	24	46	10	12
18	21	36	14	27	50	$11\frac{1}{4}$	12
20	22	38	16	30	54	$12\frac{1}{2}$	12
24	24	42	18	36	60	15	12
30	30	52	20	45	72	$18\frac{5}{8}$	15
36	33	62	22	54	84	$22\frac{1}{2}$	15
42	36	72	24	63	96	26	15

Dimensions (inches)

Stub Flanges   Adapted from American Standard for Stub Ends, B-16.9-1968.

Elbows   Adapted from American Standard Steel Butt Weld Fittings B-16.9-1958. (Long radius elbows.) Exceptions are 2" & 3" elbows where E = 2×D.

45° elbow   2", 3" and 4" sizes cannot be flanged.

Fɪɢ. 1. *Dimensions of reinforced-polyester pipe fittings.*

3.5.9   *Vacuum service*—In sizes from 2 through 18 inches reinforced-polyester pipe and fittings having an internal pressure rating of 125 psi and flanges having a rating of 25 psi are suitable for full vacuum service. Special engineering consideration is required for larger pipe sizes and for operation at temperatures above ambient atmospheric.

3.5.10   *Recommended installation practice*

3.5.10.1   *Pipe hangers and spacing*—Hangers shall be band type hangers which will support a minimum of 180° of the pipe surface. The maximum pipe hanger spacing shall be in accordance with Table 6.

3.5.10.2   *Underground installation*—Special consideration must be given to installing pipe underground. It is recommended that the manufacturer be consulted for installation procedures.

3.5.10.3   *Expansion*—Since the expansion rate of this plastic pipe is several times that of steel, proper consideration should be given to any pipe installation to accommodate the overall linear expansion.

3.5.10.4   *Bolts, Nuts and Washers*—Bolts, nuts, and washers shall be furnished by the customer. Metal washers shall be used under all nut and bolt heads. All nuts, bolts and washers shall be of materials suitable for use in the exterior environment.

3.5.10.5   *Gaskets*—Gaskets shall be furnished by the customer. Recommended gasketing materials shall be a minimum of $\frac{1}{8}$ inch in thickness with a suitable chemical resistance to the service environment. Gaskets should have a Shore A or Shore A2 Hardness of 40 to 70.

Purchasing specifications—piping   In purchasing RP polyester or epoxy pipe it is necessary to provide the vendor with purchasing specifications. The items to be covered should include:

1. Operating conditions of pipe and fittings such as concentrations of components, maximum-conditions operating pressure, and maximum pressure. Check your company security on this to make sure it can be given to the vendor without divulging proprietary information. For the most part it is not necessary to give definite process concentrations; a span of concentrations of chemical components is all that is necessary. Sometimes, in chemical processes, close component control is necessary to achieve optimum results. If so, do not give this away. Generally, however, it will be of small consequence. For example, in the coagulating baths used in the rayon industry, the following spread of concentrations is completely satisfactory for service-conditions specifications:

5–10% . . . . H_2SO_4
1– 8% . . . . $ZnSO_4$
15–25% . . . . Na_2SO_4
Saturated with H_2S and CS_2
Pressure, 100 psig
Temperature, 35–70°C

The fact that any one of these components may have to be controlled to within $\frac{1}{10}$ percent to make a competitive product would be completely lost on the fabricator.

2. In sizes 2 to 4 in. the installed cost of reinforced polyester pipe and reinforced epoxy are about the same. Where the great mechanical strength of the filament-wound epoxy is desired, it should be used in these sizes. It also should be used for the extremes of temperature conditions (above 220°F in sizes 2 to 4 in.).

3. The pipe should be specified in bills of material in 15- to 20-ft lengths or random lengths.

4. Whenever possible, specify the flanges made on at the factory, but restrict the use of flanges to a minimum and use coupled joints or butt joints for field fabrication.

5. Do not specify wax to be used in the finish coat. It will cause difficulty in making butt joints if you do.

6. Consider color-coding your pipe (especially polyester) to identify the pipe by vendor. This has many advantages in allocating responsibility.

7. If the pipe is to be submerged, make sure that the outside has a heavy gel coating or is constructed of a pipe of high-resin–low-glass content, such as $75:25$ percent. The folly of burying in corrosive chemicals a high-glass-content pipe such as 20% resin/80% glass can soon be observed in severe deterioration unless the exterior is very heavily protected with a resin-rich coat.

8. Do not, under any conditions, consider general-purpose resin in the manufacture of glass-reinforced polyester pipe for use in chemical service.

9. If the pipe is installed outdoors and no insulation is necessary, then there are two alternatives:

   a. Use ultraviolet inhibitor in the resin from which the pipe is constructed.

   b. Paint the outside of the pipe with a 3-mil epoxy gray or black paint. This will effectively stop all ultraviolet deterioration.

10. Blind flanges are available and should be used where required.

11. Consider having your piping requirements on a large job furnished by a single vendor in a competitive-bidding situation and make him responsible for the entire detailed assembly. If you do this, your fabrication cost will be at a minimum and you will have had all the advantage possible of his fabricating facilities. This is especially important if you are in a short labor market where you must make use of every hour of the fabricator's time available to you. For the typical preparation of a detailed assembly see Fig. 3-3.

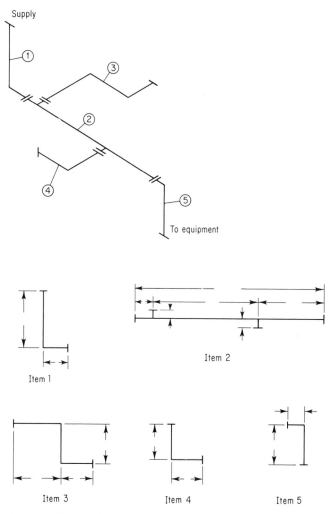

FIG. 3-3 Detailed assembly—glass-reinforced polyester pipe.

Purchasing specifications—fittings and flanges

1. It is to be understood that the basic face-to-face end-radius dimensions for reinforced polyester pipe fittings are adapted from the American Standard Steel Butt Weld Fittings. Exceptions are 2- and 3-in. elbows. Radiuses on long-radius elbows are $1\frac{1}{2}$ times fitting sizes. As long as a new system is being built, these fitting dimensions will not cause a problem. However, lined fittings such as lead-lined steel or rubber-lined steel were generally built from 150-psi ASA fittings, either

long-radius or standard.   Neither the standard fitting nor the long-radius fitting has the same *center-to-face* dimensions as those of the Commercial Standard.   Some vendors in filament-wound construction have departed from the Commercial Standard to follow 150-psi ASA standard fittings. Other vendors have adopted a compromise approach in which part of their fittings meet a 150-psi ASA standard fitting dimension but the rest of the line does not.   Interchangeability between reinforced polyester and lined-metal fittings is extremely important where a plant is turning from one type of construction to another.   The alternatives, therefore, are to rely on the few vendors who manufacture fittings to the 150-psi ASA standard or make designs sufficiently flexible to widen his choice of fitting suppliers.   Quite often the least expensive vendor does not make a 150-psi ASA fitting.   The engineering designer is therefore forced to accept the fact that standardization and economics have to be compromised.

As of this writing steps are being taken by at least one user in the field to develop a complete line of hand-laid-up reinforced polyester fittings, with the same center-to-face dimensions as the 150-psi standard fitting.   Prototypes in this area look promising.   (See Chap. 16.)

2. Have your flanges made on at the factory, where possible, and keep their use to a minimum.   In general, one pair of flanges every 40 ft

FIG. 3-4 Combining reinforced epoxy flanges with hand-laid-up polyester pipe is easy, as shown in this photo.   Burst test on joints of this sort will test right up to the bursting strength of the pipe.

in straight runs of pipe is sufficient. In some piping designs, where runs of hundreds or thousands of feet are common, the use of pipe with bell-and-spigot ends fastened together with an adhesive is a tremendous advantage. It provides the absolute minimum in joining costs and the greatest ease of erection.

3. Keep the use of hand-laid-up flanges to a minimum to hold flange costs down. Where flanges are required, use a cemented-on flange.

4. All flange drilling, regardless of pressure specification, should be 150-psi ASA.

5. Where service conditions permit, feel free to use epoxy-cement flanges on polyester pipe or vice versa. See Fig. 3-4 for an epoxy flange fastened onto a section of polyester pipe.

6. Shrinkage characteristics on the epoxy adhesives are somewhat less than the polyester adhesives. Both, however, can be used satisfactorily.

Table 3-1 indicates the approximate weight per foot in pounds for polyester piping constructed according to Commercial Standard specifications, TS-122C.

The wall thicknesses shown in Table 3 on minimum pipe-wall thick-

APPROXIMATE WEIGHT PER FOOT, LB[4]

Table 3-1

Pipe size, in.	Internal pressure rating, psig					
	25	50	75	100	125	150
2	0.8	0.8	0.8	0.8	0.8	0.8
3	1.1	1.1	1.1	1.1	1.6	1.6
4	1.5	1.5	1.5	2.1	2.1	2.1
5	2.0	2.0	2.6	2.6	2.9	2.9
6	2.4	2.4	3.1	3.1	3.5	4.8
8	3.2	4.1	4.6	4.6	6.4	7.3
10	4.0	5.1	5.7	7.9	9.0	10.2
12	4.7	6.1	9.4	10.7	12.1	15.9
14	7.2	7.9	10.9	14.1	18.4	20.5
16	8.3	8.9	14.2	18.6	21.9	27.4
18	9.3	14.0	17.9	23.4	26.0	34.7
20	10.3	15.5	19.8	27.1	36.0	38.0
24	12.3	21.8	31.0	40.4	45.0	59.5
30	16.6	28.7	42.6	56.0	73.5	87.5
36	27.6	46.0	67.5	87.5	104.0	131.0
42	32.0	59.0	88.0	125.0	151.0	187.0

nesses, in inches, are in general determined by use of the Barlow formula,

$$t = \frac{pD}{2S}$$

where t = wall thickness, in.
p = internal pressure, psi
D = inside diameter, ir
S = design stress, psi
A typical problem using the Barlow formula is worked out below.

Problem

Determine the wall thickness of an 8-in. pipe to carry a 10 percent sulfuric acid solution at a temperature of 150°F, a safety factor of 10:1, and a pressure of 130 psig.

t = wall thickness = ?

Assume

p = 130 psig
D = 8.0
S = 15,000 psig

Then

$$t = \frac{130 \times 8}{2 \times 15,000/10}$$
$$= 1,040/3,000$$
$$= 0.35 \text{ in., or rounded off to } \tfrac{3}{8} \text{ in.}$$

The rationale in using a safety factor of 10:1 in reinforced plastic pipe is traced to the long-term strength studies made with this material in service where there were corrosive solutions at elevated temperatures as high as 210°F. Losses of 20 to 50 percent of the original strength may occur over prolonged periods of time. By using a safety factor of 10:1 we are thus assured of long-term safety factors of 5:1 in nearly all cases.

3.3 VACUUM SERVICES

Polyester piping can be designed to withstand a full vacuum. It is necessary, however, to study the matter closely so that correct construc-

REINFORCED–POLYESTER PLASTIC PIPE   EXTERNAL COLLAPSING PRESSURE, PSIG[6]   Based on 20-ft Lengths

Table 3-2

Pipe dia., inches	Wall thickness, inches										
	1/8	3/16	1/4	5/16	3/8	7/16	1/2	9/16	5/8	11/16	3/4
6	12.3	38.5	100	207	378	570	800	1040	1410	1780	2200
8	5.36	17.3	44.7	94	173	264	378	515	680	865	1040
10	2.78	9.23	23.7	50	93	143	206	284	378	485	610
12	1.64	5.36	14.2	30	56	86	125	173	230	298	378
14	1.04	3.42	9	19.3	36	56	81	113	151	196	249
16	0.7	2.3	6.1	13	24.6	38	56	78	104	136	173
18	0.51	1.62	4.3	9.3	17.4	27	39.7	56	75	98	125
20	0.43	1.2	3.2	6.88	13	20.7	29.7	41.5	56	73	99
24	0.33	0.89	2.1	4.1	7.65	12	17.6	24.6	33.4	43.6	56
30	0.23	0.63	1.5	2.97	5.2	7.6	10.4	13.8	18	23	29.7
36	0.18	0.48	1.15	2.28	3.99	5.78	8	10.7	13.8	17.4	21.6
42	0.14	0.39	0.91	1.8	3.17	4.6	6.4	8.5	11	13.9	17.4
48	0.11	0.31	0.75	1.48	2.59	3.75	5.2	7	9.18	11.4	14.3
54	0.09	0.26	0.63	1.24	2.18	3.16	4.4	5.9	7.7	9.7	12
60	0.08	0.22	0.54	1.06	1.85	2.7	3.8	5.06	6.6	8.3	10.3
Flex modulus	700,000	700,000	800,000	900,000	1,000,000						

NOTE: Table gives collapse pressure in psig. Any reading above 14.7 psig will withstand a full vacuum. The addition of stiffeners every 10 ft will change the above figures. External pressure ratings are based on elastic stability. In cases where compressive strength is the limiting factor a 10:1 safety factor has been incorporated.

tion can be established, the proper wall thickness determined, and stiffener rings located when required. In addition, the engineer should watch carefully for possible vacuums caused by siphons; valves shut off in vertical lines; and pump priming. This, of course, is in addition to the pieces of process equipment which are properly designed to operate under a vacuum.

As piping or equipment size increases, it is common practice to use stiffener rings on properly designed centers to permit RP operation under high-vacuum conditions. This is particularly true on equipment above 36-in. diameter. Figure 11-4 is useful in determining the collapsing pressure with pressure considered on the sides only and the edges simply supported. On this premise the stiffener becomes an assumed edge whose design is of sufficient strength to act as a rigid member at that point. For reinforced plastics Poisson's ratio is assumed to be 0.30.[5] Most condensers, tanks, scrubbers, etc., which operate under a vacuum can be considered as round cylinders with pressure on sides and ends.

For a case history involving the design of a barometric condenser the reader is referred to Chap. 14.

It is assumed that the reader is aware that sec. 3.5.9 of the Recommended Commercial Standard endorses the use of 125-psi pipe and fittings in sizes 2 to 18 in. for full vacuum service and flanges with a rating of 25 psi or above. Normally, this blanket endorsement is good for temperatures up to 150°F. Above that temperature special considerations are necessary. To assist the reader in this, reference is made to Table 3-2, covering the external collapsing pressure in psig, based on 20-ft lengths. This table covers the common sizes from 6 to 60 in. diameter with graduated wall thicknesses of from ⅛ to ¾ in. It is based on stiffeners every 20 ft. Where stiffeners are applied on less than 20-ft spacing, a decrease in wall thickness of the pipe will result, or a higher safety factor is obtained.

Problem

To illustrate the value of stiffeners, assume we had a 36-in.-dia pipe with a ½-in. wall which is good for a collapse pressure of 8 psi with stiffeners on 20-ft centers.

Calculate the collapse pressure with stiffeners on (a) 10-ft and (b) 5-ft centers.

The formula is

$$W_c = KE \left(\frac{t}{D}\right)^3$$

(Refer to Fig. 11-4 for nomenclature and method of calculation for value of K.)

$$\frac{1}{r} = \frac{120}{18.5} = 6.5$$

$$\frac{D}{t} = \frac{37}{0.5} = 74$$

$$K = 7.2$$

$$W_c = (7.2)(1,000,000)\left(\frac{0.5}{37}\right)^3$$

$$= (7,200,000)\left(\frac{0.125}{50,600}\right)$$

$$= 17.8\ \text{lb}$$

Thus a 36-in. pipe with a ½-in. wall could stand a full vacuum with stiffeners on 10-ft centers.   By repeating the above calculations we find that, with stiffeners on 5-ft centers, the collapse pressure would be 39.5 lb.

3.4   TYPICAL SIZE DISTRIBUTION

A typical size distribution of reinforced polyester pipe covering 20,000 ft of all sizes is shown in Fig. 3-5.   These distribution ratios are based on economic factors, availability, and corrosion resistance.   They are typical distribution ratios found in a large chemical plant.

1. In the smaller sizes covered by 2, 3, and 4 in., there is little to choose, costwise, between hand-laid-up polyester pipe and the machine-made epoxy pipe.[7]   The choice will often lie in a field other than economics in this size area.   Choice of assembly methods, availability, service environment, or the engineer's personal preference may be sufficient to dictate the final choice.

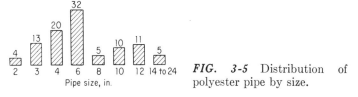

FIG. *3-5* Distribution of polyester pipe by size.

2. In sizes 6 in. and above, the present economic factors tend to favor the reinforced polyester pipe.   If the service requirements are suitable, the designer will find a complete range of polyester pipe available for conditions up to large pipe of 60-in. diameter.   In the larger sizes,

FIG. 3-6 A collection of 4- and 10-in. polyester pipe has been in acid service for over three years, with excellent performance. It installs well with conventional Y-pattern valves or rubber-lined butterfly valves.

however, above 12 in., the project should be considered on a system design basis. For a small collection of polyester pipe in sizes 4 to 10 in., see Fig. 3-6.

3.5 ESTIMATING INFORMATION AND COST TABLES

The engineer's purpose will be served best if a first-class, top-quality product is obtained at a competitive price.[1] Buy top-quality material even though it costs a bit more, and resist bargain hunting even though it is known that some reinforced plastic pipe can be bought for less. Also, constantly examine the quality of pipe and stock received and call to the vendor's attention any noticeable flaws (for the purpose of quality-control improvement). Most vendors of established reputation pride themselves on their quality-control efforts and are desirous of furnishing a consistently first-quality product. However, as in most manufacturing processes, imperfections may occasionally creep in from time to time, unknown to the manufacturer, which the user discovers on installation. Some such imperfections are:

1. Off-standard dimensions from face-to-face fittings.
2. Lack of trueness in fittings or pipe.
3. Uneven walls and voids at joints.
4. Manufacturing imperfections which may cause system leaks.
5. Wall thickness less than the SPI specifications for the standard purchased. One should be ruthless in this regard and insist upon the standard being met. The purchaser should not specify and pay for a 100-lb pipe and then accept a 50-lb pipe at his unloading dock.
6. Improperly cured resin that will not come up to the standard Barcol hardness specifications of 35 to 40 for a polyester resin.
7. Resin-starved surface with inadequate wet-out.
8. Crazed, checked, or cracked surfaces. This is particularly true of pipe having heavy-resin interior surfaces which are not reinforced.

The vendor is generally glad to replace any of these items and to institute additional controls to make sure that flaws are not repeated.

Lead or lead-lined steel pipe is used extensively in some industries. However, lead-lined pipe is largely unsatisfactory because of its high initial cost, high installation cost, and high cost of maintenance. While rubber-lined pipe costs less initially than lead pipe, it is still far more expensive than RP pipe. Also, rubber-lined pipe is difficult to repair, and even more difficult to maintain than either lead or RP pipe. However, the maintenance on rubber-lined pipe is generally less than on lead or lead-lined steel pipe.

Cost savings of reinforced plastic pipe over lead-lined steel pipe can be conservatively stated as being at least 40 percent.

A few factors indicate the relative economy of RP piping versus lead. For example, it is possible to install about $8 worth of lead piping for each lead-burner man-hour. At today's prices it is possible to fabricate and put in $14 worth of plastic piping for each pipe-fitter man-hour. The ease of installation is perfectly clear from these figures. Further, note that in relative basic costs, bonded-lead-lined steel pipe costs 70 percent more than its polyester counterpart. In general, reinforced plastic means a sizable reduction in piping costs where it replaces lead, rubber-lined steel, lead-lined steel, or stainless steel.

Competition in piping in the RP field is growing progressively more intense as better and more economical methods are developed for producing this product. This applies to both the reinforced polyesters and the epoxies. In general, the latest piping information shows that RP polyester pipe may be purchased for 20 to 30 percent less than RP epoxy, and in some cases even less than this.   In any large piping system it will generally be found that if both the polyester and epoxy are suitable for the conditions indicated, the polyester system can be installed for less money.   Certain qualifications do exist in this area, however.   Where there are manpower shortages, special consideration must be given to installation methods.   Factors to be concerned with are as follows:

1. Field installations will proceed most rapidly if the assembly forces have the minimum number of adhesive or wrapped joints to make. To this end it is quite often best engineering and soundest economy to detail the piping required, using the simple drawing methods illustrated in Fig. 3-3.   The engineer should then submit his proposed piping-detail bill of material to a number of fabricators and obtain bids.   Using this method, the engineer benefits from the fabricator's making nearly all the adhesive and wrapped joints.   This has a number of advantages:

   a. The assembly cost by the fabricator is nearly always less than that which could be accomplished by the purchaser.   The fabricator is much more proficient in this area.   Studies of comparative costs on butt-and-strap joints in the field and in the fabricator's shop show that the fabricator's price is nearly always the lower of the two.

   b. The reliability of the adhesive or butt joints made in the fabricator's shop is generally greater because this is a technique that he is using constantly.   Quite often a specially trained crew is used by the fabricator, so that joining is developed to an extremely high degree of efficiency and reliability.   In addition, the fabricator has special jigs and techniques, all of which tend to keep his costs down.   This is not meant to imply that the reliability of a plant-made adhesive or butt joint is not good. With proper training it is exceptionally good, but it is difficult to beat the day-to-day know-how and reliability of the fabricator, since he does this work on a constantly competitive basis, while in the ordinary manufacturing plant, work of this type is done on a "when required" basis, and very often on an "occasional" basis.

Figures 3-7 and 3-8 show one of the largest installations of reinforced polyester piping in the world, at the Hooker plant in Niagara Falls, New York.   More than 12,000 ft of pipe fabricated of Hetron 72 was installed, ranging in size from 2 to 18 in. in diameter.   Fabrication was by the conventional hand-lay-up technique, with a 10-mil type C glass surfacing veil on the interior, followed by two mat layers, and then alternate layers of mat and woven roving.   The system handles saturated sodium and potassium chloride brines containing chlorine (in some instances saturated with chlorine) at pH's as high as 11. Pressures are as high as 80 psi, temperatures up to 175°F, and velocities as high as 10 fps. The system has been performing quite satisfactorily for approximately five years.

2. Bolting flanges together, like assembling an erector set, is probably the simplest method of fabricating piping systems. From an assembly point of view only, the next-shortest assembly times can probably be most nearly achieved by interference-set adhesive systems, followed by the butt-and-strap joint.

Most of the high-quality corrosion-resistant polyester pipe sold today is made on mandrels by a hand-laid-up process, while the bulk of the RP epoxy piping is a machine-made pipe. Some machine-made polyester pipe is available, however.

The Society of the Plastics Industry is working aggressively to standardize the industry output. Fittings of standard dimensions and pipe manufactured to meet standard specifications are being produced subject to manufacturing limitations. As of this writing the industry

FIG. 3-7   Polyester piping.[8]

FIG. 3-8   Polyester piping.[8]

still has some distance to go.   Nevertheless, this is a step forward, and one that is most necessary.   Fittings and pipe need to be interchangeable regardless of which vendor they are purchased from, and if interchangeability is to be possible with other corrosion-resistant fittings, then all fittings must ultimately meet an ASA standard.

Assembly time, or ease of assembly, should not be confused with the final installed cost.   The easiest, fastest installation methods are generally the most expensive.   On an in-place basis the engineer must consider:

1. Manpower availability
2. Job payout time
3. Material availability
4. Labor costs
5. Material costs

In considering these factors, some odd, controversial conclusions may develop:

1. In the installation of polyester pipe, the labor cost, when compared with the total installation cost, is relatively small.   Changes in

specifications that affect the material can be money-saving.   Labor is not nearly as rewarding.

2. Training and techniques are the keys to success.

3. Where manpower is in short supply, manpower availability may dictate the use of the shortest assembly methods.

4. The shortest payout time may dictate the installation of the material which can be installed in the shortest possible time regardless of the cost.

Flanges, reducers, branches, tees, and all other fittings are available from the vendors, in standard sizes, configurations, and lengths.   Custom-designed fittings are also available to meet special requirements. RP molded flanges are now being produced by high-compression techniques to meet 150-psi ASA bolting specifications.

Table 3-3 indicates the comparative relative costs per linear foot for various materials in pipe sizes 2 to 12 in.   For some types of material, this ratio will hold substantially constant over the complete range.   The economy of other types of pipe improves as the size increases.   For example, up to 4 in. in size, Schedule 10, Type 304 stainless steel costs about the same as, or less than, 100-psi polyester.

Although Type 304 stainless steel is included in our studies, this grade of stainless steel shows comparatively little corrosion resistance to anything but the mildest substances.

In some cases, it is difficult to arrive at exact cost comparisons because of variations in vendor's pricing and because of changes made in the manufacturing specifications.   Where product lines were not complete, it was necessary to use prices from several vendors to establish continuity.

When buying reinforced polyester pipe, do not buy the product alone.[1]   While it is true that the product of any pipe manufacturer must be competitive in terms of price and quality, the buyer has a right to expect the backup of a top-flight service organization.   Technical depth is important to back up the product.   This should extend throughout the manufacturer's organization, right down to the shop personnel who fabricate the pipe, and is equally important in the supplier's field construction crews.   These crews must have a high degree of competence and a thorough knowledge of their product.

Field experiences have shown that fittings designed with smooth interior surfaces and with strength and corrosion-resistance qualities comparable with the pipe are vital to overall system integrity.   These qualities receive primary attention when fittings are selected.   As an example, major suppliers manufacture flanges that have chemical resistance and strength properties comparable with the piping systems. These flanges are designed to standard ASA dimension schedules for bolt

COMPARATIVE RELATIVE COST PER LINEAR FOOT FOR VARIOUS MATERIALS AND SIZES

Table 3-3

Description	Pipe sizes, in.							Weighted average
	2	3	4	6	8	10	12	
100 psi, 230°F, reinforced polyester...	1.0	1.0	1.0	1.0	1.0	1.0	1.0	1.0
150 psi, 230°F, reinforced polyester...	1.0	1.2	1.16	1.35	1.33	1.33	1.38	1.25
150 psi, 300°F, reinforced epoxy......	1.18	1.38	1.50	1.47	1.42	1.38	1.24	1.37
Schedule 10, Type 304 stainless steel.	0.82	1.1	0.98	1.36	1.4	1.4	1.32	1.2
Rubber-lined steel, $\frac{3}{16}$-in. rubber lining, 20-ft lengths, flanged..........	1.93	2.62	2.32	2.42	2.34	2.23	2.26	2.30
Saran-lined steel, 10-ft lengths......	1.21	1.87	1.84	2.46	2.62	....	....	2.0
Schedule 5, Type 316 stainless steel.....	1.04	1.50	1.58	1.67	1.53	1.63	1.69	1.52
Lead-lined steel, homogeneously bonded, flanged ends, 10-ft lengths	2.16	2.60	2.40	2.80	3.10	2.92	2.95	2.70
Penton-lined steel, 10-ft lengths......	3.10	4.10	4.6	6.1	6.2	....	....	4.82

circles, bolt sizes, and orientation, thus assuring proper hookup to pumps, valves, equipment, and existing systems.

The relative basic material costs for a variety of 90° elbows are shown in Table 3-4. The following conclusions apply to fittings in general:

1. It is apparent that minimum system cost can be achieved by using the contact-molded elbow with wrapped joints in sizes 6 to 12 in. There seems to be some small price advantage in cement end elbows in the 2- to 4-in. sizes. This assumes an equal degree of proficiency in making both adhesive and butt-and-strap joints. Such, however, may not be the case.

2. A good choice on a flanged system is to use the contact-molded elbow with a butt-and-strap press-molded flange, or separate flange,

COMPARATIVE RELATIVE COSTS FOR 90° ELBOWS IN VARIOUS MATERIALS AND SIZES

Table 3-4

	Pipe size, in.						Weighted average
Description	2	4	6	8	10	12	
100 psi, 230°F, chemical-grade reinforced polyester, plain ends, long radius................	1.0	1.0	1.0	1.0	1.0	1.0	1.0
150 psi, 300°F, cement end, reinforced epoxy.........	0.65	0.84	1.87	2.02	1.8	1.27	1.4
Rubber-lined cast iron, $\frac{3}{16}$-in. lining, standard radius.	1.54	1.5	1.68	1.5	1.43	1.34	1.5
100 psi, 230°F, chemical-grade reinforced polyester, press-molded, flanged end, long radius..	1.8	2.0	2.1	2.3	1.9	2.1	2.0
150 psi, 300°F, cement end, reinforced epoxy, flanged end...................	1.86	2.18	2.74	2.94	2.80	2.07	2.4
Lead-lined steel, flanged....	....	3.1	3.58	3.1	3.27	3.07	3.2
100-psi, 230°F, chemical-grade reinforced polyester, hand-laid-up flanged end, long radius..	3.56	4.0	3.8	3.5	2.7	2.5	3.4

attached. This produces a system that is generally less expensive than the conventional 150-psi epoxy-flanged elbow. See Fig. 3-9 for a good field stock of elbows, fittings, and flanges.

3. Again, the higher cost of using contact-molded stub ends as a means of providing a system of flanging (whether it involves tees, elbows, reducers, etc.) becomes evident. The use of contact-molded stub ends should be minimized in order to keep system costs down, unless they are used as a part of the vendor's bidding on a detail drawing. Contact-molded stub ends are tremendously strong and should be used where rough physical service is anticipated.

4. In fittings per se there does not appear to be a great deal of difference in price between a polyester fitting with press-molded flanges and a flanged epoxy fitting in sizes 2 to 4 in. However, in sizes 6 to 10 in., the savings in the press-molded, flanged, and reinforced polyester fittings are considerable. Corrosion-resistance characteristics of these two basic

FIG. 3-9 Stub ends, elbows, and tees make up a well-diversified stores supply of reinforced polyester fittings.

resins, however, need to be evaluated to determine the proper resin system to be used.   In severely oxidizing environments, the resin of choice will often be polyester.

5. In any machine-made polyester (or epoxy) system, cement end fittings will provide considerable economies over flanged end fittings. Couplings will provide a less expensive installation than flanges for runs of pipe.   Adhesive couplings have also been developed for hand-laid-up polyester pipe.   To achieve a minimum system cost in machine-made polyester (or epoxy) pipe, use cement couplings and cement fittings wherever possible.   Use the minimum number of flanges necessary to provide system assembly and maintenance.   For costing information on cement end fittings and couplings refer to Chap. 4.   Machine-made polyester and epoxy systems are comparable in price.

6. To provide the minimum cost in the assembly of a polyester system, use wrapped joints and plain-end fittings wherever possible. Use the minimum number of flanges necessary to permit assembly and proper maintenance.

NOTE: Strict adherence to the guidelines above, in particular, items 5 and 6, can produce real economies in system assembly.

7. Table 3-5 is a cost study of a typical polyester piping system. It shows the influences of many of the variables which go into the cost of putting together a polyester piping system.   Solely by the method of construction it is possible for the cost of the system to be doubled on an installed cost basis.   Unless we are dealing with long runs of pipe with relatively few fittings, it is doubtful that it is possible to achieve much better performance than about 70 percent wrapped joints and 30 percent flanged joints.   Use the press-molded flange where possible.   The rugged hand-laid-up stub end should be used at the points of extra strain in the system or at any flanged joint where it is necessary to use a ring gasket. Examine your proposed system design closely on the drawing board. It can pay handsome dividends.   In one large piping system constructed almost entirely of polyester, an engineer reduced the cost of the system by nearly $6,000 through the simple expedient of going back over system design with a fine-toothed comb and providing for system assembly

TYPICAL POLYESTER PIPING SYSTEM, INSTALLED COST PER FOOT[9]

Table 3-5

Description	Pipe size, in.					
	2	4	6	8	10	12
70% wrapped joints, 30% flanged joints, press-molded flange....	3.24	4.78	6.40	8.85	11.86	14.59
50% wrapped joints, 50% flanged joints, press-molded flange....	3.87	5.74	7.69	10.62	14.16	17.73
100% flanged joints, press-molded flange...............	5.20	7.74	10.17	13.49	17.82	22.26
70% wrapped joints, 30% hand-laid-up stub ends............	3.67	5.32	6.97	9.42	12.60	15.40
50% wrapped joints, 50% hand-laid-up stub ends............	5.16	7.35	9.39	12.35	16.38	20.15
100% flanged joints, hand-laid-up stub ends................	6.92	9.88	12.44	15.79	20.78	25.48

NOTES: 1. Labor rate, $3/hr.   For multiplying factors to correct to other rates, use Table 4-6.
2. Pipe, fittings, and flanges are based on 100-psi Commercial Standard.
3. These cost figures can be used for estimating purposes.   Long runs of straight pipe can be installed at a minimum cost.   Battery piping using many fittings in a close area should be estimated separately.

TYPICAL PIPING SYSTEM 6-inch Pipe

Table 3-6

Material of construction	Fabricating and installing methods	Cost per foot, labor						Material	Total	Ratio M/L	Cost index
		Fabrication		Installation		Total					
		Hours	Dollars	Hours	Dollars	Hours	Dollars	Dollars	Dollars		
Black steel, lap weld, Schedule 40	20% weld joints; 80% flanged joints	0.0486	0.1434	0.3446	1.0166	0.3932	1.1600	3.0714	4.23	2.65	0.94
	100% flanged joints	0.0811	0.2392	0.3338	0.9847	0.4149	1.2239	3.2670	4.49	2.67	1.00
Polyester-reinforced plastic	70% plain-end lock joints; 30% flanged joints	0.0378	0.1115	0.2635	0.7773	0.3013	0.8898	3.7046	4.84	4.16	1.08
	70% wrapped joints; 30% flanged joints with press-molded flanges	0.0527	0.1555	0.3223	0.9508	0.3750	1.1063	5.2940	6.40	4.78	1.43
	70% wrapped joints; 30% flanged joints with stub ends flanged	0.0730	0.2153	0.3223	0.9508	0.3953	1.1661	5.8007	6.97	4.97	1.55
	50% wrapped joints; 50% flanged joints with press-molded flanges	0.0932	0.2749	0.3230	0.9528	0.4162	1.2277	6.4655	7.69	5.27	1.71
	50% wrapped joints; 50% flanged joints with stub ends flanged	0.1540	0.4543	0.3230	0.9528	0.4770	1.4071	7.9864	9.39	5.67	2.09
	100% flanged joints with press-molded flanges	0.0811	0.2392	0.3338	0.9847	0.4149	1.2239	8.9451	10.17	7.31	2.27
	100% flanged joints with stub ends flanged	0.1622	0.4785	0.3338	0.9847	0.4960	1.4632	10.9722	12.44	7.50	2.77
Epoxy	70% adhesive joints; 30% flanged joints 50% adhesive joints;	0.1007	0.2971	0.2371	0.6994	0.3378	0.9965	7.7075	8.70	7.73	1.94

Description										
Lead-lined black steel, Schedule 40										
50% flanged joints	0.1081	0.3189	0.2743	0.8092	0.3824	1.1281	8.1846	9.31	7.25	2.07
100% flanged joints	0.0811	0.2392	0.3338	0.9847	0.4149	1.2239	9.0022	10.23	7.35	2.28
100% flanged joints, 20' max. lengths, sleeved	0.7432	2.1926	0.3338	0.9847	1.0770	3.1773	8.8253	12.00	2.78	2.67
100% flanged joints, 10' max. lengths, sleeved	0.8716	2.5712	0.3709	1.0942	1.2425	3.6654	9.0699	12.74	2.47	2.84
100% flanged joints, 17' max. lengths, homogeneous	0	0	0.3412	1.0065	0.3412	1.0065	26.8761	27.88	26.70	6.21
304L stainless steel, Schedule 5										
70% plain-end lock joints; 30% flanged joints with flanged retainer rings	0.0865	0.2552	0.2635	0.7773	0.3500	1.0325	8.9645	10.00	8.68	2.23
50% weld joints; 50% flanged joints with flanged retainer rings	0.1013	0.2988	0.3554	1.0484	0.4567	1.3472	9.1122	10.46	6.76	2.33
50% weld joints; 50% flanged joints with flanged stub ends	0.1013	0.2988	0.3554	1.0484	0.4566	1.3472	9.7050	11.05	7.20	2.46
100% flanged joints with flanged retainer rings	0.2094	0.6177	0.3338	0.9847	0.5432	1.6024	10.4095	12.01	6.50	2.67
100% flanged joints with flanged stub ends	0.2094	0.6177	0.3338	0.9847	0.5432	1.6024	11.6348	13.05	7.26	2.91
316L stainless steel, Schedule 5										
50% weld joints; 50% flanged joints with flanged retainer rings	0.1013	0.2988	0.3554	1.0484	0.4567	1.3472	12.4859	13.83	9.27	3.08
50% weld joints; 50% flanged joints with flanged stub ends	0.1013	0.2988	0.3554	1.0484	0.4567	1.3472	13.5197	14.87	10.04	3.31
100% flanged joints with flanged retainer rings	0.2094	0.6177	0.3338	0.9847	0.5432	1.6024	13.8967	15.50	8.67	3.45
100% flanged joints with flanged stub ends	0.2094	0.6177	0.3338	0.9847	0.5432	1.6024	16.0332	17.64	10.00	3.93

Labor rate $3/hr.

with the minimum number of flanged joints permissible for transportation, construction, and maintenance.   It is good business to do the Monday morning quarterbacking on the drawing board before the system is installed.   In installing polyester piping, material/labor ratios run 5:1 to 7:1.   Put another way, material is 80 to 85 percent of the installation cost and labor is 15 to 20 percent.

8. Table 3-6 provides an absolute comparison of typical piping systems versus materials of construction and various types of assembly, covering a piping system in the 6-in. pipe size.   Here is a direct comparison of steel, polyester, epoxy, lead-lined steel, 304L stainless steel, and 316L stainless steel.   This table shows how, simply by changing construction methods, it is possible to effect real system economies.   Also note the wide variance in material/labor ratio.   Then observe the cost index for an absolute comparison of one type of construction with another.   Additional detailed figures give the labor fabrication time and installation time on a per-foot basis.

3.6   PRICE RANGES—POLYESTER PIPE AND FITTINGS

To give some idea of the cost of polyester pipe and fittings, typical price ranges are shown in Table 3-7, through the common sizes of 2 to 12 in., and in some cases larger sizes, where data are commonly available.

The actual price covering the immediate purchase of these materials is governed not only by the usual economic factors of supply and demand but by other variables such as:

1. Size of the immediate order
2. Yearly contractual requirement, which establishes the discount
3. The variance which exists from vendor to vendor
4. Pipe specification being purchased

For estimating purposes the reader may use the midpoint of the range as an approximate purchase price for 100-psi specification items.

3.7   REPAIR TECHNIQUES APPLIED
TO A DAMAGED PIPE[7]

The types of physical damage which may occur to a reinforced polyester piping system may in general be classified as follows:

1. Physical damage through striking the pipe with a hard object. This mechanical damage permits the corrosive liquid to attack the glass

PRICE RANGES OF POLYESTER PIPE AND FITTINGS, 100-PSI SPECIFICATION

Table 3-7

Pipe size, in.	Pipe cost per foot	90° elbow		45° elbow		Tee		Flange		
		Plain end	Flanged	Plain end	Flanged	Plain end	Flanged	Press type	Hand-laid-up (stub ends)	Blind hand-laid-up
2	$ 1.60–$ 2.20	$12–$ 15	$ 55–$ 60	$ 4–$ 8	$ 49–$ 53	$ 17–$ 27	$ 76–$ 80	$ 9–$10	$16–$20	$17–$ 32
3	2.00– 2.75	13– 16	62– 67	5– 9	55– 59	19– 27	90– 94	10– 12	20– 24	20– 36
4	2.65– 3.50	17– 20	75– 80	7– 10	67– 70	23– 35	100– 105	13– 15	23– 31	24– 42
6	3.50– 4.75	20– 25	95– 110	12– 14	87– 89	31– 38	134– 160	18– 22	28– 41	28– 58
8	5.40– 6.70	32– 35	125– 140	20– 24	115– 120	48– 50	192– 200	22– 25	34– 50	33– 80
10	8.15– 9.15	48– 55	160– 175	29– 35	140– 145	62– 66	225– 240	25– 30	43– 60	43– 93
12	10.50– 12.30	65– 70	185– 200	41– 47	160– 170	88– 90	268– 280	27– 32	54– 66	53– 110
14	15.10– 16.80	100– 110	....	55– 60	....	120– 130	....	65– 70		
16	20.25– 24.00	140– 150	....	75– 80	....	160– 170	....	75– 80		
18	23.00– 30.00	170– 180	....	90– 100	....	170– 180	....	75– 80		
20	27.00– 34.00									
24	40.00– 47.00									

NOTE: The above approximate costs for flanged 90° elbows, 45° elbows, and tees are based on using flanges made up in accordance with the current Commercial Standard (TS-122C, dated Sept. 18, 1968). By comparing the cost of press-type flanges with hand-laid-up stub ends, it is possible to gauge the economies attainable by using press-type flanges fastened onto plain-end fittings with a good polyester adhesive. In this manner flanged-fitting costs can be considerably reduced. For example, a 3-in. 90° flanged elbow manufactured in accordance with the Commercial Standard costs, approximately, $62 to $67. This same elbow can be furnished for approximately $50, using press-type flanges. Likewise, an 8-in. 45° elbow costing $115 to $120 with hand-laid-up flanges can be bought for approximately $85 using press-molded flanges. Large tees are even more economical, using press-type flanges. For example, a 12-in. flanged tee, using hand-laid-up flanges, would cost $268 to $280. Using press-molded flanges, such a tee could be furnished for approximately $190 to $200. The user also has the option of substituting epoxy flanges, which are basically a press-molded flange and are less expensive than polyester flanges in the smaller sizes, up to 8 in. Press-molded flanges may also be used to prepare stub ends if desired. Also, a press-molded flange may be used directly as a fitting flange for minimum-cost construction. The price range on blind flanges of the hand-laid-up type is exceedingly wide. Cheaper methods may be used, such as a cast-iron flange protected with a piece of polyester sheet, sheet rubber, or lead blank.

fiber. The responsibility for installing a first-class product begins with the vendor but extends to the truckers, the warehouse handlers, and the field installing personnel, to ensure that the pipe which goes into service is in the same good condition as when newly manufactured.

2. Improper anchoring and supporting.

3. Intense hammering or pressure surges.

4. Unstable flow, which, combined with supporting conditions, results in an angular twisting moment in the pipe. This may cause the pipe to split.

Reinforced plastic pipe is easy to repair and modify, which makes it most desirable as a material of construction. All that need be done is to saw through the pipe (using ordinary hand tools), insert a new section or fitting, and apply new wrapped joints. See Figs. 10-1 and 10-2. Most such changes can be made in a very short time, depending on capability and training. Major modifications to any system, though taking longer, reduce downtime to less than one-third that required for other piping materials.

Occasionally it is necessary to repair a system on the run, for example, where weeping or a small crack may appear, generally caused by some mechanical damage. Several methods are possible:

• Make up a hot mix, using a generous supply of catalyst to set the polyester resin in a few minutes. Dry the area and make a hot patch using glass mat. Such a repaired pipe can be used in 30 min.

• The process liquid can be shut off only long enough to dry the area with a cloth. Wrap pipe with an adhesive tape. Make a hot patch with polyester resin and glass cloth over the tape. One repair of a split pipe was done in this way, with the process pump off only 7 min.

• Leaking epoxy coupling joints, which may occur for a variety of reasons, can also be repaired satisfactorily with the wrapped-joint technique. The secret is to get the area dry, then build a polyester wrapped joint right over the leaking epoxy coupling. There is a notion current in the industry that a polyester wrapped joint cannot be made on an epoxy pipe. It is nothing more than a fable. If you merely roughen and dry the surface of the epoxy pipe, a wrapped joint made from a polyester resin and cloth will do the job very well. This is recommended as a repair method only, but it will work.

Some of the RP manufacturers sell repair kits especially designed for repairing small breaks or leaks. Generally, this amounts to a combination of rubber plugs or mats, an adhesive, and a pipe saddle held in place by several clamps or bands.

3.8 INSULATION VALUE

Reinforced plastic pipe has unusually low thermal conductivity. In medium-temperature applications (0 to 200°F), no insulation is needed for RP pipe. This represents a significant savings on capital costs and maintenance. For example, 1½ in. of insulation applied to 6-in. metal pipe costs approximately \$3/linear ft. Another important factor related to RP pipe's low thermal conductivity is condensation. Condensation is held to a minimum, or is at least retarded, by RP pipe. Again, in this area, RP polyester piping is generally somewhat thicker than RP epoxy piping, owing to the method of manufacturing. Under these conditions, RP polyester piping is more resistant to sweating than RP epoxy pipe, although both have approximately the same thermal conductivity per inch of thickness (1.8–2.0 Btu/(ft²)(hr)°F/in.).

There are certain areas, however, in which insulation should be considered. These are:

1. Cold applications in the range of 32 to 50°F in which sweating of the pipe may occur on a hot summer day. Although this sweating is of no consequence as far as the piping is concerned, the drip on the floor and equipment below may be damaging.

2. In some lower-temperature processes where the delta T between the refrigeration medium and the process itself is relatively small, the heat loss may be an objectionable amount. This is true especially if large surface areas are involved, and is more particularly true in tanks than in piping. A heat balance of each individual process in this type of application becomes a necessity, so that the engineer may then judge whether the economic value of the insulation is a necessity for process operation.

3. Insulation as necessary on RP pipe may be required to prevent freezing in outdoor installations.

3.9 RELIABILITY—EXPECTED SERVICE LIFE

Based on a wide range of experience, the initial reliability in reinforced plastic systems has been good. It is difficult to speak categorically of a service life without tying one's self to the service conditions. Not only that, but the engineer must continually look at alternatives to see where the targets lie. Underground cast-iron sewers may boast a service life of 50 to 80 years. The reinforced plastics industry is so young that few installations have a service life longer than 10 to 15 years, and the majority of them probably less than 5 years. This does not mean that the material will not last a lot longer. It simply means that the industry is still relatively young and the service period has not yet had the time to accumulate.

In many exceptionally difficult applications in the chemical industry service life may be measured in weeks or months for some of the best metals available at a practical price, but in these tough applications reinforced plastics performed exceptionally well, outlasting their metal counterparts manyfold. The following case histories illustrate this.

1. A major installation of polyester pipe was made in 1960. This was a 24-in. sewer line installed as a replacement system for a like-diameter lead-lined steel system that had required considerable maintenance. The new RP pipe has not had a leak or required any repair whatsoever in 7 years. Removing the lead-lined steel system which had been fabricated in 7-ft lengths required a crane plus the sweating efforts of a hard-working crew of four men. Installing the reinforced polyester replacement required two men and went in place in one-third the time. Figure 3-10 is an illustration of this installation.

FIG. 3-10   A 24-in. reinforced polyester sewer line has been carrying hot acid-waste liquors. Plant acidic effluent is carried underneath a north-south railroad line for 150 ft, replacing lead-lined steel pipe. Operation has been trouble-free.

FIG. 3-11 10- and 12-in. reinforced polyester return lines carry rayon spin bath at velocities up to 8 fps. These lines have been in continuous service for 4 years. Although installation has performed well, present practice would dictate the use of fewer flanged joints and a subsequent reduction in installation cost. Use the minimum number of flanges to get the job done.

2. The use of reinforced polyester piping in acid-slurry service has been demonstrated over many years. Here bitingly sharp crystals of Glauber's salt ($Na_2SO_4 \cdot 10H_2O$), suspended in a solution of 6 to 12 percent sulfuric acid, are pumped continuously through RP systems. The polyester piping, with its abrasive handling characteristics, has been amply demonstrated in service for the last 5 years, outlasting its corrosion-resistant metallic ancestors by over 600 percent and still performing effectively.

3. In a redo of two large rayon-spin-bath systems, hundreds of feet of reinforced polyester piping in 10- and 12-in. sizes was employed on pumped return lines. Designed to carry hot acid solutions at velocities up to 8 fps, these lines have been in continuous service for 4 years. This piping replaced a lead-lined steel system which had failed after 7 years service. See Fig. 3-11.

3.10 HEAD LOSS VERSUS FLOW RATE

One of the outstanding advantages in the use of reinforced plastic pipe is its glasslike interior surface, which minimizes resistance to fluid flow. At least in theory, the pipe interior surface should remain smooth throughout its entire service life.   Figure 3-12 shows a head loss versus the flow rate for reinforced plastic pipe, whether epoxy or polyester, in sizes 2 to 24 in., as calculated by the Fanning equation.   To those engineers who are more familiar with the widely used Williams and Hazen formula, it will readily be recognized that the developed C value, which accounts for surface roughness, is exceptionally high, being in the 150+ range. New cast-iron or wrought-iron pipe is generally considered to have a clean, new-pipe C value of 130.   However, there is a C value of 100 commonly used for design purposes.   This service coefficient represents a substantial multiplier correction of 65 to 70 percent.   When using cast-iron or steel pipe between new and service conditions, long-term buildup on the pipe wall, or tuberculation, can easily provide sufficient frictional resistance to flow to justify such a correction.   If, however, the engineer is dealing with a system in which internal deposits are not anticipated on the reinforced plastic pipe, then, obviously, the long-term value of reinforced plastic pipe, in preference to its steel or cast-iron counterpart, far transcends the initial cost comparison.   For example, the friction loss in 1,000 ft of 6-in. polyester pipe at a flow of 1,000 gpm would be approximately 42 ft of head.   With the installation of cast iron at a C value of 100, the anticipated pressure loss would be 121 ft.

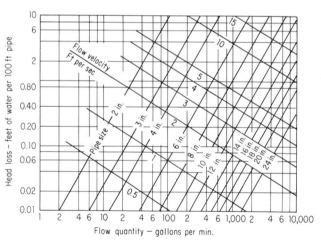

FIG. 3-12 Head loss versus flow rate.

Our power requirements, based on friction alone, would be nearly three times as much, using the cast-iron pipe over a long period of time after tuberculation had occurred.

It is quite often possible to reduce the size requirement one step in going from cast iron or steel to a reinforced plastic pipe where tuberculation is a problem.

In many systems the buildup will occur regardless of the pipe wall in question. It may be slower on a smooth, slick surface, but the probability is that eventually it will occur. Statistical studies on record have indicated that the continued slickness and resistance to buildup of reinforced plastic pipe can demonstrably manufacture a better product. In some cases in the textile industry quality is significantly better in certain applications over its metallic ancestor. In other cases the reduction in physical defects has been dramatic, amounting to 75 to 80 percent. Not only that, but when reinforced plastic pipe is chemically cleaned, the resultant effort produces a system in "as new" condition, with no physical damage to the piping system itself.

POLYESTER PIPE CAPACITY, GAL/LINEAR FT

Table 3-8

Inside diameter, in.	Gallons per foot
2	0.16
3	0.36
4	0.65
6	1.47
8	2.61
10	4.08
12	5.88
14	8.00
16	10.45
18	13.23
20	16.32
24	23.52

Figure 3-13 is an annotated graph of the conventional C value versus the material of construction.[10] From this graph it is relatively easy to see the difference in flow characteristics versus the material involved.

3.11 LIGHT STABILITY—ULTRAVIOLET

An item of special importance is the necessity for providing and specifying light stability in the ultraviolet range. This is essential especially where outdoor installation is contemplated. Unless this is provided, in a matter

150   New reinforced plastic pipe

140   Transite, fiber

130   Copper, brass, lead, tin, glass, seamless steel

120

110   Wood stave

100   Commonly used value for design purposes with
       solids buildup

90

80

70

60   Corrugated steel

Relative frictional coefficient – C value

FIG. 3-13   Relative frictional coefficient of various piping materials. (Ref. 10, p. 28.)

of a year or two the pipe will suffer severe degradation on the exterior surface from ultraviolet, and the glass filament will fray out, so that the purchaser will become considerably alarmed at the appearance presented. This is more alarming than it is dangerous.   After the initial fraying out, the phenomenon will stop far short of having any particular effect on the safe carry capacity of the pipe.   However, many vendors prevent this from occurring by adding an ultraviolet inhibitor to the resin, which is an effective "sun-tan lotion."   Experience has shown that pipe installed with this ultraviolet inhibitor in the resin shows no degradation over a considerable number of years' exposure to the sunlight.   Even storage outdoors is perfectly safe.   Other vendors provide a 3-mil coat of black epoxy paint, which will also effectively screen out the ultraviolet rays. It is, of course, axiomatic that any other type of structure to be installed outdoors and made of reinforced plastics should also be furnished with the ultraviolet (UV) inhibitor. Commonly, a maximum of 0.25 percent absorber is specified.[11] It should be specified to be incorporated in all pipe or structures used in outdoor installation.

3.12 RELEASE FILM AND ITS DETECTION

In the construction of some of the larger sizes of polyester pipe, 4 in. and above, and generally in tanks and any shapes requiring a release mechanism, the mold may be wrapped in Mylar (polyester film), polyethylene paper, or cardboard. The vendor will usually take every care to ensure that none of this mold release, or molding material, is shipped with the finished product. However, when this does occur, the Mylar film or polyethylene paper or cardboard is stripped from the interior of the pipe or tank as the warm solution hits it. As the system begins to circulate, this loose material clogs up orifices, valves, elbows, and the system in general. For this reason it is always wise to circulate warm water through a new system for a complete test-out before the RP system is placed on stream. This is a precaution to "de-bug" the system of a defect that can spell economic loss.

Although a close inspection of the material prior to assembly will be beneficial, a Mylar film is difficult for the uninitiated to detect because it is completely transparent. A polyethylene film is somewhat simpler to detect.

The conclusions, then, to be drawn are as follows:

1. It is essential to check your system closely prior to startup to make sure that bits and pieces of film have not been left, inadvertently, in the manufactured part.

2. It is advisable to allow a short time for the circulation of water through the system before giving the signal to go.

3.13 ADVANTAGES OF HAND–LAID–UP POLYESTER PIPING SYSTEMS OVER MACHINE–MADE POLYESTER OR EPOXY SYSTEMS—A REVIEW

• The machine-made filament-wound system is dependent upon its inner corrosion barrier for complete performance. When this is lost, the system is compromised quickly, and failure occurs. In the hand-laid-up polyester system the corrosion barrier is very deep, since all sections built of random mat are part of the corrosion barrier. Experience in depth in corrosion environments supports this observation.

• At least in theory, the machine-made pipe should have fewer defects because it should lend itself to better quality control, rather than relying on the craftsman's skill. In practice, however, this is not found to be the case. Defects in machine-made manufactured pipe are many times greater than in the hand-laid-up pipe. Porous elbows, leaking pipe, damaged inner liners, resin-starved inner liners, flanges and fittings not true, all are found in systems supposedly well automated, fully inspected, and fully tested. Experience has indicated that the hand-laid-up product, by its very method of manufacture, receives much closer personal attention prior to shipment.

• Surprisingly, many of the machine-made epoxy piping systems have yet to incorporate an effective UV shield in their manufacture, so that "blooming" of the pipe exterior is quite evident after several years' service in an outdoor installation. Painting of the pipe is done on request by some vendors, at an additional charge. A few vendors offer a painted pipe at no extra charge. All this is easily accomplished in the polyesters by the addition of a UV inhibitor, so that outdoor storage or service for many years shows no ill effects.

• The best corrosion-resistant structures can be designed using a discontinuous reinforcing and with glass/resin ratios of approximately 25:75. The corrosion resistance is in the resin; the strength is in the glass. When inner liners are breeched and the backup systems are running 75% glass/25% resin, the useful end to the life of the lay-up is not far off. For chemical resistance in severe corrosion conditions where the polyesters are suitable, a hand-laid-up system should be the choice.

• The machine-made filament-wound systems appear to be much more liable to transportation and handling damage. Inner liners become starred and cracked, especially those provided without suitable reinforcing. Such liners may be damaged without visible change to the outside of the pipe. It is naturally desirable to get the pipe into operation in first-class condition, but the consequences of not doing so will be infinitely worse with a filament-wound pipe than with a hand-laid-up pipe.

• Many installations require submergence of the RP pipe in corrosive solutions. Under such conditions the filament-wound pipe fares very poorly since the corrosion barrier is on the inside only. Not so the hand-laid-up pipe—it is corrosion-resistant inside and out. It has stood the test many times in severely corrosive installations. Submerged sparging units of hand-laid-up pipe have performed most effectively at temperatures of 200 to 210°F.

• Perhaps the union of the best qualities of hand-laid-up piping and machine-made filament-wound piping would be beneficial. Such a pipe might be built of:

A corrosion barrier of 40 to 60 mils reinforced with C glass or asbestos
A secondary corrosion barrier at least 90 mils thick consisting of two layers of 1½-oz mat
Filament winding for the required structural strength for 150-psig service topped off with a 10:1 safety factor
An outer layer of 1½-oz mat
A finished hot coat to which a UV inhibitor has been added

Thus, by combining the good points of each system, we should come up with durability, high impact strength, and great corrosion resistance.

REFERENCES

[1] Mallinson, J. H.: Reinforced Plastic Pipe: A User's Experience, *Chemical Engineering*, Dec. 20, 1965.

[2] Killeen, N.: International Pipe & Ceramics, Arcade, N.Y., personal communication, July 21, 1965.

[3] Recommended Product Standard for Custom Contact Molded Reinforced Polyester Chemical Resistant Process Equipment, TS-122C, Sept. 18, 1968.

[4] Steelman, C., Heil Process Equipment Co., Cleveland, personal communication, June 26, 1967.

[5] O'Leesky, S., and G. Mohr: Handbook of Reinforced Plastics of the SPI, p. 542, Reinhold Publishing Corporation, New York, 1964.

[6] Steelman, C., Heil Process Equipment Co., Cleveland, personal communication, July 7, 1967.

[7] Mallinson, J. H.: Plastic Pipe: Its Performance and Limitations, *Chemical Engineering*, Feb. 14, 1966.

[8] Szymanski, W. A., Hooker Chemical Corp., Durez Div., North Tonawanda, N.Y., personal communication, Jan. 5, 1967.

[9] Loftin, E. E., FMC Corporation, American Viscose Div., Front Royal, Va., personal communication, July 10, 1967.

[10] Cameron Hydraulic Data, *Compressed Air*, Phillipsburg, N.J. (Ingersoll-Rand Co., New York).

[11] Duracor Reinforced Plastic Pipe, The Ceilcote Company, *Bulletin* 10-F, Berea, Ohio, rev. March, 1966.

4
Epoxy Piping

4.1 INTRODUCTION

We now come to our second system of reinforced plastic pipe, the epoxies. In terms of dollar volume reinforced epoxies far exceed custom-contact-molded polyester piping dollar value. Epoxies predominate in the filament-wound-pipe business, which in 1966 did a volume of approximately 18 million dollars. Two manufacturers accounted for approximately 50 percent of the business. Custom-contact-molded polyesters was a 6-million-dollar industry in 1966. Although the filament-wound construction is largely epoxy and the volume of filament-wound piping is three times the size of hand-laid-up pipe, this represents only part of the story. The statistics are greatly colored by the fact that large quantities of epoxy pipe are used in oil-field applications and predominantly in the smaller sizes. The viewpoint, therefore, of piping usage in the chemical industry is somewhat different. Both epoxies and the polyesters are used in chemical processes, depending upon the service conditions involved. Machine-made epoxy pipe, from the overall viewpoint, appears to predominate in the smaller sizes. It is widely available on a commercial basis in sizes up to 12 in. and available from a smaller number of vendors in sizes up to 24 in. One large chemical plant in the last seven years showed a usage of 35,000 ft of filament-wound epoxy pipe and about 15,000 ft of hand-laid-up polyester pipe. Filament-wound epoxy predominated in the 1- to 4-in. sizes. At 6 in. there was a good mix of both, depending on the particular installation in question. In sizes 8 in. in diameter and extending to 24 in., the piping was mostly hand-laid-up polyester. Generally, where both the epoxies and polyesters are applicable for the service conditions, the choice lies in the economic field. At 6 in. and above, polyester piping is the economic choice. In the oil fields large quantities of straight pipe are used with relatively few fittings, while in chemical plants large amounts of fittings are used, so that methods of assembly and fitting costs become of paramount importance in overall system economics.

The largest user by type of process is at present the petroleum industry, where several million feet of glass-fiber-reinforced piping systems has been installed in the oil fields in sizes 2 to 8 in., at pressures up to 1,200 psi and temperatures up to 150°F. These oil-field installations date from 1957, or perhaps a little earlier.[1] Epoxy down-well tubing has been used to solve other petroleum corrosion problems.

Glass-reinforced epoxy pipe is also being used by gas utility companies as distribution piping, to minimize gas leakage. Several hundred thousand feet of this pipe has been used where corrosive soil conditions worked against the use of steel or copper.[1]

The glass-fiber-reinforced epoxy pipe is also used in a wide variety of chemical industries in handling acids, bases, and many chemicals. It has been approved for use by saline-water groups for potable water and meets FDA requirements for food processing.[1]

Reinforced epoxy pipe is being built by a number of different methods of construction. There is no SPI or Commercial Standard as a guide to follow. Nevertheless, the engineer should not regard the lack of standards in this field as a reason for not using filament-wound epoxy pipe. Granted that standards are a recognized customer need, such standards may also serve as a straightjacket, tending to limit improvement. While standards can be an asset to both purchaser and fabricator, the more advanced consumer is quite often in a position to improve upon the standard to suit his own particular needs.

The reinforced epoxy piping industry today may well never be covered by a fabrication standard, although *certainly, we should expect performance standards to be developed ultimately.* Attempts at preparing standards for reinforced epoxy pipe tend to result in standards which are proprietary in nature, since methods of construction vary widely.

The reader is referred to Appendix C for a Proposed Standard on Filament Wound FRP Piping.[2]

We should also point out that, by varying resin, hardener, and conditions of curing, it is possible to attain a wide range of epoxy-resin systems, so that speaking in general terms of the chemical resistance of "epoxies" may be as erroneous as speaking of corrosion resistance of "metals."[1]

4.2 EPOXY PIPE—BASIC CONSTRUCTION

The raw materials which make up epoxy pipe are as follows:[1]

Glass fibers   Normally, two types of glass fibers are used, an E (electrical), which possesses enormous strength, and a C (chemical), which possesses a higher corrosion resistance.   E glass has an average ultimate tensile of 450,000 psi, which is five to eight times stronger than steel, but has a tensile modulus of $10\frac{1}{2}$ million psi, which is similar to aluminum and about one-third that of steel.   The E grade has little chemical resistance and is used primarily for structural purposes.   The C grade is much more resistant to acids.   Both grades are normally not affected by salt solutions.

Resin   To provide the chemical resistance in the pipe an epoxy resin is used which is a condensation of bisphenol A and epichlorohydrin. The resins may be juggled to provide different molecular weights. When a hardener is added, cross-linking occurs, and a solid high-molecular thermoset plastic material is produced.   The epoxy resins possess good chemical resistance, low thermal conductivity, and low electrical resistance.   The epoxy resin can be further tailored to introduce halogens into the molecule, so that a fire-retardant product is produced.   By varying the hardener, molecular weights, and cure conditions, the physical characteristics of the resin can be suited to any particular service.   For this reason care should be taken in generalizing on "epoxies" since we are using a generic term in much the same manner as when we speak of "liquids," "metals," or "gases."

The interior of the pipe should be a resin-rich surface of at least 20 mils, reinforced with a C glass, asbestos, or an "organic veil."   Some liners made for heavy-duty service are 60 mils.   (Practices in the industry vary, and all these liners are not reinforced.)   In reinforced epoxy piping, asbestos is a popular interior reinforcing.   In this interior resin-rich surface the resin constitutes upward of 90 percent of the weight, and the C glass or veil, the balance.   Since the C glass is much more chemically inert than the E glass, it will offer superior chemical resistance. However, in addition to this, the C glass reinforcement greatly reduces the chances of resin fracture under impact, cutting, or thermal shock. Not only that, but the use of a surfacing veil will greatly add to the abrasion-resistant properties of the laminate.   The construction of the pipe may then vary.   The remainder of the pipe may be a filament-wound E grade glass saturated with a suitable resin.   To obtain optimum performance in this area a high-performance coupling agent which bonds together the resin and glass is generally used.   (Silanes are often used as coupling agents.)   In other piping designs of this type, additional mat is added to the interior wall prior to the use of the filament-wound exterior.   Over the filament-wound exterior an additional resin-rich surface may be provided to furnish the high-performance exterior.   It is obvious that, with the number of variables in this type of pipe, one could expect to see marketed many variations, designed to fit a particular set of service conditions or with claims of possessing markedly better operating performance. Glass helix angles may be varied; liner reinforcements may be changed; liner thickness may be altered; filament winding may be changed to sock-type construction; outside protection may incorporate a UV shield, a resin-rich surface, or in some cases simply a coat of black paint. It is apparent from the great number of possibilities which exist that many different types and grades of pipe can be created. From a list of those available in the marketplace, this wide choice is perfectly apparent. While it is easy to list the variables that can affect pipe construction, the physical characteristics which the user desires are easier to classify. Normally, in the chemical plant, desirable characteristics are as follows:

1. A corrosionproof interior with a smooth inner surface.
2. A pipe which possesses considerable strength to resist physical damage. This is often a special requisite where resistance to hammering and vibration is necessary.
3. A resistance to failure from ultraviolet degradation when used or stored outside. A 3-mil coat of black epoxy or furane paint is the best recommendation for guarding against ultraviolet degradation in epoxies. UV inhibitors are not effective in epoxy pipe.
4. A good method of system assembly to provide installation in the shortest possible time.
5. Maximum reliability.
6. Ease of repair.

Normally, the high stress failure of glass-fiber-reinforced epoxy pipe is found to be due to loss of adhesion of the glass fiber. It is almost idle to speculate on stronger reinforcing mechanisms when what is actually required is an increased adhesion between the resin and the glass fiber. Magnification cross-section studies of pipe-wall failures due to overstress repeatedly show the failure to lie in this area. While an increase of 100 percent in reinforcing strength would be good, the reinforcing plastics designer would much rather have 50 percent increase in his coupling strength.

Strangely enough, whether the attack is mechanical or chemical, the brunt of the attack is borne by the glass-resin interface. Since this area is vital to the entire structure of the material, any decay results in a weakening of the entire laminate. This is one of the basic reasons why all glass-fiber reinforcement should be provided with excellent protection against corrosive liquids or fumes—a resin-rich surface is essential, both outside and in. The preservation of this resin-rich interface, whether by C glass reinforcement or an organic veil, is therefore

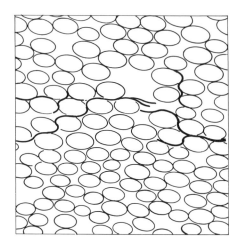

FIG. 4-1 Mechanism of failure in glass-fiber-reinforced epoxy pipe.[3]

a basic necessity. Its use provides good shrinkage control plus the sealing of structural reinforcement.

For the mechanism of failure of glass-fiber-reinforced epoxy pipe see Fig. 4-1.

4.3 EPOXY PIPING—COMMON CHARACTERISTICS

All good epoxy piping systems, at this writing, seem to have the following points in common:

1. They are a machine-made pipe, with an upside temperature limitation of approximately 300°F. Pressure ratings hold up well at elevated temperatures. Figure 4-2 shows the performance of 2-in. Chemline pipe over a range of −300 to +300°F. It will be readily observed that the physicals hold up remarkably on this well-constructed pipe.

2. With but few exceptions they possess extremely high impact resistance.

3. They are a filament-wound or sock-wound construction.

4. They possess a resin-rich inner liner for corrosion control. This smooth inner liner also provides a high flow factor and low gas permeability.

5. Failure, if it occurs, will generally be of a weeping type, permitting repairs to be made on regular maintenance tours, rather than catastrophic.

6. The main body of the pipe is of E glass construction, suitably wound or interwoven to give exceptionally high wall strength.

7. In general, they are finished on the outside with a pigmented

finish or some method of preventing ultraviolet degradation of the lining exterior.

8. Of emphatic importance: they are generally more expensive than a hand-laid-up polyester in sizes 6 in. and above.

9. Epoxy piping is assembled by adhesive systems rather than butt-and-strap joints, which are characteristic of polyester systems.

10. The thermal conductivity of epoxy piping is similar to that of the reinforced polyester systems. However, since, normally, the thickness of the epoxy piping is less than the polyester systems, heat loss from an epoxy system may be greater.

11. Edgewise porosity is the greatest enemy of filament-wound material. This point cannot be emphasized too strongly.

The base knowledge to permit use of heat-cured epoxy filament systems has been adequate since 1954.[2] Glass finishes compatible with the epoxies have been available for some time, although the techniques to eliminate edgewise porosity have been known and in use since 1958. The knowledge and practice does not seem to be universal. Edgewise porosity can result in 50 percent loss on the original design value of 80,000-psi tensile in a short period of time. Case histories are at hand

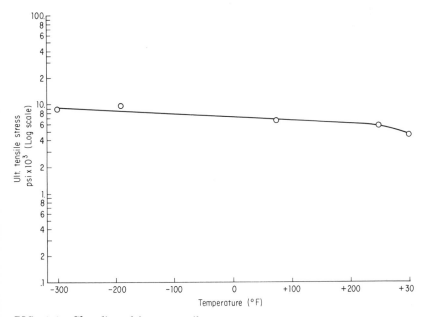

FIG. 4-2 Chemline ultimate tensile stress versus temperature, 2-in. IPS specimen.[4]

showing complete failures due to edgewise porosity in 90 days. It is known, however, that glass-reinforced epoxy piping structures which have been produced from known resins, glass finishes, and winding technique have given definite results and are completely predictable. These three variables, resins, finishes, and winding techniques, have been so well established that, by knowing any two of them, edgewise porosity can be used as a meaningful control to accurately measure the third.[2]

In epoxy liners 15 to 25 mils is construed as being useful for light duty and 40 to 60 mils is graded as heavy-duty liner. Now the testing of a liner is something that cannot be accomplished in a short time. One typical lining failure not discovered in short-term testing is blistering. This phenomenon is evidenced by blisterlike bubbles in light-duty liners subjected to high temperatures (let us say, 190 to 210°F) over a considerable period of time. Frequently, blistering will be picked up only in the long-term tests, running a year or more. This will limit service life. Normally, liners 40 to 60 mils in thickness, if well constructed, will be free of blistering. Such liners, however, should be reinforced with a C grade glass or some type of asbestos reinforcing, or checking or cracking of the liner may cause ultimate structural failure in a much shorter time. The liner should, of course, be an integral part of the structure, and definitely should not depend on an adhesive joining with a filament winding.

4.4 EPOXY FITTINGS—A GENERAL VIEW

Fittings are generally constructed to be sold with each type of epoxy pipe. These are usually of two types:

Filament-wound fitting   This type has physical characteristics comparable with the pipe. The filament-wound fitting is generally wound over a chemical-resistant inner core, so that the filament winding is fully protected. Filament winding itself has relatively poor resistance to corrosive chemicals and must be protected at all times. The construction of the inner core, that is, the chemical-resistant part of the fitting, is therefore of extreme importance. Filament-wound fittings are not universally built to a standard dimension, although some vendors are standardizing on the ASA long-radius dimensions or the ASA standard dimensions. In some cases a portion of a fitting line will meet these dimensions and the remaining portion of the line will not. This limits interchangeability between fitting lines unless both have been built to an ASA standard dimension. It is thus quite probable that a 6-in. 90° elbow from one vendor cannot be interchanged with a 6-in. 90° elbow from another vendor. These items need to be carefully considered. There will always be pressure from the individual vendor to use his line in chemical-plant work. Obviously, there must be a limit to this proliferation or the user will speedily discover that his spare-parts maintenance becomes a real problem.

While some of the filament-wound fittings are smoothly contoured, others are of mitered construction. As of this writing the mitered joints seem to be a source of weakness from which fitting failure may occur, especially in borderline conditions. Experience indicates the filament-wound smoothly contoured fitting to be more desirable for rugged service than a mitered fitting.

The flange adhesion on a filament-wound fitting may be a point of concern. As long as the inner liner of the fitting maintains its integrity, all will go well, but when lost, failure may well be in the flange area. This produces an attack on the filament overlay, so that fitting failure may result from the destruction of the filament-wound overlay at the flange area, and the fitting literally pulls apart, even though the flange adhesive joint is completely sound. This weakness has also been observed in reducers, where the extent of reduction is on the order of 2:1, such as an 8 X 4 or 6 X 3 reducer. Work is going ahead in this area to provide maximum reliability for long-term service.

Epoxy resin molded fitting or flange   These will normally have a heavier wall because of the reduced physicals, although chemical resistance will be the same as the parent material. Fittings of this type are commonly referred to as "gunk" (premix) molded fittings. Molding pressure varies from relatively low up to as high as 1,000 psi. There is no standardized design for these fittings, and they vary from vendor to vendor. Some of those making flanged fittings are attempting to hold to the ASA standard radius fitting so that it will be interchangeable with cast-iron or other lined fittings.

The filler in a gunk system is normally extremely short glass fibers, although again, this varies from vendor to vendor. Some vendors use a fiber as short as $\frac{1}{32}$ in., while others use a mixed-length filler, running from $\frac{1}{2}$ to $\frac{3}{4}$ in. in length. Longer reinforcing fibers appear to provide better physicals and resistance to cracking or breakage. Fortunately, advances are being made in this field at an appreciable rate, so that today's molded fitting provides considerably better performance than a fitting of five years ago. For a modern epoxy molded fitting see Fig. 4-3.

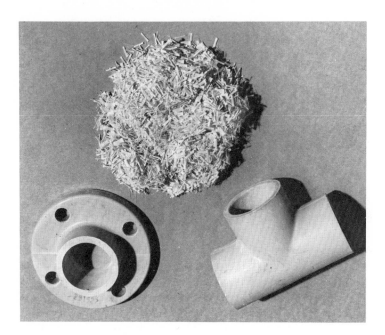

FIG. 4-3   High-strength glass-fiber-reinforced (60 percent glass) epoxy molding compounds are used to mold the pipe fittings in the foreground. The 2-in. tee on the right has an average burst pressure of 2,500 psi and a working pressure of 450 psi.[1]

4.5   MANUFACTURING METHODS—PIPE AND FITTINGS

Two of the more popular methods for manufacturing reinforced epoxy pipe are the following.

Bondstrand manufacturing process[5]

Epoxy pipe.   This is an unusual pipe in that the asbestos-filled interior liner is made with either polyester or epoxy resin and the wall structure is made with epoxy resin. The steps in the process are as follows:

1. Crocidolite asbestos mat, precut to the required dimension, is laid flat on a table and uniformly impregnated with a chemically resistant epoxy resin.
2. The steel mandrel is lowered onto the mat, then rolled forward. As the mandrel moves, it picks up the impregnated mat.

3. The mat around the mandrel is next uniformly compressed with a layer of glass roving and resin.

4. The mandrel is then moved to a filament-winding machine, where the epoxy wall is applied by the "wet-winding process."

In this process glass filaments in the form of roving are impregnated with an epoxy resin by passing through a resin bath. The impregnated roving is continuously applied to the mandrel in the form of a flat band. A moving carriage travels back and forth along a rotating mandrel. This carriage carries the resin bath and roving, thereby delivering the glass band to the mandrel. The band is applied to the mandrel in a precise helical angle calculated to give maximum strength.

Upon completion of the windings the mandrel and pipe are placed on a conveyor which runs through a curing oven. Here the pipe is raised to a temperature of 300°F. The pipe is completely cured as it leaves the oven. It remains only to remove the pipe from the mandrel and trim the ends to complete the process.

Fittings. Some fittings, such as tapered reducing bodies, may be filament-wound in the usual manner by applying glass and resin to a rotating mandrel. In the manufacture of more complicated shapes, such as tees and ells, a partial-lay-up technique is employed. The usual reinforcement of glass cloth or woven roving, however, is not used for the lay-up material. In order to maintain the high degree of strength afforded by the filament-winding process, a resin-impregnated mat is made by filament winding on a large mandrel. When the required thickness is on the mandrel, the mat is slit and unwrapped to form a flat sheet. This is then cut in appropriate shapes and layed up on a fitting mandrel over the liner material. The mandrel with the lay-up is over-wound with helical windings by placing the wrapped mandrel in a filament-winding machine. The uncured part is conveyed through an oven, reaching a cure temperature of 300°F.

Lower-pressure fittings may also be manufactured from asbestos- or glass-filled premix molding material, using either compression or transfer molding techniques. This type of fitting is in common use today.

Fibercast manufacturing process[6]   Production of thermosetting pipe by centrifugal casting, the equipment used to produce centrifugally cast pipe at Fibercast Company, consists of three major components:

1. A high-speed rotating cylindrical tube called the "mold tube," whose inside diameter conforms to the finished-product outside diameter

2. A heating jacket surrounding the rotating mold tube which controls the cure temperature

3. A mechanism for ejecting the cured pipe from the mold tube

The temperature of the mold tube can be set and accurately controlled up to 300°F. The speed at which the mold revolves can be varied by means of changing gears in the drive unit. The minimum rpm of the mold is generally considered to be the one at which complete wet-out of the glass is achieved, along with void-free pipe.

The materials that go into the pipe can be broken down into two basic categories:

1. Structural supporting material, or reinforcement. The reinforcement can be composed of glass, asbestos, sisal, and synthetic organic and metallic materials in the form of woven cloths, mats, braids, or random or directionally laid fibers. The most commonly used reinforcement for making centrifugally cast pipe is fiber glass, in its various forms.

2. The second material in centrifugally cast pipe is the thermosetting resin.

In the processing of centrifugally cast pipe, the reinforcement is first loaded into the mold tube; the preweighed or metered resin charge is then injected into the rotating mold. The resin charge is sufficient to more than cover the reinforcement, or structural-supporting fibers. The most commonly used thermosetting resins are epoxies, vinyl esters, and polyesters.

After a preset cure time, the rotating mold stops, at which time the pipe is pulled from the mold by means of mechanical jaws.

The advantages and disadvantages of centrifugal casting are as follows:

The resin-rich liner has a smooth surface which allows for excellent fluid-flow characteristics; also, the resin-rich liner gives maximum protection of the glass to chemical attack and abrasive wear. In addition, the smooth ID allows maximum protection against solids buildup. If solids buildup does occur, the smooth resin liner allows ease of solids removal.

It produces a pipe of smooth and uniform OD, which makes for more efficient fabrication in the field and superior appearance. The smooth-OD pipe also has improved electrical properties.

It usually has a lower glass content than filament-wound pipe; the advantage of this is increased chemical resistance and resistance to weeping. However, a disadvantage of the low glass content is the lower internal-pressure ultimates obtained at equal wall thicknesses in comparison with filament-wound pipe.

4.6 HELPFUL ENGINEERING HINTS IN INSTALLING EPOXY PIPING

1. Connections to pumps or equipment that involves vibration, shock, or mechanical movement should include flexible connections or expansion joints. This will remove the strain on pipe and fittings. A bellows-type expansion joint, such as a Teflon joint, is preferred.

2. Use full-faced gaskets in all installations. Where a reinforced epoxy flange is being bolted to a raised-face flange, a filler ring of hard material should be used to obtain an even bolt pressure and square alignment. See Fig. 4-4.

3. Torquing pressures of 35 to 50 lb are commonly recommended for reinforced epoxy flanges. Tracing of reinforced epoxy pipe to prevent freezing in outside lines is common. Tracing temperatures should not exceed 260°F, and the spiral method should be used. Do not trace one side of epoxy piping only, since an uneven expansion and possible bowing will occur. Consider well the thermal conductivity of the pipe in question.

4. Many other additional hints in joining RP pipe, particularly epoxy, will be found in Chap. 7.

5. For supporting and anchoring epoxy pipe the reader is referred to Chap. 8.

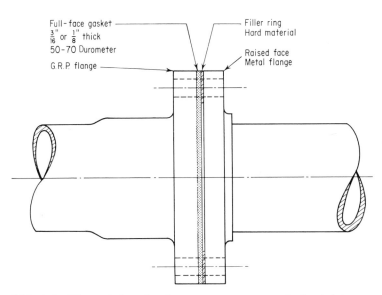

Full-face gasket
$\frac{3}{16}$" or $\frac{1}{8}$" thick
50-70 Durometer
G.R.P. flange

Filler ring
Hard material

Raised face
Metal flange

FIG. 4-4 Glass-reinforced polyester or epoxy to steel pipe joint.

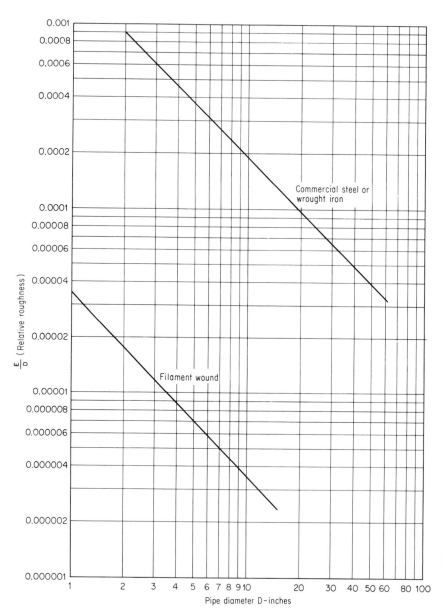

FIG. 4-5 Relative roughness, E/D, versus pipe diameter for filament-wound, commercial steel, or wrought-iron pipe.

6. The relative roughness, E/D, as a function of pipe diameter, in inches, for filament-wound (epoxy or polyester), commercial steel, or wrought-iron pipe is graphically illustrated in Fig. 4-5.[7] Because of this difference in relative roughness and because reinforced plastic pipe is free from tuberculation, which so commonly affects metallic pipe, it is quite often possible to use pipe one size smaller to do the job, or if installed the same size to obtain savings in pumping costs over the life of the installation.

7. For calculations of the common head loss versus flow rate the reader is referred to Fig. 3-12, which defines head loss as a function of flow quantity. In using Fig. 3-12 it should be borne in mind that the ID of standard polyester pipe conforms to the pipe size. Not so with all epoxy piping. Epoxy ID's are generally slightly larger than the corresponding nominal pipe sizes. There is so much variation in the inside diameters of epoxy piping that the reader must identify the exact ID of the pipe in question and then use Fig. 3-12 for a suitable interpolation. The Williams and Hazen coefficients for epoxy and polyester pipe are approximately the same.

8. "Quick-thread,"[8] or "fast-thread," piping systems have been developed for use in mildly corrosive service with epoxy piping. This type of assembly is commonly in use in oil-field applications, where it is resistant to the corrosive effects of crude oil, saltwater, natural gas, soils, and H_2S. The high resistance to scale buildup provided by the smooth inner surface substantially reduces paraffin deposits in 95 percent of the installations where it had formerly been a problem. Cost studies have shown that, on a basis of installed costs, including total material costs, stringing and laying, and ditching and backfilling, epoxy fiber-glass line pipe can be installed for as much as 21 percent less than comparable lined and coated steel pipe. Under these conditions the fiber-glass epoxy pipe has outlasted its predecessors manyfold. Threaded epoxy line pipe may fill a need in chemical-plant requirements. Where service conditions are severe, consultation with the manufacturer on the proposed use should be solicited.

4.7 REPAIR OF EPOXY PIPING

Many of the vendors will furnish, at nominal cost, pipe-repair kits containing all the components needed to make a patch, should a pipe be locally damaged. They are intended for use in a temporary repair job to minimize downtime. In general, they consist of:

1. A quick preparation of the surface to be repaired
2. The adhesive and catalyst
3. A section of pipe to embed in the adhesive

In this method of repair acceleration can be accomplished by heating the part. Again, the engineer should make sure that where heating is used it is done in a safe manner. The enemies of a good repair job are cold and dampness. Both must be dispelled if success is to be achieved. In Chap. 7, detailed instructions for fabrication under low-temperature conditions are to be found. Basically, this involves the application of heat to the area being cured. The joint or adhesive should be brought up to a temperature of approximately 180° over a 5-min period, kept there for another 10 min, and then slowly increased to 200 to 230°F for another 10 min. Watch out for bubbling. If it occurs, stop heating and reapply. Let cool to 100°F before handling or using.[9] Another helpful step, in cold weather, is to heat the adhesive on a hot plate to about 120°F, then remove it and add the catalyst. Warm, dry, clean surfaces and working conditions are the key to good repairs.

The author has also often used a polyester patch, successfully, to repair an epoxy pipe. This was done, literally, by making a butt-and-strap joint over the damaged section. First, of course, a section was sanded to obtain good adherence. Since the polyester shrinks on curing, the repair bonds tightly, and is a very effective method. The secret is to obtain a clean, dry, roughened surface for bonding.

4.8 REINFORCED EPOXY PIPE—EQUIVALENT LENGTHS FOR FITTINGS[10]

Here we are referring to the lengths of straight pipe that have the same resistance to fluid flow as the common fittings listed in Table 4-1. Equivalent lengths are in feet.

EQUIVALENT FRICTION LOSSES IN PIPE FITTINGS

Table 4-1

				Pipe size, in.				
Fitting	$1\frac{1}{2}$	2	3	4	6	8	10	12
90° standard elbow......	4	5	8	10	15	19	24	28
45° standard elbow......	3	3	4	6	8	10	13	15
Standard tee (flow through run).....	3	4	5	7	10	13	16	19
Standard tee (flow through branch)..	9	10	15	20	29	38	47	55

4.9 INSTALLATION COSTS OF EPOXY PIPING SYSTEMS

To provide the engineer or estimator with some rough guidelines for estimating installation costs in chemical service, a number of tables have been prepared, and are discussed in detail below. They must be used with the usual degree of caution, but should permit the establishment of preliminary figures on the cost of installing an epoxy system.

Table 4-2 This table shows the price ranges at which epoxy pipe and fittings with 150-psi rating can be purchased in the market today. Over a period of years the competitive position of the user in the marketplace has improved in the epoxy field. Although the table is largely self-explanatory, there are the usual factors which influence the purchasing position of the user in his attempt to obtain the best system at a reasonable price. Some of these factors are:

  1. Volume discounts.
  2. The use of socket- and coupling-type designs as against flanged systems. Flanged systems are inherently more expensive. The designer should use only such flanges in system design as are necessary for assembly and maintenance.
  3. Normally, installation can be without insulation unless temperatures are above 212°F, or where safety factors may be involved or outdoors in severe cold, where freezing may be involved.
  4. Price variations within the industry on reinforced epoxy pipe and fittings at this time are considerable. The higher prices are justified by vendors claiming to offer better pipe, but experience has indicated that satisfactory material may be purchased that will provide long-range performance in the price structure indicated. Where a question exists, the estimator may feel confident in using the mid-average of the bracket for satisfactory offerings.

Table 4-3 Since no two piping systems are ever exactly the same, the estimating engineer needs a reference source from which manpower requirements can be determined. Table 4-3 is the result of a study of a considerable number of field installations in order to establish average labor units. Taking the number of joints and fittings from the drawing, the estimator is then able to determine the manpower requirements for the job. By multiplying these manpower requirements by the labor rate, an approximate labor figure will be achieved.

Table 4-4 Quite obviously, the installed cost of any epoxy piping system has a large number of variable parameters, so that Table 4-4 is satis-

PRICE RANGES—EPOXY PIPE AND FITTINGS, 150-PSI RATING

Table 4-2

Pipe size, in.	Pipe cost per foot	90° elbow Socket type	90° elbow Flanged	45° elbow Socket type	45° elbow Flanged	Tee Socket type	Tee Flanged	Cement coupling	Flange	Blind flange	Reducer by larger size Socket type	Reducer by larger size Flanged
2	$1.80-$2.45 $2.20 avg	$8.05-$10.10	$22-$29	$8.05-$10.10	$22.25-$29.40	$9.40-$11.75	$31-$39	$2.20-$2.75	$3.60-$4.50	$5.00-$6.50	....	$24-$25
3	$2.45-$3.45 $2.80 avg	$11.05-$13.60	$28-$35	$11.05-$13.60	$28.00-$36.00	$13.70-$17.05	$39-$49	$3.20-$4.05	$4.30-$5.35	$6.40-$7.70	$24-$28	$28-$30
4	$4.30-$5.20 $4.75 avg	$13.65-$17.15	$35-$45	$13.65-$17.15	$36.00-$45.00	$17.25-$21.25	$48-$60	$4.40-$5.50	$6.25-$7.85	$9.00-$9.75	$38-$39	$37-$40
6	$5.25-$8.70 $6.40 avg	$38.75-$51.50	$56-$73	$38.74-$51.50	$56.00-$73.00	$50.00-$62.00	$75-$93	$7.00-$8.70	$8.45-$10.75	$14.75-$18.50	$49-$54	$44-$54
8	$7.75-$11.20 $8.65 avg	$56.00-$72.00	$76-$106	$56.00-$72.00	$76.00-$105.00	$71.00-$88.00	$109-$135	$11.70-$19.50	$13.30-$16.80	$21.60-$26.80	$68-$72	$75-$82
10	$9.90-$14.45 $11.50 avg	$68.00-$103.00	$114-$161	$68.00-$103.00	$114.00-$161.00	$106.00-$130.00	$140-$207	$16.50-$19.50	$21.00-$30.00	$33.00-$40.50	$74-$76	$92-$98
12	$12.00-$15.40 $13.50 avg	$82.00-$100.00	$132-$170	$82.00-$100.00	$132.00-$170.00	$123.00-$150.00	$173-$237	$22.00-$24.50	$31.00-$38.00	$44.00-$54.00	$90-$96	$119-$127

NOTE: There are other special fittings available, not listed here, such as crosses, reducing flanges, reducing tees, reducing bushings, saddles, laterals, U turns, traps, and pipe plugs. Many special grades of epoxy pipe and tubing are also available.

PIPING FABRICATION AND INSTALLATION[11]
EPOXY PIPE

Table 4-3

	Hours per joint by size, in.						
	2	3	4	.6	8	10	12
Adhesive couplings							
1. Shop installation............	0.3	0.3	0.4	0.7	0.9	1.1	1.3
2. Field installation............	0.5	0.5	0.6	1.0	1.2	1.5	1.7
Flanged joints							
3. Shop fabrication							
a. Install flanges............	0.6	0.7	0.8	1.2	1.4	1.6	1.8
b. Bolt up flange joint.......	0.2	0.2	0.3	0.4	0.5	0.7	0.8
4. Install flanged joint—field...	0.4	0.6	0.8	1.1	1.5	1.8	2.1
5. Permanent hanging at 10-ft height per linear foot of pipe..	0.15	0.15	0.15	0.2	0.2	0.25	0.25
6. Average per linear foot for total job (combination of above).....................	0.22	0.23	0.25	0.35	0.39	0.47	0.51

NOTE: All times include layout, measuring, travel, materials procurement, and all other auxiliary work. Item 4 includes only those joints not bolted up in shop. Item 6 is an approximate man-hour figure which may be used for total installation of a job. Recognize it as only approximate, but good enough to serve.

TYPICAL EPOXY PIPING SYSTEM,
INSTALLED COST PER FOOT[11]

Table 4-4

	Pipe size, in.					
	2	4	6	8	10	12
70% adhesive joints, 30% flanged joints..........	3.23	5.74	7.04	11.84	16.23	19.24
50% adhesive joints, 50% flanged joints..........	3.50	6.21	9.31	12.75	18.07	21.54
100% flanged joints.......	4.17	7.25	10.23	13.96	21.28	25.52

NOTE: The above total installation cost is based on a labor rate of $3/hr. To convert to other rates, refer to Table 4-6 for the appropriate conversion factor.

factory for preliminary estimating work only.   It is based on:

1. The purchase of epoxy pipe and fittings at the midpoint of the piping range and in the lower third of the fitting range.
2. A labor performance essentially as shown in Table 4-3.
3. No difficult hanging or relocation work.   It is presumed that the installation will be approximately 10 ft above the floor and that conventional beam clamps, rods, and saddle-type hangers will be used.
4. Labor costs in these calculations at $3/hr.

Using the same system design, the study went on to vary the method of construction to determine the effect on the final cost.   The study showed that adhesive joints represent the minimum-cost approach and that the system design should be such that flanged joints should be used only to the extent necessary for assembly and maintenance.   As the number of flanged joints in a system is increased, the cost of a system goes up.

Based on this study, labor and material costs as a function of the piping size were distributed as shown in Table 4-5.

Table 4-5   It is evident from the cost study that material/labor ratios are exceedingly high.   This is especially significant in high-cost labor areas.   Even if labor costs were escalated to the $4 to $4.50 area, the material/labor ratio would be very favorable, especially when compared with other types of materials used in corrosion-resistant piping.   The reader is referred to Table 3-6, which shows typical installed costs for various types of piping and the various material/labor ratios which can be achieved.   It will also be observed, from Table 4-5, that material/labor ratios are partially a function of piping size.   This is particularly true in working reinforced plastic material because the ease of fabrication and

DISTRIBUTION OF LABOR AND MATERIAL COSTS
AS A FUNCTION OF LABOR COSTS

Table 4-5

Pipe size, in.	Labor, %	Material, %	Total, %	Ratio M/L
2	19	81	100	4.25
4	12	88	100	7.35
6	12	88	100	7.35
8	10	90	100	9.0
10	8	92	100	11.5
12	8	92	100	11.5

the lightness of the material minimize handling costs and thus contribute to a better material/labor ratio as the sizes increase.

Table 4-6   This table was prepared to permit simple correction of Table 4-4 with changes in labor rates.

Problem

Let us suppose that we wish to correct Table 4-4 in 10-in. pipe for a $5/hr labor factor to be installed by a competent contractor.   If our installation were 50 percent adhesive joints and 50 percent flanged joints, the estimated installed cost would be

18.07 × 1.05 = $18.97 per foot

Note that the purchasing practices on reinforced plastic epoxy pipe have a greater bearing on the installed cost than the labor rate.

MULTIPLYING FACTORS FOR TABLE 4-4 BASED ON VARIABLE LABOR RATES

Table 4-6

Pipe size, in.	Labor rates per hour				
	$2	$3	$4	$5	$6
2	0.94	1.00	1.06	1.13	1.19
4	0.96	1.00	1.04	1.08	1.12
6	0.96	1.00	1.04	1.08	1.12
8	0.97	1.00	1.03	1.07	1.10
10	0.97	1.00	1.03	1.05	1.08
12	0.97	1.00	1.03	1.05	1.08

Distribution of installed footage by pipe size   A case-history study of a large chemical plant covering epoxy piping installed over a 10-year period revealed a suspected size distribution conforming to economic studies.   See Fig. 4-5A.   Most popular sizes were 2, 3, and 4 in., which represented a good economic selection in view of the competition.   Some 88 percent of all the reinforced piping installed was in the 2-, 3-, and 4-in. sizes.   Only 12 percent represented 6- and 8-in.   Not shown on the graph, being actually less than 1 percent, was the 10-in. size.   Since this survey was made, some quantities of 1-in. were being used.

4.10 BRIEF CASE HISTORIES OF EPOXY PIPING

Heater discharge piping   In the rayon industry it is customary practice to provide a hot dilute sulfuric acid solution as a "stretch" liquor for additional yarn strength.   For reasons of economy this solution is con-

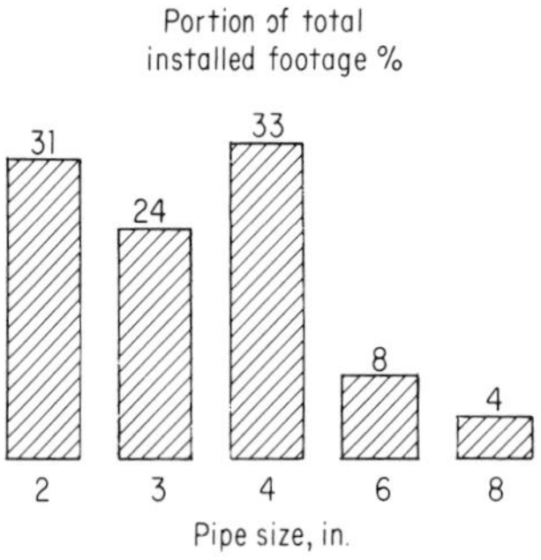

Portion of total
installed footage %

31   33   24   8   4

2   3   4   6   8

Pipe size, in.

FIG. 4-5A   Portion of total installed   footage,   percent, epoxy piping.

FIG. 4-6   Filament-wound reinforced epoxy piping handles 200°F condensate successfully.

tinually recirculated with small amounts of water and sulfuric acid makeup. By hot we mean above 90°C. Normally, the heat is provided by an injection-type heater. Now, it is difficult to imagine a more corrosive environment than the short section of piping leading from this heater as the collapsing steam bubbles subject the piping walls to a continuous hydraulic hammering. Often the life of lead-lined steel pipe would be measured in terms of weeks or perhaps a few months. In this same application filament-wound reinforced epoxy piping has run 1 to 2 years. When the salesman arrives with a new line of pipe, this is the ground on which it is tested. If it survives this test, we can be sure it will safely serve for most other uses.

Condensate lines   Condensate can, in practice, prove one of the most pesky corrosive solutions. High temperatures and borderline pH can be extremely troublesome. Figure 4-6 shows a filament-wound RP condensate piping arrangement which has been in service for 4 years and which has appeared frequently in magazine releases by various companies. Under similar conditions it will outlast copper, wrought iron, steel, or lead-lined steel.

FIG. 4-7   Reinforced epoxy battery piping is used in this fully automated ion-exchange system to effect a saving of $7,000 over rubber-lined steel.

An ion-exchange plant   This plant was constructed for the recovery of a heavy metal from an acid-waste stream.   The heavy metal was present in concentrations of 400 to 600 ppm, in a hot solution of 20 percent sulfuric acid.   Feed to the ion-exchange system was designed to operate under considerable pressure.   This fully automated system used a glass-reinforced epoxy piping system as a material of choice, and is shown in Fig. 4-7.   Replacing rubber-lined steel with epoxy battery piping, covering three ion exchangers and two small tanks, resulted in a saving of $7,000.

Additive system—central mixing room   Many additive and finish systems are a "natural" for reinforced epoxy pipe.   Figures 4-8 to 4-10 show three views of reinforced epoxy pipe and fittings carrying paper additives from a central mixing room to a number of paper machines, at a large plant located in Louisiana.   This system is reported to be the world's largest central mixing and distribution system.   Installation was by a

FIG. 4-8   Additive-system manifold.

FIG. 4-9 Additive-system piping between central mixing room and paper machines.

contractor. Note the judicious assembly and socket-type fittings, using adhesives, with flanges kept to a minimum for maintenance and assembly. Also note the use of nipples and saddles for easy wide entrance into the RP epoxy piping.

4.11 ADVANTAGES OF MACHINE–MADE EPOXY PIPING SYSTEMS OVER HAND–LAID–UP POLYESTER SYSTEMS

1. The tremendous built-in strength capabilities of the machine-made (either filament-wound or sock-type construction) systems are indeed impressive. These systems are capable of resisting great abuse from internal hammering and shock, which would destroy a random-mat pipe. Catastrophic failures in these machine-made piping systems have a very low incidence rate. Typical failure when it occurs is one of weeping, which may be adequately repaired or replaced in normal maintenance

FIG. 4-10   Distribution piping associated with paper machines.

tours.   Where a piping system is subjected to hammering, coupled with
intense corrosive conditions at high temperatures, the material of choice,
provided the chemical service conditions are suitable, will generally be a
machine-made epoxy pipe.

2. Normally, the assembly options available on machine-made epoxy
pipe (either filament-wound or sock-type construction) are fairly wide.
Adhesive systems are common, and assembly can be accomplished with
good speed.   Where assembly time is of the essence, the coupling socket-
type-fitting adhesive system will provide a rapid rate.   Adhesive socket-
and-bell construction is also available to permit good assembly time.

3. The general resistance of epoxies to alkaline systems is somewhat
superior to the polyesters, so that, where exposure to high-alkaline conditions is prevalent, the material of choice may very well be an epoxy.
Since there are no hand-laid-up epoxy systems available, it follows that
the engineer must automatically choose a machine-made epoxy system.

4. The outside diameter of the machine-made epoxy piping system
is rigorously controlled.   This produces a pipe with a pleasing external

appearance and one with which the craftsman, previously associated with stainless steel or steel pipe, is familiar. Some of the epoxy-pipe vendors have standardized on the steel pipe, Schedule 40, OD as a standard dimension. In fact, a great deal of reinforced epoxy pipe engineering has gone into identifying the RP epoxy pipe with a steel pipe in such matters as outside diameter, supporting distances, and fitting face-to-face dimensions. All this is intended to keep the engineer on familiar ground in effecting his transition from steel or stainless-steel piping to the light, tough reinforced epoxy piping. Also, the coefficient of linear thermal expansion is quite often about the same as that of steel.

5. Composite designed flanges made for use with epoxy piping have successfully been used with ring gaskets for a number of years with good results. This is contrary to the manufacturer's instructions, which specify full-faced gaskets (practically a universal specification with RP piping). Many of the epoxy fittings manufactured today are of the premixed type, which will not perform satisfactorily with ring gaskets.

6. Certain sizes are manufactured in epoxy piping which are not generally available in the hand-laid-up polyester piping. These sizes are 1, 1½, and 2½ in. They can, of course, be made to order in polyester pipe, but are generally constructed as special items at a premium price. These smaller sizes of epoxy piping permit a wider adoption of this type of material and render them suitable for compressed-air lines in corrosive areas, electrical conduit, sampling lines, distribution nozzles, bleed lines, and a host of other applications which require small corrosion-resistant piping of the RP type. They are available with a complete line of fittings, and also screwed ends, in addition to the adhesive type of fittings. Various grades of this small piping are offered to permit the user to achieve real economies.

REFERENCES

[1] Kelly, Mark E., Jr., and Richard Hof: Glass Fiber Reinforced Epoxy Piping Systems, *Materials Protection*, vol. 4, no. 10, pp. 50–53, October, 1965.

[2] Boggs, H. D., Amercoat Corp., Brea, Calif., personal communication, Nov. 7, 1966.

[3] Kelly, Mark E., A. O. Smith Corp., Smith Plastics Div., Little Rock, Ark., personal communication, Oct. 7, 1966.

[4] Fiser, J. T.: The Effects of Temperature from −300°F. to +300°F. on the Tensile Properties of Red Thread, Green Thread, and Chemline, A. O. Smith Corp. *Technical Report*, Little Rock, Ark., Sept. 17, 1965.

[5] Information on manufacturing process furnished by C. G. Munger, Amercoat Corp., Brea, Calif.

[6] Information on manufacturing process furnished by Fibercast Company, Sand Springs, Okla., July 14, 1967.

⁷ Munger, C. G., Amercoat Corp., Brea, Calif., personal communication, November, 1966.

⁸ Como, Frank: A Survey of Fiber Glass Epoxy Lined Pipe Service, *Materials Protection*, February, 1968.

⁹ Amercoat Corp., Bondstrand *Bulletin* R3-64, Brea, Calif., 1964.

¹⁰ Information furnished by A. O. Smith Corp., Smith Plastics Div., Little Rock, Ark., 1964.

¹¹ Loftin, E. E., FMC Corporation, American Viscose Div., Front Royal, Va., personal communication, July 10, 1967.

5

Composite Piping, Tanks, and Structures

5.1 INTRODUCTION

The field of composite materials seems to be almost limitless. Here the engineer has an opportunity to tailor-make the particular application, be it piping, tanks for static storage, or combination RP structures, such as stacks, towers, ducts, vents, fans, trays, hoods, etc.

By judiciously choosing the material in contact with the corrosive, and subsequently overlaying it with an RP material, the application range of the substrate and overlay can be improved through synergistic action to produce a piping system, tank, or other structure which neither alone could achieve. Composite thermoplastic and thermoset construction has been more widely used in Europe than in the United States, but its usage in this country is increasing. Composite structures, however, are not limited to thermoplastics and thermosets; they can also be used in thermoset-and-thermoset construction. The design of a mixing chamber constructed of PVC and polyester composite material is shown in Fig. 5-1, and a number of these structures are shown in operation in Fig. 5-2.

Some of the other advantages to be found in composite tanks are better suitability for the corrosion environment of the inner liner and added improvements, as follows:

1. Strength by the addition of fibrous-glass reinforcing to thermoset materials.

2. Chemical resistance, tailor-made by a combination of thermoplastic and thermoset materials, reinforced with fibrous material.

3. Equipment economies through composite glass-resin materials such as isophthalic and bisphenol resins.

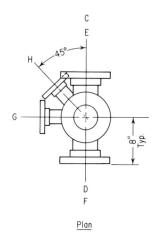

Plan

Nozzle schedule			
Nozzle	Size	Type	Remarks
A	4"	Flanged	Vent
B	6"	"	Outlet
C	4"	"	Mixed acid
D	4"	"	Blend acid
E	4"	"	Mixed acid
F	4"	"	Filtered acid
G	2"	"	93% H_2SO_4
H	2"	"	Liquid sulfate solution

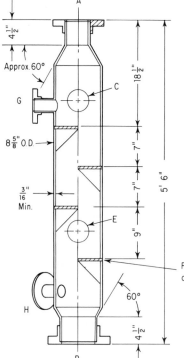

Material: High temp. PVC or polypropylene overwound with $\frac{1}{4}$" reinforced polyester. Overwind extends to flange hub.

Four semicircular baffles $\frac{1}{2}$" thick with $\frac{1}{2}$" thick 45° gussets at center cut slots in wall to suit; heat fuse baffles to wall.

FIG. 5-1   Cross-sectional view of composite mixing chamber.

FIG.5-2 Multiple-composite mixing-chamber installation. Eighteen systems for spin-bath control are centralized in a large control room. Here baths are continuously adjusted, using various ingredients to maintain specifications. Corrosive chemicals are blended in mixing chambers to ensure homogeneity before being added to the process.

4. Dimensional stability through composites of fibrous-glass-reinforced polyesters and epoxies.

5. Abrasion resistance by the addition of fine silica to a resin-glass matrix.

6. Resistance to ultraviolet (sunlight) degradation by the addition of UV inhibitors to the resin-glass matrix.

7. Both chemical and abrasion resistance through the use of organic veiling in resinous materials.

8. Fire retardancy by the addition of antimony trioxide with the resin.

Basically, composite materials are as old as nature itself. Bone, bamboo, horn, and hair are composite materials.

Composite materials have become so loosely defined as to mean a change in lay-up across the tank wall. For example, the following type of tank has been designated as a composite structure:[1,2]

1. An interior high-chemical-resistance resin reinforced with glass surfacing mat, 5 to 10 percent by weight.

2. A layer of resin-impregnated chopped-strand glass, 30 percent by weight. This is a second-line corrosion barrier.

3. A third layer of filament-wound material to carry the bulk of the wall strength and running 60 to 75 percent glass.

4. A final outside layer reinforced with surfacing mat and UV inhibitor. *There has thus been created a new four-ply composite structure, each ply of which is built specifically to perform a certain function. It has much to recommend it.*

Composite tanks are available in sizes up to 12 ft in diameter and heights to 30 ft.

Structural overlays may be done in filament-wound epoxy fiber glass or filament-wound polyester fiber glass. Structural overlays are also available in the hand-laid-up polyesters, the internal liner being specifically tailored for the application. By these techniques it is possible to handle solvents, 93 percent sulfuric acid, hydrofluoric acid, etc. Indeed, it is difficult not to become enthusiastic about composite materials. Some of the more interesting uses of them are examined in the remainder of this chapter.

5.2 ABS (ACRYLONITRILE–BUTADIENE–STYRENE) AND RP POLYMER[3]

The wide corrosion-resistance properties of ABS polymer pipe have been known and utilized for nearly twenty years. It has good chemical and abrasion resistance over a variety of chemicals, such as chlorine gas, dilute sulfuric acid, and ferrous sulfate solution. Other chemical solutions to which it is resistant are calcium chloride, calcium hydroxide, citric acid, copper sulfate, distilled water, ferric chloride, Kraft liquor, sodium chloride, and 50 percent sulfuric acid. Nearly all these chemicals show resistance to ABS up to 160°F. The conventional plastic, however, has a long-term hoop stress of only about 1,200 psi, although stronger modifications have been developed which extend it up to 3,200 psi. It would naturally follow that a prestressed filament-wound overlay of the ABS would produce a lightweight, strong, corrosion-resistant pipe which would extend the capabilities of the ABS thermoplastic resin. This piping, manufactured under the name of Fiberplus,* is currently available in sizes 4 to 15 in. in diameter. It is reported to excel in applications demanding abrasion resistance, and has other uses, such as air ducts, food-processing piping, gas mains, sewer piping, transportation of sand and gravel, water mains, etc. The piping is available in pressure ratings from 75 to 500 psig. The pipe is naturally a nonconductor, not

---

* Registered trademark of Fiberplus Pipe, Inc., Costa Mesa, Calif.

PHYSICAL DATA AND COST OF ABS AND POLYESTER COMPOSITE PIPE, CLASS II FIBER-PLUS PIPE[3]   Nominal Length 40 Ft

Table 5-1

Nominal size, in.	Nominal ID, in.	Nominal OD, in.	Nominal wall, in.	Weight per linear foot, lb	Working psi, lb	Cost per foot (list), dollars
4	4.075	4.403	0.164	1.92	300	2.77
6	6.076	6.408	0.166	2.89	300	4.13
8	8.000	8.348	0.174	3.80	225	5.12
10	10.000	10.388	0.194	5.25	225	7.10
12	12.000	12.428	0.214	6.69	150	9.04
15	15.000	15.488	0.244	10.90	150	15.20

subject to electrolysis, and possesses the usual plus factors of low heat conductivity, light weight, and low installation costs, commonly found in plastic piping.   To give an idea of piping costs of this material, Table 5-1 shows the cost of 150- to 300-psig working-pressure pipe and fittings. The ABS inner core is not resistant to solvents such as acetone, aniline, benzene, gasoline, or other highly oxidizing acids, such as 50 percent chromic acid, 40 percent nitric acid, 93 percent sulfuric acid.

A unique wrapping method has been prepared for this type of pipe which exposes the ABS pipe and then literally applies a fiber-glass wrap through a bag molding kit around the outside joints of the two pipes. Methods have also been developed coupling Fiberplus to steel, and mating steel flanges to Fiberplus flanges.   The pipe is normally sold in lengths up to 40 ft.

5.3 BISPHENOL AND ISOPHTHALIC FIBROUS-GLASS SYSTEMS[4]

In the cost structures the high-performance chlorinated polyester and bisphenol polyesters have the same price relationship.   Isophthalic resins may be purchased for approximately two-thirds the cost of the bisphenol resins.   General-purpose polyesters may be purchased for approximately half the price of high-performance bisphenol resins.   It naturally follows that composite structures could therefore be built with combinations of bisphenol and isophthalic, chlorinated polyester and isophthalic, or bisphenol and general-purpose, etc.   Any of these combinations, adequately engineered, should produce a satisfactory lay-up and result in substantial cost savings. As an example, a tank constructed using the dual-laminate approach would have a high-performance bisphenol polyester interior of 60 to 120 mils thickness, followed by the body of the tank of an isophthalic overwind. The tank would then be finished using a high-performance bisphenol resin on the outside surfacing mat. Such a composite construction as this may reduce tank costs by 10 to 20 percent over a construction of all-high-performance bisphenol or chlorinated polyester. It is being done by a number of companies. In evaluating bidding on polyester tanks, bear in mind that a low bid can be achieved through such construction. If necessary, have all the bidders rebid on the composite construction to achieve the minimum bidding prices.

The isophthalics are used more often than the general-purpose resins in combination with high-performance chemical resins. This is done because the isophthalics possess a much greater degree of chemical resistance than the general-purpose resins, and yet provide some savings in costs due to lower resin costs. In dual-laminate construction a tank which would be completely constructed of a high-performance bisphenol resin costing $5,000 could probably be built in composite lay-up for $4,000 to $4,500. The purchaser should realize, however, that if the inner corrosion barrier is breached, the entire structure may be in serious difficulty unless prompt repair measures are taken with the high-performance resin.

5.4 EPOXY AND RP POLYESTER

Some types of centrifugally cast polyester pipe are quite often loaded with carbon black. These centrifugally cast pipes possessing no internal glass reinforcement have a high degree of corrosion resistance in the areas where epoxy resins are adaptable. However, they lack the shatter and impact resistance associated with glass reinforcement. Some of these centrifugally cast pipe are overlaid with a filament-wound reinforced polyester, or if desired, a filament-wound epoxy, to give maximum strength in the areas where they may be subjected to impact and vibration. A stainless-steel wire net may also be embedded in the outer $\frac{1}{8}$ in. of a composite lay-up pipe to provide added shock resistance.

5.5 FURANE AND REINFORCED POLYESTER[5,6]

The use of furane and polyester composites affords the engineer another means of problem solving. The furanes themselves will not support combustion, having a tunnel-test rating of less than 20; thus they are rated as noninflammable. In weight we are dealing with a material approximately one-fifth that of conventional iron or steel fittings. Thermal conductivity is slightly higher than the polyesters, but not appreciably so. Furane-lined piping, tanks, and RP structures are widely available from a variety of fabricators. The strong point of the furane systems is their resistance to solvents such as acetone, ethyl alcohol, benzene, carbon tetrachloride, carbon disulfide, chloroform, the fatty acids, methyl ethyl ketone, toluene, and xylene, many of which would quickly impair the polyester or epoxy resins.

While the furanes excel in solvent resistance, there are some areas in which they are not as good as the high-chemical-resistance polyesters. A few of these areas are wet and dry chlorine gas; chromic acid plating solutions; hypochlorous acid; some of the nitrate solutions such as lead, nickel, and zinc; brine solutions saturated with chlorine; sodium hypochlorite solutions; trichloroacetic acid; etc.

FIG. 5-3 Furane and polyester composite storage tank. The excellent corrosion resistance of the furanes is supplemented with fiber-glass-reinforced polyester overlays to form a composite tank which has resisted dilute sulfuric acid solutions near their boiling point for many years.

For the most part, however, the furane-lined RP structures have a very broad spectrum of corrosion resistance in both acids and alkalis. The furanes themselves do not possess the rugged physical strength of a glass-reinforced structure, although corrosion-resistant piping in low-pressure supply and drain-line work may be purchased, built completely from a furane resin. In addition, some tanks and structures have been made solely from furane resins. However, using the furane as the inner lining of a composite structure, with the fibrous-glass polyester overlay either hand-laid-up or filament-wound, effectively unites the best qualities of both systems in many areas of application. See Fig. 5-3 for a furane-lined polyester tank which has been in a severe corrosive environment of 5 percent sulfuric acid at 200°F for 5 years with completely satisfactory performance. Note the metal backup on the heads, to improve strength. Advances in tank design on this type of structure now make these metal backup heads unnecessary, and thus reduce tank costs.

5.6 GLASS AND RP POLYESTER[7]

One of the major suppliers of glass process equipment has combined the excellent corrosion resistance of borosilicate glass with the tough filament-wound laminate of a fiber-glass polyester resin. The filament-wound polyester greatly increases the impact resistance of the system, and has added a threefold increase in pressure ratings, up to 150 psig, and temperatures to 350°F. The outer filament winding, even under a heavy blow, maintains system integrity and prevents catastrophic failure. Pipe may weep or drip, but there is always time to replace the damaged section in an orderly fashion. Although such accidents are not common, they do occur. By the use of the overlaid reinforcement, the unique corrosion properties of glass have been greatly expanded.

Cost studies on an installed cost basis indicate this composite piping to be comparable (within ±10 percent) with some aluminum systems, Schedule 10 Monel-welded, Schedule 10 nickel, Saran-lined steel, Schedule 40 in 316 stainless steel butt-welded, and rubber-lined steel. It offers considerable cost economies when compared with Hastelloy C, Inconel, Teflon-lined steel, and glass-lined steel.

5.7 POLYFLUOROCARBON AND REINFORCED POLYESTER[8]

Open-topped chemical tanks made from a polyfluorocarbon lining 8 to 20 mils thick are being marketed in Japan. Such a lining is preformed on a wooden mold. Following this a fibrous-glass-reinforced polyester is spread on to a thickness of 120 to 160 mils to provide additional strength and rigidity. Stainless rims and bars are used for the additional reinforcement of the tank body.

Such tanks are being marketed for use in chrome-plating operations and are claimed to be satisfactory in a temperature range of -180 to $+400°F$. Suggested applications would be those in which the polyfluorocarbon lining would show satisfactory resistance.

5.8 POLYPROPYLENE LINED—FIBER-GLASS-REINFORCED POLYESTER[9]

Here a special preparation of the polypropylene sheet provides an effective bond between the two materials with a minimum bonding shear of 500 psi. Distortion temperature of the combination is higher than the polypropylene alone. Such applications are well suited for food handling and food processing. The fiber-glass side of the laminate affords an excellent opportunity for high-strength bonding of structural additions. Polypropylene lining is normally 80 to 125 mils. Fiber-glass reinforcement is added as required and is increased in high-stress areas.

There are techniques to ensure perfect joints and linings through spark testing. Normally, it is desirable to weld the lining joints on both sides since this produces the strongest possible construction. Welding of the polypropylene is usually hot-gas welding, and weld strengths of up to 90 percent are readily achieved. The use of adhesives with polypropylene gives poor results and is to be avoided. The epoxy and polyester adhesives may, however, be used in repairing the laminated structure. Repair will be required only if there is a weld failure or mechanical damage. Repair of mechanical damage is accomplished by cutting out the damaged area until all delamination is removed. The polypropylene repair section should then be welded into the original polypropylene. The repair reinforcement is then overlapped over the damaged section and tied into the original overlay reinforcement.

Some tests have indicated that the use of polypropylene materials as a reinforcing medium provides improved impact resistance. This leads to the observation that a desirable composite could be developed in which polypropylene fiber was used as a reinforcing material at impact points, while glass was used for the remainder of the structure to provide the suitable physical characteristics. For example, a pipe whose inner wall was reinforced with polypropylene on which an overlay of glass was provided would have interesting possibilities.

Equipment of this composite design has been used in the United Kingdom under the trademark Celmar* since 1961, and more recently in the United States under the trademark Prolite.[10]† Indeed, one of the largest duct systems engineered to date is the composite polypropylene and reinforced polyester system installed in England. This system comprised some 2,000 ft of 11-ft-dia duct.[11]

Maximum stress due to thermal expansion in this composite laminate will occur in the 140 to 158°F region. Where stiffeners are required in the design and development of cylindrical tanks, the same type of stiffener as is used in all other RP tank work is quite adequate. Stiffening, of course, is done in the RP laminate using steel strips, glass rope, shaped foam, or shaped balsa wood, such shapes generally being squares, triangles, or half-rounds.

In building certain items such as tank lids, baffles, or fan blades, it is good to have a sandwich construction consisting of polypropylene, RP, and polypropylene so as to obtain the highest corrosion resistance on all surfaces.

For the complete range of applications of this particular system the engineer should consult qualified fabricators.

5.9 PVC AND REINFORCED POLYESTER COMPOSITE

One of the best of the corrosion-resistant materials is PVC. Unfortunately, it is fairly weak, mechanically. Its union, however, with reinforced polyester in the form of a composite structure has successfully solved many difficult corrosion problems. Proprietary methods have been developed by a number of fabricators to literally chemically weld the reinforced polyester and the PVC into a tough continuous structure having the enhanced chemical resistance of the PVC and at the same time expanding the pressure (Figs. 5-1 and 5-2) and temperature capability because of the reinforced polyester overlay. Again, techniques of this type can solve many of the difficult corrosion problems so prevalent in the industry. The thickness of the PVC and the fibrous-glass polyester overlay must be engineered to suit the service conditions. In piping hookups it is quite common to use a PVC Schedule 40 or Schedule 80 pipe and simply overlay the pipe with at least ¼ in. of RP polyester. In dealing with pipe or any circular structure there need be no chemical bond between the two layers since a compression bond will form due to the normal shrinkage characteristics of the polyester on curing. This compression bond is very satisfactory. Such designs have been used exten-

---

* Trademark of British Celanese, Ltd., Coventry, England.

† Trademark of Rowland Products, Inc., Kensington, Conn.

sively for chlorine absorption towers[12] and 93 percent sulfuric acid mixing chambers, where the acid is continually diluted in conjunction with other chemicals from 93 percent H_2SO_4 to levels of 5 to 20 percent concentration. A maximum temperature limitation of 180°F is suggested for satisfactory performance. High temperatures and stress in the PVC may result in an embrittlement-type failure of the PVC.

There are case histories where the attempted mixing of two liquids in a tank attacked the tank severely; but when the mixing had been done prior to pouring the fluid into the tank, there was no effect whatever. The dilution of 93 percent sulfuric acid with water is a good case history. Dilution should definitely occur prior to adding the solution to a tank in a suitably designed mixing chamber. A prototype of such a mixing chamber is shown in Fig. 5-1. A solution of 93 percent sulfuric acid will destroy any of the polyesters or epoxies in a short time, but diluted below 40 percent represents no hazard. The 93 percent sulfuric acid destroys through oxidation and dehydration. Keep the number of baffles in the mixing chamber to at least four, and make the mixing-chamber diameter twice the size of the largest entering nozzle. Inadequate mixing-chamber design will quite often result in the destruction of the RP pipe below the chamber. A good mixing-chamber design is therefore essential.

Another example is the use of benzene added in a tank with an acid solution. Adequately mixed, this will present no problem. Unmixed, the benzene will attack the polyester.

Other aids to bonding may be mechanical roughening of the PVC surface prior to making the overlay. Still other techniques employ a mildly solvent solution on the PVC so that the subsequent polyester overlay becomes literally chemically bonded to the PVC. Proprietary methods have developed these bonding techniques to exceptionally high values.

In PVC fibrous-glass overlaid structures, PVC flanges are normally used, and the overlay extended right up to the base of the flange. Flange faces are serrated, and full-faced gaskets are always used.

5.10 RP POLYESTER AND RP EPOXY[13]

Composite piping systems are being manufactured specifically designed for transporting many oxidizing materials which conventional RP epoxy pipe cannot handle. Such piping is made of a reinforced polyester inner wall, which is then overlaid with a fiber-glass epoxy outer wall. The type of production being a filament winding, a structure is achieved that gives high strength, light weight, and resistance to the usual wide range of chemicals compatible with the polyesters. In addition, the low thermal expansion of the epoxy governs. A polyester or epoxy adhesive system may be used in making up cemented joints. A complete line of fittings, couplings, and flanges is available in this system. The cost is about the same as filament-wound epoxy pipe.

REFERENCES

[1] Big Upsurge Coming in RP Tanks, *Modern Plastics*, May, 1967.

[2] Isham, A. B.: Design of Fiberglass Reinforced Plastic Chemical Storage Tanks, Twenty-first Annual Meeting of the Reinforced Plastics Division of the SPI, 1966.

[3] Fiberplus Pipe, Inc., *Bulletin*, Costa Mesa, Calif., 1967.

[4] Whiteman, R. L., Carolina Fiberglass Corp., Wilson, N.C., personal communication, June 3, 1966.

[5] Beetle Plastics, Inc., *Bulletin* P-100, Fall River, Mass.

[6] Amercoat Corp. *Bulletin* R 6/66, Brea, Calif., 1966.

[7] "Process Equipment," Corning Glass Works, Corning, N.Y., 1966 (brochure).

[8] "Chementator," p. 76. Reprinted with permission from *Chemical Engineering*, May 23, Copyright 1966, McGraw-Hill, Inc.

[9] Polypropylene Lined Bonate Products, Beetle Plastics, Inc., *Bulletin* PPB-1266, Fall River, Mass., 1966.

[10] Prolite Technical Manual, Rowland Products, Inc., *Bulletin*, Kensington, Conn., 1966.

[11] "Plastic Fabrication," Courtaulds Engineering, Ltd., Coventry, England (brochure).

[12] PVC/Reinforced Polyester Composite Defeats Caustic Cl_2, *Chemical Processing for Operating Management*, December, 1966, pp. 111 and 112.

[13] Bondstrand Technical Data, Amercoat Corp. *Bulletin* B104, R 11/63, Brea, Calif., November, 1963.

6

A Reinforced
Plastic Workshop
in the Plant

In dealing with the operational requirements of any trade, those responsible are acutely concerned with providing the optimum conditions for the work to be done.   Whether it be pipe fitting, spray painting, lead burning, or reinforced plastics, the need is for a shop conducive to safe working conditions, plus suitable jigs and tools.

Figure 6-1 illustrates the basic components involved in a reinforced plastic shop.   Such a shop can safely and adequately permit two to ten men to work in this area, depending upon the type of work in which they are engaged.   It is presumed that they would share common equipment, such as power saws and drill presses, with other shops.   They should also have with each bench:

1. A disk sander
2. A sabre saw
3. An assortment of aluminum spiral rollers consisting of:
    a. $\frac{3}{8}$-in.-radius corner roller
    b. 1-in.-dia $\times$ 3-in. roller, general-purpose for inside corners, $\frac{5}{8}$-in. radius in small flat areas
    c. 2-in.-dia $\times$ 3-in.-long roller for laminating on flat or curved surfaces
    d. 2- $\times$ 7-in. roller to be used for medium and large flat surfaces

To this list add assorted felt rollers and 2-in. paint brushes with stiff bristles.   The cutting table should be fixed with rolls of 10- and 20-mil C glass surfacing mat, $1\frac{1}{2}$-oz E chopped-strand mat, 24-oz woven roving, and a 36-in. roll of Avisco cellophane.   The cellophane (or Mylar) can be used continually as a work surface.   Assorted rolls of 10-mil C glass in size 6 in. and rolls of $1\frac{1}{2}$-oz E glass in 3, 4, and 6-in. widths should be available in the shop for prompt use.

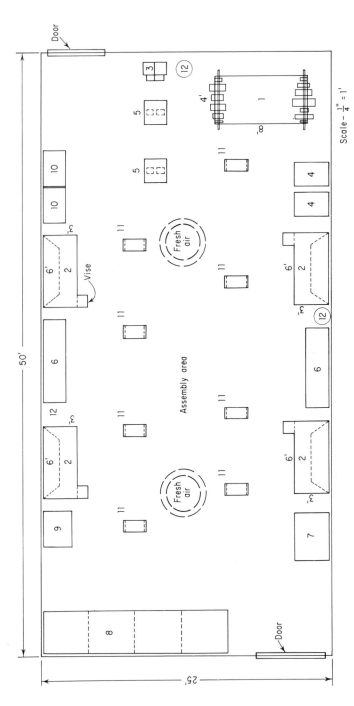

FIG. 6-1 A practical layout for a reinforced plastics shop in an industrial plant. This layout represents a consolidation of the best ideas from a number of shops. (1) 4- × 8-ft cutting table with fixtures for assorted rolls of glass at each end. Woven roving should be on the right, facing the table; (2) 3- × 6-ft workbenches with hood, exhaust fan, and vise. (3) Power sander with a coarse wheel or open-grit sandpaper (tapering tool) with exhaust; (4) Two 55-gal drums of resin (unpromoted), one fire-retardant, the other high-performance chemically resistant (an alternative is a storage cabinet containing pints, quarts, or gallons of a resin assortment for the various types of work); (5) Stands with casters and rollers to permit easy handling of pipe when used with the tapering sander; (6) Open shelving for storage of epoxy and polyester adhesives and assorted small tools; (7) Open shelving for jigs and molds; (8) Open shelving for flanges and fittings; (9) A 9-ft³ refrigerator for catalyst with all sources of ignition removed. Label: "For storage of catalyst only. Do not store cobalt naphthenate in this refrigerator.";

In a small shop, resin may be purchased, prepromoted in pints, quarts, or gallons. In a shop where considerable usage occurs, 55-gal drums of unpromoted resin may be found to be more useful. Make sure the shelf life of this resin is known, since it varies from resin to resin. *The longest shelf life is obtained by purchasing the unpromoted resin and storing it in a cool room.* If an unpromoted resin is bought, it is common practice to store in 55-gal drums, then to draw 5-gal pails of resin from the drum and promote it with a 6 percent cobalt naphthenate solution, which is done by intimately mixing the 6 percent cobalt naphthenate with the resin. The shop can then draw on this promoted resin and use it as required. The catalyst should be bought in small quantities, such as individual $\frac{1}{2}$- to 2-oz bottles, each containing about 12 to 60 cm³ of MEKPO, and limiting the storage in the refrigerator (all sources of ignition removed) to perhaps 200 oz.

SAFETY PRECAUTIONS. *Caution* must be observed when using the 6 percent cobalt naphthenate and MEKPO system. Under no circumstances should the two chemicals be mixed in the same container or poured into the resin at the same time. When mixed together, these two chemicals can be explosively reactive; when handled with care as directed, no danger will ever be encountered. For this reason storage of the cobalt naphthenate solution and the MEKPO are always in separate cabinets and measured in separate graduates.

Other shops purchase their MEKPO in approximate pint polyethylene bottles containing 1 lb. When carried to the job site outside the shop, the catalyst is transferred to a polyethylene laboratory wash bottle approximately 8 oz in size. A 10-cm³ graduate is then used to measure the MEKPO from the wash bottle. From the graduate it is added to the resin in the quantity desired. It is not permissible to return the MEKPO to the parent bottle once it is withdrawn. *Contamination is the worst enemy of safety and must be carefully avoided.*

Each 3 × 6 worktable is equipped with its own hood, exhaust system, and dust collector, the hood being approximately 3 ft long and the exhaust some 600 cfm. This permits buffing, grinding, and making RP sanded joints in a safe, comfortable atmosphere. Each bench should be equipped with a convenience electric outlet for utilizing electric tools such as buffers, sanders, sabre saws, etc. (Some shops prefer air tools, for safety reasons, because of the styrene fumes which may be involved.)

The cutting table can be further improved by providing a rotating knife with a slot cut in the table, and can be automated to permit the knife to make the complete cut and return to the operator (similar to portable-type knives used in the textile industry). However, most shops will probably be content with a heavy pair of scissors or shoe knife. A measuring rule embedded in the table surface is also helpful.

A roll of pressure-sensitive tape 1 or 1½ in. wide will also be found helpful on interior surfaces, to hold aligned pieces of pipe or ductwork in place, when accessibility permits.

A small supply of acetone will also be found useful for cleaning aluminum and felt rollers periodically, to minimize glass pickup. Make sure, however, that the roller is completely self-draining and is essentially dry when used again on the resin and mat. Do not let resin and solvent accumulate inside the barrel since it will dribble back out and lead to unsatisfactory work. Some plants may prefer to use methyl ethyl ketone or styrene, but medical approval should first be obtained.

Mold storage must be provided.

Either inside the shop or immediately adjacent to the shop there should be a small stockroom carrying pipe, fittings, and flanges. As the plant develops experience in various types of reinforced plastic pipe, it is only natural that one or two varieties will become favorites, very likely because of ease of assembly. At that time specific tools should be purchased to facilitate installation of these particular piping systems. Field tapering tools, which are in effect a "centerless grinding operation," permit rapid preparation of pipe ends prior to cementing or wrapping.

Good housekeeping practices in a shop of this sort are essential. Metal waste cans should be emptied at the close of each day's work, and the shop should be swept thoroughly daily. Styrene checks in the shop should be made at least monthly on a routine basis, and more often if an ambitious work program is under way.

The use of reinforced plastics in a shop such as this is limited only by the imagination and ambition of the operator. The operator, however, should confine himself to the intent and purpose of such a shop, which is to permit installation and repair of reinforced plastic equipment. The shop should not attempt to compete on a mass-production basis with the outside fabricator, although the temptation to do so at times is great.

From time to time, however, there will be certain jobs which, in any medium-sized chemical plant, can keep a shop of this type extremely busy. These involve:

1. Erecting piping systems, which will probably be the main function of the shop. Such erection can become quite complex at times and involves the making of many wrapped joints and strapping in fittings, flanges, etc. If the shop performed no other function than this, it would pay for itself.

2. Erection of duct systems. In this case be extremely careful to use a chlorinated polyester resin to which 5 percent antimony trioxide has been added. Both the duct and wrapped joints should be made of this material. This provides maximum fire retardancy.

3. Making tank covers, which can be constructed from sheets of Haysite* material, suitably reinforced with angle or channel.

4. Constructing small tanks and screen boxes (also Haysite).

5. Making odd reducers and shapes, specifically to fit some peculiar condition in the plant.

6. Making stub ends. In flanging systems, it will be found that the best flanging system produced is the so-called stub-end flange. With a few simple molds it is possible for the shop to make their own stub ends at a saving. It requires an extremely competent shop to do this, but several years' practice will produce such competency.

7. Making repetitive shapes. Each plant will have some particular small repetitive shapes which they will find they can make at a saving. For example, one plant found it to their advantage to make reinforced plastic gutters because the conventional gutters corroded in a very short time. Another plant made machine guards of RP material because of corrosive conditions. Still another had a need for odd-sized reducers. Another, for elevator buckets. The building of small prototype parts from reinforced plastics for corrosion-resistant work is common with a small shop.

Obviously, in a shop of this type the size can be scaled down to meet the individual requirements of the plant, and may consist of a worktable and a cutting table, the remaining material being stored in other areas and brought to the shop as required. This may especially be true if a number of crafts are all using the material, such as painters, pipe fitters, and sheet metal and general mechanics. Unless close control is maintained, "kangaroo" shops will spring up in a number of areas, each one catering to the craft involved, as being convenient and acceptable to the work area. Such shops generally leave a lot to be desired in the way of labor utilization. The highest labor efficiency can be achieved only by doing the bulk of the work required in the shop. *Shop time is more productive than field time.* Every job should be engineered to do the maximum amount of work in the shop and the minimum amount of installation time in the field. Along this route lies the objective of least costs. Data have consistently shown that field joints will take anywhere from two to four times as long to fabricate as the same joint made under ideal conditions in the shop. There are two basic places to do work for the least amount of money. One of them is the fabricator's shop, and the second is a well-equipped plant shop. Avoid making wrapped joints 20 ft away from the floor, teetering on the end of a ladder.

The use of jigs and fasteners is also of great importance in achieving low-cost assembly. An adjustable jig is shown in Fig. 6-2.

* Registered trademark of Haysite Corp., Erie, Pa.

Make sure that, at each door and within the shop, NO SMOKING, LIGHTS, and OPEN FLAMES signs are prominently displayed.

About 2,500 cfm of fresh air should be brought into the area. In the design proposed this would provide a complete air change every 6 min,

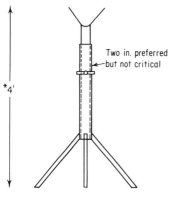

Two in. preferred but not critical

*4'

*When moveable insert is down.

FIG. 6-2 Adjustable jig for use in shop work. At least four of these should be available for use as item 11 in Fig. 6-1.

which would maintain completely satisfactory working conditions at all times. The exhauster at each table will tend to continually pull the fumes away from the applicator and maintain very satisfactory conditions in the working areas of fibrous glass, styrene, and acetone.

It is recognized, in the final analysis, that each shop must solve its own individual ventilation problems. What has been said here is intended as a guide only.

7

Joining Reinforced
Plastic Pipe

7.1 METHODS OF JOINING REINFORCED PLASTIC PIPE

To be successful in joining reinforced plastic pipe it is necessary to master the technique, training, and tools and then adhere to specifications. The joint can be stronger than the pipe itself or it can be the weakest link of the chain. (Indeed, some competent designers suggest that the wrapped joint be made weaker than the pipe. The reasoning behind this is that if pressures are generated which will rupture the pipe, there is a good chance that the entire pipeline will be crazed. If the strap joint is slightly weaker than the pressure rating of the pipe, the strap joint will serve as a safety valve and prevent damage to the piping structure. In the author's experience this would be applicable only to systems subjected to violent hammering or shock.) A number of methods are commonly used in the fabrication of reinforced plastic pipe. Joint reliability by any of these methods can be essentially high if the four items mentioned above are successfully applied.

There are three basic methods by which reinforced plastic pipe may be joined.

Butt joints—butt-and-strap technique   This type of joint is made by butting together two pieces of pipe and overwrapping the joint with successive layers of resin-saturated mat, or mat and roving, out to the proper widths and lengths. This is the standard method for fabricating polyester pipe and is commonly used for joining straight runs, elbows, tees, and reducers. The following description is extracted from the Commercial Standard:

> *Butt Joints*—This type of joint shall be considered the standard means of joining pipe sections and pipe to fittings. . . . All pipe 20 inches in

diameter and larger shall be overlayed both inside and outside. Pipe less than 20 inches in diameter shall be overlayed outside only, unless the joint is readily accessible. . . .

Commonly, the axial stress in the joint is the same as in the pipe itself so that laminate construction can be the same as the pipe being joined.

The width of the strapping material is of great importance because it is, actually, the shear length of the pipe joint and must be long enough to withstand shear stress. Table 7-1A indicates the different types of shear stress commonly encountered in laminate piping.

In strap joints the material shrinking around the pipe as it cures makes a very tight joint which, when based on 100-psig design and a 10:1 safety factor, will test above 1,200 psi on a complete section. Practical experience indicates that hand-laid-up strap joints may begin to leak at 80 to 90 percent of their theoretical pipe strength. When joints, however, are made on flat surfaces, we cannot avail ourselves of the shrinking effect of the polyester, which adds to the shear strength. A joint on a flat surface may therefore fail at 75 percent of the design ultimate.

To calculate the shear surface length for pipe, the engineer should consider the total force exerted on a blind flange and divide by the shear strength of the joint according to the example below for an 8-in.-dia pipe, 100-psig design, with 10:1 safety factor:

$$\text{Shear length} = \frac{\text{area} \times \text{ultimate pressure}}{\text{circumference} \times \text{shear strength}} \quad \text{Ref. 1}$$
$$= \frac{(\pi D^2/4) \times 1{,}000 \text{ psi}}{\pi D \times 1{,}000 \text{ psi}} = \frac{\pi \times 8^2 \times 1{,}000 \text{ psi}}{4 \times \pi \times 8 \times 1{,}000 \text{ psi}}$$
$$= 2 \text{ in.}$$

TYPES OF LAMINATE SHEAR STRESSES[1]

Table 7-1A

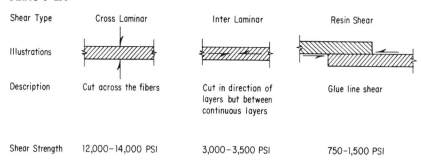

Shear Type	Cross Laminar	Inter Laminar	Resin Shear
Illustrations			
Description	Cut across the fibers	Cut in direction of layers but between continuous layers	Glue line shear
Shear Strength	12,000–14,000 PSI	3,000–3,500 PSI	750–1,500 PSI

Now the shear length is equal to half the strap width, so that minimum width of strapping of 4 in. would be required. Normally, fabricated reinforced plastic pipe is built with an additional safety factor, so that the butt-joint minimum total width as shown in Table 7-1B would be 5 in. Reference should be made to Table 7-1B, which gives the minimum total width in inches of overlays in each butt joint. If reference is made to the piping specification, minimum width of overlay can be readily determined. Since the 100-psi specification is relatively common in this type of work, the pipe size is correlated with the table.

Although the butt joint is considered the standard method of fabricating piping sections and fittings, it should be pointed out that this is a mechanical, or adhesive bond, not a chemical bond. The success of any mechanical bond depends upon the surface being free of contaminating materials. Complete detailed preparations for making a joint are available from any reputable vendor and need not be repeated here, except to outline in general the necessary steps. These are:

1. Prepare an absolutely clean surface.

2. Roughen the surface with a file, hand sander, or, best of all, a power sander.

3. Make sure the ends of the pipe which are cut on the job site are cut straight.

4. Align the two sections of pipe so that they are perfectly square. If a number of joints are going to be made, a few simple jigs to hold and align the pipe will pay off handsomely.

BUTT–JOINT MINIMUM TOTAL WIDTH, INCHES, OF OVERLAYS[1a]

Table 7-1B

Pipe-wall thickness, in.	Minimum of overlay width, in.	Pipe size, 100-psi specification, in.
$\frac{3}{16}$	3	2, 3
$\frac{1}{4}$	4	4, 6
$\frac{5}{16}$	5	8
$\frac{3}{8}$	6	10
$\frac{7}{16}$	7	12
$\frac{1}{2}$	8	14
$\frac{9}{16}$	9	16
$\frac{5}{8}$	10	18
$\frac{11}{16}$	11	20
$\frac{3}{4}$	12	

5. All raw edges of the pipe should be coated with resin to prevent edge penetration of the corrosive fluid.

6. Many fabricators tab the pipe ends with several plastic patches to hold the alignment until the final joint can be made.   These tabs can be made of 2-in. square mat 2 layers thick.   As an alternative, pressure-sensitive tape may be used as simplified tabs, and will work very successfully.

The strapping material should be laid out in sequence as shown in Table 7-2.   Two types of material are normally used for making wrapped joints, namely, a chopped-strand glass fiber with a weight of $1\frac{1}{2}$ oz/ft² and 24-oz woven roving, which provides greater strength in the laminate, improves impact resistance, and gives a high strength in the axial direction. Make sure that the length of mat or roving is long enough to surround the pipe completely, plus an overlay of approximately 2 in.   To make a smooth joint it is common to use a number of different strap widths, starting with the narrowest strap next to the pipe and the widest strap on the outside of the joint.   The proper training and techniques are necessary to make an acceptable joint.   Figure 7-1 indicates the relative positions of the saturated layers of mat.

Single-layer application   This method of butt-and-strap joining is commonly used by the beginner who is unsure of himself.   It consists, simply, in saturating and applying individual layers of reinforcing materials until the desired thickness is reached.   Although it is the choice of the beginner, it is not the choice of the experienced applicator, because of the time it takes to make a joint and, normally, the need for a longer pot

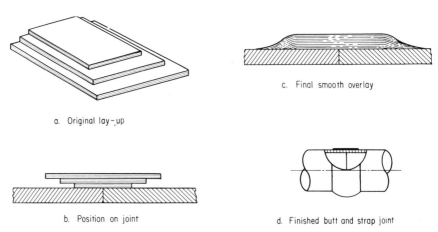

a. Original lay-up

b. Position on joint

c. Final smooth overlay

d. Finished butt and strap joint

FIG. 7-1   Successive steps in lay-up of a butt joint on polyester piping.

CONSTRUCTION SPECIFICATION AND MATERIAL REQUIREMENTS PER JOINT FOR 100-PSI PIPE[2]

Table 7-2

					Pipe size, in.						
	2	3	4	6	8	10	12	14	16	18	20
Pipe wall thickness, in.	3/16	3/16	1/4	1/4	5/16	3/8	7/16	1/2	9/16	5/8	11/16
Layers of 1½-oz mat	4	4	5	5	6	6	7	9	10	12	13
Layer sequence:											
3-in. mat width	1-2	1-2	1-2	1-2	1-2	1-2	1-2-3	1-2-3	1-2-3-4	1-2-3-4	1-2-3-4-5
4-in. mat width	3-4	3-4	3-4-6	3-4-6	3-4	3-4	4-6	4-5-6	5-6-8	5-6-7-8-10	6-7-8-10
6-in. mat width	…	…	…	…	6-7	6-8	8-9	8-10-11	10-11 12	12-13 14	12-13-14-15
8-in. mat width	…	…	…	…	…	…	…	…	…	…	…
Layers of 24-oz woven roving	…	…	1	1	1	2	2	2	2	2	2
Layer sequence:											
4-in. width	…	…	5	5	5	5	5	7	7-9	9-11	9-11
6-in. width	…	…	…	…	…	7	7	9	…	…	…
Mat required per joint:											
3-in. linear ft	1.4	2.0	2.7	3.8	5.0	6.2	11.2	13.0	19.7	22.4	30.8
4-in. linear ft	1.4	2.0	3.9	5.7	5.0	6.2	7.5	13.0	14.9	27.8	18.7
6-in. linear ft	…	…	…	…	5.0	6.2	7.5	13.0	14.9	16.7	30.8
8-in. linear ft	…	…	…	…	…	…	…	…	…	…	…
Woven roving required per joint:											
4-in. linear ft	…	…	1.4	1.9	2.5	3.1	3.8	4.4	9.9	11.3	12.4
6-in. linear ft	…	…	…	…	…	3.1	3.8	4.4	…	…	…
Resin required per joint, pt or lb	0.3	0.42	0.75	1.25	2.25	3.3	4.3	6.3	10.4	14.0	16.7
Catalyst required per joint (MEKPO), cm³, based on 6 cm³/pt of resin	1.8	2.5	4.5	7.5	13.5	20	26	38	63	85	100

life on the resin to complete it. The danger is that the resin will become oversaturated and may sag, with a poor cure resulting or leaking of the joint, due to sagging.

Of even greater concern to the cost-conscious engineer is the relative high cost of the single-layer method of application. Joining times are multiplied by factors of 2 or 3 as the tyro fumbles with the technique. Bad habits are sometimes hard to break, and this method of single-layer application may continue, to the pronounced economic disadvantage of the butt-joint method.

Multiple-layer application   This is the butt-and-strap joining method used by the professional and experienced applicator. It provides the shortest time per joint in both mat saturation and application and makes removal of entrapped air easier. In this method, the layers of mat are presaturated on a worktable covered with cardboard, film, or waxpaper. In a typical $\frac{7}{16}$-in. joint the three bottom layers are 6 in. wide, the middle three layers are 4 in. wide, and the top three layers are 3 in. wide. The secret is to saturate each layer only partially before the next layer is added. There will be some drainthrough from the upper layers to the lower layers, which will provide complete saturation.

Other steps are as follows:

1. If the mat is rolled from the center to the edges, it will eliminate air. Common paint rollers about 3 in. wide are satisfactory for the job. Spiral-grooved aluminum rollers are also excellent. These are shown in Fig. 7-2. The spiral aluminum rollers are good bubble poppers. A spiral-grooved roller will also tend to push the resin to the outside, which may reduce the thickness of the strap joint. The thinner laminate will have a higher glass content and may be stronger, but since the corrosion resistance lies in the resin, we should aim at making our resin content as high as possible.

Another method commonly used is to rapidly dab the air pockets with a stiff-bristled 2-in.-wide paint brush and push them to the edges.

Either of these methods will create a dense, tough laminate and give a joint equal in strength to the pipe itself.

2. Every effort should be made to do the work in a temperature range of 65 to 85°F. Temperatures below this will slow the cure, and additional catalyst will be required.

3. Avoid excess humidity. Do not attempt to make wrapped joints in the rain or snow or in a high-humidity area. If the mat gets wet or damp, it may cause a white, or milky, joint. When in doubt, wipe the laminate with an acetone solvent or any solvent that will absorb moisture.

4. Do not make a single wrapped joint over $\frac{3}{8}$ in. thick. Make successive lay-ups, cutting the mat in half for each lay-up. Do not apply

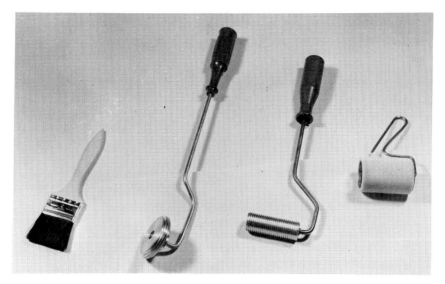

FIG. 7-2 Different tools for rolling out mat. (*a*) A stiff-bristled brush; (*b*) a ½-in. aluminum roller for the fillets; (*c*) a 3- or 5-in. serrated aluminum roller for use as a bubble popper; (*d*) a common paint roller.

the second lay-up until the first lay-up has kicked off and begins to cool. Surface preparation for the second lay-up is unnecessary.

5. Completely saturate the outer laminate to provide a good close finish. The finished joint should look good and feel good. If the joint is outdoors, add wax and 2 percent CAB-O-SIL* to the final coat. This will serve as an effective UV inhibitor.

Adhesive joints   Adhesive joints are commonly used in reinforced epoxies (or polyesters). Making an adhesive joint varies from system to system. In general, the steps are as follows:

1. Mix two components, the cement and the catalyst.

2. Use a workable temperature. The higher the temperature, the shorter the pot life after the catalyst is added.

3. The container for mixing the cement should be used only once.

4. The adhesive may be furnished in paste or liquid form.

5. If possible, make your adhesive joints at 60°F or above, and make sure that your adhesive components and pipe and fittings are at 60°F or preheated to 60°F before joining.

6. Apply mixed cement to surfaces to be joined, using one-third of

* Registered trademark of Cabot Corp.

the required amount in the female opening and two-thirds on the male end of the pipe. Be sure that cement is applied to the cut end of pipe.

7. Assemble fitting to pipe and rotate at least 180° to distribute the cement and eliminate air pockets.

8. Wipe a fillet of cement around the fitting and remove the excess.

9. To hasten the hardening of an *epoxy* adhesive joint:

   a. Use a torch to apply heat evenly to the entire area.

   b. Apply heat slowly, taking at least 5 min per connection until the area reaches a temperature near, but not exceeding, 180°F.

   c. Keep the joint at this elevated temperature for another 10 min or until the adhesive fillet ceases to feel tacky.

   d. Slowly increase the temperature to 200 to 230°F for another 10 min. If bubbling occurs at the edge of the joint, stop applying heat for a few minutes, then reapply.

   e. Let the joint cool to 100°F before handling.

The application of heat to an adhesive joint is not necessary for either corrosion resistance or strength. The use of heat serves one purpose only, namely, the reduction of curing time.

In the preparation of an epoxy joint, the pipe and fittings should be clean, sanded, and ready to assemble before the adhesive components are mixed, since their pot life is only 15 to 30 min.

The vendor generally furnishes the adhesive in kits. Only one kit should be mixed at a time. Each kit contains sufficient adhesive to complete a number of connections, depending on the size of the joint.

By all means follow the vendor's instructions on the preparation of the adhesive; that is indispensable to obtaining a good joint. Be especially careful to mix or shake the components thoroughly prior to usage. On such elementary details hangs the success or failure of RP joining methods. Entire systems have been in difficulty because these simple instructions were not followed. Make sure the adhesive used is resistant to the solution in the line.

Packaged polyester adhesives are also available where such an adhesive is necessary for compatibility with the piping system. Polyester adhesive requirements may be estimated from the following table:[3]

Pipe diameter	2	4	6	8	10	12
Pounds per joint	¼	½	¾	1¼	2	2¾

The catalyst methyl ethyl ketone peroxide is added to the polyester adhesive in an amount equal to 12 cm^3/lb.

To those interested in preparing their own polyester adhesive, the following formulation[4] is recommended as having been used successfully, developing a very high strength:

FORMULATION:

100	parts	Hetron 197
100	parts	Hetron 32A
16.5	parts	⅛″ Ferro Cat-X Chopped Glass
49.5	parts	Cabot Lite F-1*
37.0	parts	A-A Mica
2.0	parts	Triton-X-100
4.0	parts	Cab-O-Sil M-5*
5.0	parts	Union Carbide-Vinyl Silane A-172
1.0	parts	6% Cobalt Naphthenate Solution

* Registered trademark of Cabot Corp.

MIXING PROCEDURE:

Equipment—Hobart Mixer or equivalent.

1. Charge Hetron 197, Hetron 32A and 6% Cobalt Naphthenate Solution, to mixing vessel. Mix completely.

2. With continuous mixing add the Triton-X-100 and Silane A-172. Mix completely.

3. Add Cab-O-Sil M-5 slowly with mixing till uniform.

4. Add the A-A Mica slowly with continuous mixing. Mix until uniform.

5. Add the Cabot Lite F-1 slowly and mix till uniform.

6. Add the ⅛″ Ferro Cat-X glass and mix 10–20 minutes till uniform.

CAB-O-SIL may be used to adjust the viscosity of the adhesive. Adjustment should be to a paste form. The selection of the adhesive, be it epoxy or polyester, is definitely a part of the engineer's responsibility. Epoxy adhesives show less shrinkage characteristics than polyesters, but provided chemical compatibility exists, both can produce highly satisfactory joints.

Flanged joints—flanging system   Those familiar with flanging systems in metal piping will find its counterpart in reinforced plastic piping.

Details of various flanging systems are discussed in a later part of this chapter. In polyester work the practice is to keep the use of flanges to a minimum, the butt joint being used as a standard means for joining sections of pipe. Certainly, some flanges are necessary for system maintenance and fabrication. Various attempts have been made to design flanges possessing certain attributes. Some of the more common flanges on the market today are:

1. Flat-face flange

2. Serrated-face flange with serrations molded into the face approximately ⅛ in. apart

3. Vanstone-type, or raised-face, flange
4. Flange with an angular taper on the face to provide maximum compression at the flange ID with minimum bolt torquing force

7.2 COMPARATIVE JOINING COSTS

The comparative joining costs of reinforced plastic joints may be a subject for some deliberation.   Obviously, it will vary from shop to field and with the efficiency of the labor and the type of joint being applied.   Table 7-3 represents an attempt to place approximate quantitative estimates on joining costs.   Field joining costs are an average of many different installations, and might be said to be a line 10 ft off the ground.   If the mechanic is doing it early in the day, one cost will be achieved, and at "a quarter to four," another set of costs will be developed.   The study of joining costs can be most frustrating to the engineer.   He may find, for

COMPARATIVE RP JOINING COSTS, LABOR PLUS MATERIAL

Table 7-3

	Pipe size, in.						
	2	3	4	6	8	10	12
Butt-and-strap joint:							
Shop	$5–$6	$6–$7	$6–$7	$10–$12	$12–$15	$15–$18	$20–$25
Field	$10	$12	$12	$18	$23	$25	$35
Adhesive coupling joints:							
Shop	$5	$7	$8	$12	$19	$35	$45
Field	$8	$12	$12	$18	$28	$47	$60
Flanged joints (cemented), field	$13	$18	$23	$31	$47	$61	$98
Flanged joints, polyester stub ends, field	$48	$62	$76	$112	$137	$168	$196

NOTES: 1. Wrapped-joint costs are typical vendor's costs and will be approximated in any good shop.
2. Coupling costs include coupling, adhesive, and joint labor.
3. Flanged-joint costs include flanges, adhesive, making adhesive joint, gaskets, fasteners, and bolting up.
4. Hand-laid-up stub-end-flange costs include flanges, wrapped joints, gaskets, fasteners, and bolting up.
5. Labor is based on a shop rate of $3/hr.

example, that he cannot meet the costs shown in Table 7-3. If so, the answers may generally be found in two areas:

1. A vendor selling the product will be glad to furnish wrapped or adhesive joints at the costs shown in Table 7-3.

2. The field prices given can be achieved through the use of the proper technique and adherence to specifications, coupled with a sufficient amount of experience. Inadequate experience will invariably result in exceeding the figures shown.

From Table 7-3 our conclusions are as follows:

1. The polyester wrapped joint and the epoxy coupling joint in sizes 2 to 6 in. are not measurably different in cost. In the sizes 8 in. and above, the polyester butt joint is comparably cheaper.

2. The epoxy-cement flanged joint is considerably more expensive than either the butt joint or the coupling joint in any size measured. Normally, it costs three times as much.

3. By far the most expensive joining methods are the polyester stub-end flanges, which average nine times the cost of a wrapped joint. It is thus good piping economy to assemble the system with the minimum number of flanged joints which will permit good field assembly and maintenance. The stub-end polyester flanges, however, possess enormous strength, and where a particularly rough application is being engineered, they should receive serious consideration. They also are reasonably safe to use with ring gaskets, whereas press-molded flanges operate best with full-faced gaskets. The press-molded flanges approximate the epoxy-cement flange joining system in relative price.

Field crews too often tend to become impatient with the cure time for some of the polyester resins when making wrapped joints, or cements when making adhesive joints. In the case of polyesters, curing times may be accelerated by the use of paradimethyl aniline. The addition of 10 drops of dimethyl aniline to a pint of resin or cement will give a set time of about 20 min at 75°F room temperature. Dimethyl aniline is often referred to as DMA. However, beware of such abbreviations in specifying this chemical since DMA also means dimethylamine, a totally different chemical, which has no effect on cure time. Communications are important—do not be misled.

7.3 BURST TESTS ON JOINTS

A great deal of work has been done on destructive tests of both wrapped joints made by the butt-joint technique and the adhesive joints. Both joints when properly made leave little to be desired.

The butt joint made with the wrapped technique will test right up to the destructive pressure of the pipe. When this is not achieved, there is something wrong with the joint (which will be dealt with in a later section of this chapter). For example, hydraulic burst tests repeatedly show failure of the 150-psig polyester pipe at pressures of 1,600 psig with no joint distress. Tests on a butt joint in a section of 100-psig plastic pipe showed a burst at 1,100 psig before failure occurred. Also, a flange set on a 4-in. epoxy pipe took more than 40,000 lb of force to pull it off. Such is the strength of modern adhesive that if a joint is properly made, any failure that occurs may very well be cracking of the flange or dismemberment of the pipe. Under test conditions at high test pressure, gasket blowout may occur prior to other failures. One of the bombers of our modern air force was largely assembled with epoxy cement.

7.4 BOLT AND GASKETING SPECIFICATIONS

The flanged-joint assembly of reinforced plastic pipe is commonly recommended to permit:

1. Suitable erection in sizes which can be moved into the job site.
2. Disassembly for maintenance or inspection purposes.

It should be well understood that flanged-joint assembly costs anywhere from three to nine times as much as butt-joint or adhesive-coupling assembly.

Flanges, nevertheless, are an absolute necessity at many points of operating systems. All flanges for reinforced plastic pipe are provided with standard ASA 150-psi schedule drilling.

The Commercial Standard of the SPI indicates:

> *Bolts, Nuts and Washers*—(To be furnished by customer.) Metal washers shall be used under all nut and bolt heads. All nuts, bolts and washers shall be of materials suitable for use in the service environment.
>
> *Gaskets*—(To be furnished by customer.) Recommended gasketing materials shall be a minimum of ⅛ inch thickness with suitable chemical resistance to the service environment. Gaskets should have a Shore A or Shore A2 Hardness of 40 to 70.

A red sheet rubber or Neoprene full-faced gasket will be found to be satisfactory in most applications. Field experience indicates that the use of ring gaskets on press-molded flanges will result in a higher failure rate than the use of full-faced gaskets. Failure commonly occurs at the neck of the flange, where cracking is observed. When the use of ring

MAXIMUM BOLT TORQUE FORCE,* FT-LB OF TORQUE[5]

Table 7-4

Pipe size, in.	Internal pipe pressure rating, psig, standard					
	25	50	75	100	125	150
2	25	25	25	25	25	25
3	25	25	25	25	25	25
4	25	25	25	25	25	25
6	25	25	25	25	35	40
8	25	25	30	40	50	60
10	25	25	30	40	50	70
12	25	25	35	45	60	80
14	25	30	40	60	75	100
16	25	30	50	70	80	
18	30	35	50	80	100	
20	30	35	60	90		
24	35	40	70			

* The indicated torque on bolts is required to seal step-face flanges in pressure pipe using red-sheet-rubber gaskets of 70 Durometer. Gaskets of 40 Durometer should not be used in pressure service above 50 psi.

gaskets is a necessity, flanges of the highest strength, such as hand-laid-up polyester, will be found to be beneficial. In-practice flange failures, which are observed in the cracking in the neck of the flange, may occur almost immediately upon installation or take several years. The use of full-faced gaskets will normally prevent failures of this type. Where neither red rubber nor Neoprene proves satisfactory, the engineer should consult the fabricator for recommendations.

The maximum bolt torque force to be used is shown in Table 7-4.

All flanged joints should be drawn up by gradually tightening and alternating bolts so that the maximum pressure is not achieved on one or several bolts while others have no pressure applied.

Gaskets   Some epoxy manufacturers recommend the use of a gasket material in the 40 to 50 Durometer range. Many engineers prefer a 70 Durometer in preference to a 40 Durometer gasket because the 40 Durometer gasket will tend to deform and spread, quite often into the pipeline, and thus obstruct smooth flow. Higher torques are required to hold 40 Durometer gaskets. Avoid a soft gasket in polyester flange service.

Specify gasket ID and OD in accordance with Table 7-5.

STANDARD GASKETING DIMENSIONS

Table 7-5

Pipe size, in.	Gasket ID, in.	Full-face gasket OD, in.
1	$1\frac{5}{16}$	$4\frac{1}{4}$
$1\frac{1}{2}$	$1\frac{29}{32}$	5
2	$2\frac{3}{8}$	6
$2\frac{1}{2}$	$2\frac{7}{8}$	7
3	$3\frac{1}{2}$	$7\frac{1}{2}$
4	$4\frac{1}{2}$	9
6	$6\frac{5}{8}$	11
8	$8\frac{5}{8}$	$13\frac{1}{2}$
10	$10\frac{3}{4}$	16
12	$12\frac{3}{4}$	19
14	14	21
16	16	$23\frac{1}{2}$
18	18	25
20	20	$27\frac{1}{2}$

7.5 WRAPPED JOINTS—POLYESTER—MANPOWER REQUIREMENTS BY SIZES

It is of interest to the engineer to have established guidelines on shop and field installations covering the approximate time necessary to make the standard butt joints. Tables 7-6 and 7-7 indicate shop allowance by qualified personnel; Tables 7-6 and 7-8 include field allowances. These tables cover lining, jigging, sanding, and making the joint. Experience indicates that schedules can be met, but also that in some organizations meeting them may be a problem. Obviously, to achieve the maximum savings in a reinforced plastic pipe program, it is necessary to get the labor requirements on wrapped joints into the areas established below. While an experienced man in the shop may complete a 6-in. wrapped joint in 45 min, the same joint in the field, where conditions may not be as good, takes $1\frac{1}{2}$ hr. Inexperienced personnel may take considerably longer. For example, one study of 16 joints made in 6-in. pipe indicated an average of 3 man-hours per joint. Where the allotted time is beyond that indicated in the table (Table 7-6), the engineer can be reasonably certain that he is dealing with poor work indexes or with a deficiency in tools, methods, training, or specifications.

WRAPPED JOINTS—POLYESTER, TOTAL MAN-HOUR REQUIREMENTS BY SIZES

Table 7-6

Pipe size, in.	Shop	Field
2	0.7	1
3	0.7	1
4	0.7	1
6	0.8	1.5
8	0.9	2
10	1.1	2.5
12	1.5	3
16	2	4
20	2.5	5
24	3	6
28	3.5	7
32	4	8
36	4.5	9
40	5	10
44	5.5	11
48	6	12
54	7	14
60	7.5	15

7.6 ESTIMATING DATA ON HANGING, BOLTING, AND DRILLING RP PIPE

Table 7-7

Hangers......... Normal, to steel, 1 hr each at 1:1 ratio, total 2 man-hours
Normal, to concrete, 2 hr each at 1:1 ratio, total 4 man-hours
Coated, material cost $6 each
Bolting.......... 12 bolts per $\frac{1}{2}$ hr at 1:1 ratio, total 1 man-hour
Drilling......... Under $1\frac{1}{2}$-in.-dia, at $0.50/hole
Over $1\frac{1}{2}$-in.-dia, at $1/hole

7.7 ADHESIVE JOINTS—EPOXY—MANPOWER REQUIREMENTS BY SIZES

Table 7-8 was prepared to furnish the engineer with man-hour requirements in the fabrication and installation of epoxy pipe joints. In using

PIPING FABRICATION AND INSTALLATION, EPOXY JOINTS[6]

Table 7-8

			Pipe size, in.				
	2	3	4	6	8	10	12
			Hours per joint				
Adhesive couplings							
1. Shop Fabrication............	0.3	0.3	0.4	0.7	0.9	1.1	1.3
2. Field Installation............	0.2	0.2	0.2	0.3	0.3	0.4	0.4
Total....................	0.5	0.5	0.6	1.0	1.2	1.5	1.7
Flanged joints							
3. Shop Fabrication							
a. Install flanges only.......	0.6	0.7	0.8	1.2	1.4	1.6	1.8
b. Bolt up in shop...........	0.2	0.2	0.3	0.4	0.5	0.7	0.8
4. Field Installation............	0.4	0.6	0.8	1.1	1.5	1.8	2.1
			Hours per linear foot				
5. Permanent hanging at 10-ft height................	0.15	0.15	0.15	0.20	0.20	0.25	0.25
6. Estimating average per linear foot (combination of above, total job)..........	0.22	0.23	0.25	0.35	0.39	0.47	0.51

NOTES: All times include layout, measuring, travel, materials procurement, and other auxiliary work.

Item 2.   Includes making all necessary tie joints in the field plus placing in position and hanging all fabricated joints.   Labor cost shown is to be multiplied by all joints in the system.

Item 3.   All flanges installed in shop, part of which can be bolted up in shop.

Item 4.   Includes only those joints not bolted up in shop.

Item 6.   This is an approximate man-hour figure which may be used for the total installation of a job.   Recognize it is approximate but good enough to serve.

the table, keep in mind that an adhesive coupling has two ends.   The table may therefore be used for adhesive couplings or equally well for 45° and 90° elbows.   Allowances should be multiplied by 1.5 in dealing with tees.

For field installations an arbitrary assumption has been made on permanent hanging 10 ft above the floor.   The table is largely self-explanatory.   Several typical problems have been worked out to illustrate the use of the table.

Sample problems to illustrate the use of Table 7-8

Problem 1

A system of 4-in. pipe requires making 20 adhesive coupling joints and hanging 100 ft of pipe as assembled in the field. How many man-hours will be required for this work?

20 couplings $\times$ 0.6 hr/coupling = 12 hr
Hanging 100 ft of 4-in. pipe at .15 hr/ft = 15 hr
Total installation time = 27 hr

Problem 2

Calculate the time to assemble and install 20 8-in. flanged joints and 100 ft of 8-in. pipe in a process room 10 ft above the floor. Assume that all flanges can be made on in the shop, and further, that 12 joints are to be bolted together in the shop and 8 joints in the field on final erection.

20 flanged joints made in the shop at 1.4 hr/joint   = 28
12 joints bolted together in the shop at 0.5 hr/joint = 6
8 joints bolted up in the field at 1.5 hr/joint   = 12
Permanent hanging = 100 ft at 0.2 hr/ft   = 20
Total = 66 hr

Problem 3

Assume a job requiring 200 ft of 10-in. pipe and 300 ft of 12-in. pipe. Estimate the total man-hours required for complete installation.

200 $\times$ 0.47 =   94 man-hours
300 $\times$ 0.51 = 153 man-hours
Total = 247 man-hours   *Answer*

The economics of polyester versus epoxy pipe system is a subject of much controversy. A practical illustration of the breakpoint in the installation of these two materials is shown in Table 7-9. Basically, the conclusions that can be drawn from this table are as follows:

1. There is essentially no difference between the installed price on polyester versus epoxy systems in sizes 2 to 4 in.

PRICE COMPARISON, POLYESTER VERSUS EPOXY, TYPICAL PIPING SYSTEM[7]

Table 7-9

	4 in.			6 in.			8 in.		
	Polyester	*Epoxy*	*Ratio epoxy/ polyester*	*Polyester*	*Epoxy*	*Ratio epoxy/ polyester*	*Polyester*	*Epoxy*	*Ratio epoxy/ polyester*
					Pipe size				
Total material cost.........	$658	$710	1.08	$849	$1,087	1.28	$1,032	$1,604	1.55
Total labor cost.........	126	106	0.84	164	151	0.92	183	166	0.91
Total materials and labor cost......	784	816	1.03	1,013	1,238	1.22	1,215	1,770	1.45
Total cost per foot.........	5.30	5.50	1.03	6.85	8.37	1.22	8.22	11.95	1.45

NOTE: This is a typical system consisting of:
1. 150 ft of piping
2. Four 90° plain-end elbows
3. Five flanges
4. Four shop-wrapped or adhesive joints
5. Seven field-wrapped or adhesive joints
6. Five flanges put on in the shop
7. Five flanges bolted on in the field
8. Hanging and supporting of 150 ft of piping

2. The breakpoint in the installation of polyester versus epoxy systems occurs at 6 in. At this point the epoxy system costs approximately 20 to 25 percent more on a total installed cost basis.

3. In sizes 8 in. and above the total installed cost of epoxy versus polyester systems may be 40 to 50 percent above its polyester counterpart. Bear in mind that we are comparing 100-psig polyester systems with the standard manufactured 150-psig epoxy system. Polyester systems may be purchased in various incremental design specifications, varying from 25 to 50, 75, 100, 125, and 150 psig design. The standard which many of the fabricators stock, however, is a 100-psig design, which fits the needs of most users. The standard epoxy specification is 150 psig. The user should be aware of this difference. In each case design factors of safety vary with pipe sizes. Commonly, it is at least 10:1.

7.8 COMBINING POLYESTER AND EPOXY PIPE

Do not for a minute harbor the notion that you are committed to either epoxy or polyester pipe. If the chemical requirements can be met by both systems, you have the option of combining polyester and epoxy pipe and fittings.

For example, many find it an advantage to use polyester pipe and epoxy flanges to achieve the most economical system. This will generally work well, and it is possible to combine this with an epoxy or polyester adhesive. There is nothing particularly sacred about the combination. See Fig. 3-4 for an example of the use of an epoxy flange on a polyester pipe.

Another combination which works well is the use of 150-psi ASA epoxy fittings in a polyester piping system. Although not necessarily the most economical combination, it does have the advantage of standardizing dimensions when replacing metal fittings.

In general, the RP polyester piping may be purchased for 20 to 30 percent less than RP epoxy, and in some cases for even less. If both systems are suitable for the conditions indicated, the polyester system can generally be installed at lower cost. However, each purchaser has to consider the capability of his own crew. Quite often he will find that an adhesive joint is easier to master than a wrapped joint. If such is the case, then, economywise, he would quite often be better off mixing the systems.

In general, epoxies show more resistance on the alkaline side and, because of their helical-wound or hoop-type construction, develop enormous strength in the pipe wall (a few polyesters are also helical-wound).

Consider this, then, for installations which have surging or hammering flows that would destroy a random-mat pipe.

The examples of flexibility cited above, while not common, are sometimes required to meet peculiar labor conditions or target dates with limited material availability.

7.9   FLANGING SYSTEMS

Polyester flanges   Standard flange dimensions are applicable regardless of type of construction, that is, press-molded, premixed, stub-end, or combination flanges.

Press-molded flanges. This type of flange is commonly made by a matched-steel-die molding process and incorporates a chopped-strand fiber-glass reinforcement with special polyester resins and inert fillers to provide void-free flanges of good physical strength and stable chemical resistance. Tensiles in the order of 13,000 psi and flexural strengths of 27,000 psi with modulus of elasticity at 1.0×10^6 are common in matched-die molding done under extremely high pressure. Compressive strengths of 25,000 to 35,000 psi are also achieved. Lowest unit costs are obtained with this type of flange. It is sometimes sold with a short length of pipe as a stub end.

Premixed flanges. The premixed flange is rarely sold on a commercial basis because of its lower physical properties. In conception it is made in simple molds from a premixed material of chopped-glass fibers and polyester resins. Little if any pressure is applied. Flanges of this type are subject to air occlusion. Physical properties may be considerably less than in the matched-die high-pressure molding. Premixed flanges are often used for experimental work and may also be used to make odd shapes or parts where only a few parts are involved.

Stub-end flanges. The stub-end flange is one of the work horses of the trade. It is constructed of random-mat cloth saturated with resin and made up in successive layers. In larger sizes woven roving is interspersed with the random mat to form in a mold a flange with integral stub. Since a flange of this type is completely hand-laid-up, its cost is considerably more than a press-molded flange. It does, however, possess very high physicals and fatigue resistance. It is especially resistant to cracking in the neck area of the flange. Although nearly all manufacturers recommend its use with full-faced gaskets, the stub-end flange has been used successfully in many cases without them. This is probably the toughest type of construction in the industry, although present costs may limit its application. It should be considered for best performance.

Combination flanges.[8]* A combination flange consists of epoxy-impregnated roving wound between two press-mold epoxy-impregnated mats. One mat is epoxy-impregnated asbestos, which forms the gasket face of the flange. This provides the same chemical resistance on this face as in the interior of the pipe—a necessity, in view of the fact that this face is partially exposed to the contained fluids during service. The other mat is a filament-wound epoxy-impregnated mat and forms the other face of the flange. This provides the strength of filament-wound structures radially, which the central layers of roving require. Dynel cloth provides a resin-rich layer for bonding and additional resistance to chemicals in very severe applications. (See Fig. 7-3 for a typical design.)

The gasket face extends past the socket periphery to form a stop for the pipe during assembly. A slope is provided in the socket to aid adhesive flow during assembly.

This laminar construction results in a tough, strong flange, almost impossible to damage in handling or service. The limiting property is in the adhesive bond. The face and bolt-hole dimensions are according to ASA configuration.

* This description is based on a personal communication from C. G. Munger, Amercoat Corp., Brea, Calif.

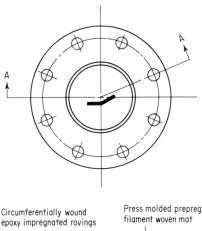

Circumferentially wound epoxy impregnated rovings

Press molded prepreg filament woven mat

Press molded prepreg asbestos mat

Impregnated Dynel cloth liner

A−A

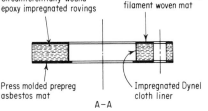

FIG. 7-3 Amercoat Corp. combination flange.

POLYESTER FLANGE DATA Flanges Are Standard up to 12 in. and 100-psi Rating[9]

Table 7-10

(*A*) *Pipe ID*	(*B*) *Flange OD**	(*C*) *Bolt circle**	(*D*) *No. of holes**	(*E*) *Diameter of holes**	*Bolt diameter**	*Length of bolt*	*Diameter of washer*	(*F*) *Flange thickness**	(*G*) *Shear surface*	*Common stub length*
1	4 1/4	3 1/8	4	5/8	1/2	2 1/4	1 3/8		Varies with manufacturer. Should be at least four times flange thickness or equal to the shear surface of the same size butt-and-strap joint.	6
2	6	4 3/4	4	3/4	5/8	2 1/4	1 3/4	9/16		6
3	7 1/2	6	4	3/4	5/8	2 1/2	1 3/4	11/16		6
4	9	7 1/2	8	3/4	5/8	2 3/4	1 3/4	13/16		6
6	11	9 1/2	8	7/8	3/4	3	2	7/8		8
8	13 1/2	11 3/4	8	7/8	3/4	3 1/4	2	1		8
10	16	14 1/4	12	1	7/8	3 3/4	2 1/4	1 3/16		10
12	19	17	12	1	7/8	4 1/4	2 1/4	1 7/16		10
14	21	18 3/4	12	1 1/8	1	4 1/2	2 1/2	1 1/2		12
16	23 1/2	21 1/4	16	1 1/8	1	4 3/4	2 1/2	1 5/8		12
18	25	22 3/4	16	1 1/4	1 1/8	5	2 3/4	1 3/4		12
20	27 1/2	25	20	1 1/4	1 1/8	5 1/4	2 3/4	1 7/8		12
24	32	29 1/2	20	1 3/8	1 1/4	Note 2	Note 3			12

NOTES: 1. Column letters *A* to *G* are for use with Fig. 7-4.
2. Bolt length = 2 × flange thickness + bolt diameter + 3/8 in.
3. Use U.S. Standard round washers.

* Conforms to proposed Commercial Standard specifications.

Conclusion. The economics of joining systems justifies minimizing the use of flanges and giving preference to the butt joint.

Table 7-10   Unfortunately, polyester flange design has not been completely standardized within the industry. Although the dimensions shown in Table 7-10 are common to many vendors, some fabricators have flanging dimensions different from these. Nearly all, however, conform to the 150-psi ASA standard bolt-up specifications. Being built to 150-psi ASA standards, these flanges will make up to valves and any other steel flanges to which the line is joined. Variances exist in flange OD washer diameters, flange thicknesses, flange hubs, and stub length. Table 7-10 has been field-tested and will produce satisfactory conditions. Many of the specifications given will meet the proposed SPI Commercial Standard and are based on a safety factor of 8:1 and a flexural strength of 20,000 psi.

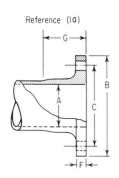

Reference (10)

FIG. 7-4 Typical flange design—polyester. A = pipe size; B = flange diameter; C = bolt-circle diameter; D = number of holes; E = hole diameter; F = flange thickness; G = shear surface.

At least one vendor makes a flange OD which is ½ in. larger than standard ASA 150-psi specifications, which are the SPI Commercial Standard. This extra diameter is said to provide more material between the OD and the bolt hole and makes a stronger flange. Many of the fabricators following the Commercial Standard use ASA 150-psi standard specifications.

Take special care to ensure that the U.S. Standard round washers do not have to be cut to prevent cocking. The flange should be suitably built to prevent this. Otherwise, washer seats will have to be flatted.

Polyester flanges—field experience   The reader is referred to pages 58–60 for the Commercial Standard specifications on flanges for reinforced polyester pressure pipe. Table 5, page 58, applies particularly to flat-faced flanges with full-faced gaskets of 70 Durometer. Various other flange designs have been worked out, including step-faced flanges, concentric ring embossments, etc., all of which are designed to provide greater sealing at lower bolt torques.

In the au hor's experience some special flange designs are capable of satisfactory operation at pressures approaching the bursting strength of the pipe. Many flat-faced designs, however, will begin to leak at approximately 400 psig. This, however, is no great hardship as the Commercial Standard specifically states that the flanging system need be adequate for only twice the design operating pressure of the piping system. Thus, the flanged joints of a 100-psig system should be capable of 200 psig pressure without leaking; a 150-psig system would require flanges tight at 300 psig, etc. These criteria are not difficult to meet. It is obvious that special consideration is needed in flanging systems where pipe operation above 200 psig is proposed. Wrapped joints provide higher operating pressure systems unless special pressure designs are used. A young employee with only the briefest training can produce wrapped joints which will test satisfactorily at 800 psi.

The most commonly observed difficulty in flanging systems is cracking in the neck area of the flange or breaking the glue line if an adhesive flange has been used. This is especially true where soft gaskets of a 40 Durometer are used instead of the harder 70 Durometer gaskets. It is also observed where ring gaskets are used, contrary to the suppliers' recommendations, instead of full-faced gaskets. As the bolts are pulled up, the lever arm of the flange exerts a high moment in the neck area and cracking in the neck may occur. Full-faced gaskets of 70 Durometer will minimize such occurrences. At the same time overtorquing of flanges may accentuate the problem.

The author should emphasize that if flanging systems are installed according to instructions these difficulties will be avoided. Nevertheless, the hand-laid-up polyester stub end shows the widest degree of latitude and greatest resistance to this type of problem. It is a truism that the end user can do more things wrong with a polyester stub end and still obtain a satisfactory job than with other flanging types. Where cracking in the neck area of polyester stub ends is observed it is generally due to either very severe overtorquing or a poor laminate in the neck area. Air occlusion through improper rollout in the neck area can produce a severe reduction in physicals which contributes to the problem. However, under these conditions, there is more of a tendency to weep than anything else and repairs can be made on a scheduled maintenance tour.

Another type of flanging system which is occasionally used by some fabricators is the backup ring flange. If external corrosion is not a problem some economies can be achieved by using a steel backup split-ring flange specially coated for mild corrosive exposure.

Epoxy flanges   At present there are no standard epoxy flanges. The only item which has been standardized in the reinforced epoxy industry is the hole spacing and sizing, which is generally 150-psi ASA. Flange thicknesses vary, and type of construction varies widely, from press-molded types to composite types, where a mixture of molding compound and filament-wound surfacing material is used. Shear surface varies from vendor to vendor. This is not meant to imply, however, that satisfactory flanges for reinforced epoxy pipe have not been developed. On the contrary, some of them provide very satisfactory and durable services. The strength of the adhesive joint is enormous and is well suited for chemical-process service. The engineer in this area, however, does not have an SPI Commercial Standard to refer to, and must be guided by his own judgment, test program, and success in related industries.

Even in polyester systems the engineer will quite often find it to his advantage to use epoxy blind flanges where such a requirement exists, providing the epoxy is compatible with the chemical service involved. One large consumer of polyester piping uses the epoxy blind flanges without exception in sizes up to 12 in. A comparison of the relative costs on pages 83 (polyester) and 110 (epoxy) indicates the considerable savings which may be achieved by using epoxy blind flanges. These savings are particularly large in sizes up to 8 in.

Flat-faced epoxy flanges are also available with American Standard or iron pipe threads in addition to the conventional cement socket. Where threaded connections are used Teflon paste should be considered to act as the thread lubricant and pipe sealant. This Teflon paste possesses good chemical resistance and is a high-pressure nonhardening sealant which remains permanently flexible.

7.10 WHY JOINTS FAIL

Occasionally, adhesive joints assembled by competent mechanics come apart, and there is much speculation as to what has gone wrong. The following discussion may provide solutions to the problem.

Poor adhesive wet-out has been demonstrated to be one of the prime causes of poor joints (in the case of the polyesters, it may be improved by using a primer such as Ceilcote P-370 or styrene). Field experience has indicated that a sandblasted surface when primed with the same polyester resin used for making wrapped joints will aid in solving polyester adhesive problems. Quite often sandblasting the inside of the flange hub exposes glass, and the adhesive does not do a good job of wetting the glass. Occasionally, the angular space between the flange hub and the pipe will be greater than $\frac{1}{8}$ in. In such a case a wrap of one or two layers of glass cloth on the pipe will save the situation. As added insurance in some cases, the outside of the flange hub has been sandblasted, and wrap-joint construction made in addition to the normal flange adhesive joint. (This, perhaps, is gilding the lily.) A little extra labor is worthwhile to assure a sound joint and one that is chemically clean.

Be sure to have a rough surface profile when making the joint. Avoid improper preparation, leaving dust, dirt, or oil on the sanded surface. Keep hands off the sanded surface. Also, beware of wax, paint, or glass on either the inside of the flange or the pipe. Any such carelessness will surely cause joint failure. Do not let oil or wax get onto the sanding disk since it may be transferred to the already ground and cleaned surface.

Maintain air temperature above 60°F. (Store your adhesive in a normally heated room.) The problem that can be caused by working in a cold air and on cold surfaces is not difficult to foresee. Adhesive bonds are temperature-sensitive. Their strengths are measured in the thousands of pounds per square inch when properly made, but if made in a cold room (below 60°F) a 6-in. flange glued to a flat surface can be lifted off with little effort. For joint reliability, either adhesive or wrapped, warm working conditions and surfaces are a necessity.

Improper mixing of the adhesive (guessing at the amount of catalyst instead of measuring it) is another cause of trouble. The consequences were clearly illustrated by one installed epoxy system, where joint reliability was severely lowered because of spotty adhesive mixing.

Avoid using old, partially set up, or partially evaporated adhesive. The bond line must not be too thick or too thin. Try to control it to a thickness of $\frac{1}{16}$ to $\frac{1}{8}$ in. A thinner bond may give wet-out trouble. A thicker bond may cause shrinkage problems. A bond line of $\frac{1}{4}$ in. is entirely too thick, and with a polyester adhesive, volumetric shrinkage problems may result. That is less likely to occur with an epoxy adhesive, but it has been observed to occur in bond lines this thick.

After applying adhesive to the outside of the pipe and the inside of the flange socket, smooth out all air bubbles and pockets with a spatula; then push the pipe slowly into the flange bell until it sets firmly into the flange lip. When you have done this, rotate the flange 180°. This will provide further insurance against pockets and thin spots.

It has proved most frustrating to attempt to use flanges or other materials which have been manufactured using a zinc stearate or silicone releasing agent. Gluing to this material is almost impossible. These releasing agents will penetrate well into the material, so that removing them by any common means is impractical. Nearly every reputable vendor will shy away from the use of such releasing agents on reinforced plastic work. Occasionally, however, flanges or structural shapes manufactured with this kind of agent are sold and result in unsatisfactory performance.

Refrain from using glued-on flanges in fabricating short lengths of pipe which are to be installed between sections of steel, lead-lined-steel, or rubber-lined-steel adjacent sections. The short section of reinforced plastic pipe will tend to act as an expansion joint and will quite often be unequal to the strain imposed upon it. Failure may be in the piping or at the joint.

Do not use glued-on flanges where the installation of the pipe is going to be difficult, for example, between unyielding sections of lined-steel headers or piping. Here the pipe fitter is likely to resort to wedges, with considerable physical abuse of the reinforced plastic in the installation. The risk of breaking the glue line on the flanges is very great. Heavy steel lines have a habit of springing, so that the original dimension does not hold. The strongest possible assembly in today's technology is obtained with hand-laid-up stub ends. In literally thousands of these which the author has observed being installed, the failure rate, even under the most difficult conditions, was virtually nonexistent. Where the going is rough, specify hand-laid-up flanged stub ends on polyester pipe.

If, however, a glued-on flange has been used and the glue line is broken, all is not lost. If the leak is small and can be suitably led off to waste, the hairline break may act as a filter, and the leak will decrease daily. This has been the experience in solutions where suspended solids existed. It will vary with the amount of solids in suspension.

If the pipe fitter has to use wedges to drive the joint apart, a halt should be called to the installation promptly. It is preferable to jack the line apart and permit the reinforced plastic spooled section to slide in between taped gasketed surfaces. Release the jack, and bolt up.

Be sure that your piping specifications are standard, making an adequate joint with the flanging or coupling system you intend to use. Occasionally, through inexperience, the engineer will call for a non-standard specification. This could spell jointing problems, but they can still be avoided if the fabricator and the purchaser know how to "communicate" with each other.

Flanged pipe stubs using press-molded flanges or assemblies should be pressure-tested at the factory or in the field at a pressure equal to twice the actual operating pressure. Double-testing is good insurance, that is, shop-testing of the stubs and subsequent testing of the entire assembly after it has been put together. This may sound like a duplication of effort, but in one field installation covering about 1,500 ft of piping it was found that, due to a manufacturing deficiency, every single one of the elbows leaked. This is a rare exception, but it can happen.

REFERENCES

[1] Fonda, A. F., The Ceilcote Company, Cleveland, personal communication, May 11, 1967.

[1a] Recommended Product Standard for Custom Contact Molded Reinforced Polyester Chemical Resistant Process Equipment, TS-122C, Sept. 18, 1968.

[2] Construction Specifications and Material Requirements per Joint for 100 PSI Pipe, The Ceilcote Company, *Bulletin* C-884-R, Cleveland, Ohio, p. 18, March, 1966.

[3] *Ibid.*, p. 13.

[4] Annis, M. C., and W. A. Szymanski, Hooker Chemical Corp., Durez Div., North Tonawanda, N.Y., personal communication, Dec. 22, 1965.

[5] Ref. 2, p. 18.

[6] Loftin, E. E., FMC Corporation, Front Royal, Va., personal communication, Dec. 20, 1966.

[7] Loftin, E. E., FMC Corporation, Front Royal, Va., personal communication, Jan. 9, 1967.

[8] Munger, C. G., Amercoat Corp., Brea, Calif., personal communication, Dec. 29, 1966.

8

Supporting
and Anchoring
Reinforced Plastic Pipe

8.1 INTRODUCTION

The support of plastic piping systems is completely different from that of any other and requires a special approach to the entire procedure. The low modulus of elasticity of polyester pipe puts it in a separate category from the hanging and supporting of steel pipe. This is another area in which the user's design concepts need to be reoriented.

In Fig. 8-1 are included a number of sketches illustrating some of the supporting and anchoring details involved in installations of RP piping systems. Much can be said about the ease of handling and supporting RP pipe versus rubber-lined, stainless, or lead-lined steel pipe. The lighter pipe is, of course, much simpler to handle and install. Note the wide-band support method universally used with RP pipe to minimize localized stresses.

The support and anchoring of reinforced plastic piping systems should be engineered to provide a firm support at points of contact while holding deflection to a minimum. A good combination of flexibility is afforded by using bends between anchor points plus the proper hangers to support the pipe. See Fig. 8-2 for additional supporting, anchoring, and guiding suggestions.

All valves should be supported independently of the piping system. See Fig. 8-3 for a typical valve support. The valve, if adequately supported, may be used as an anchor point.

Be especially careful when passing through a masonry wall to sleeve and cushion the passage with a short section of insulation material so that the RP line is not damaged by striking rough masonry surfaces.

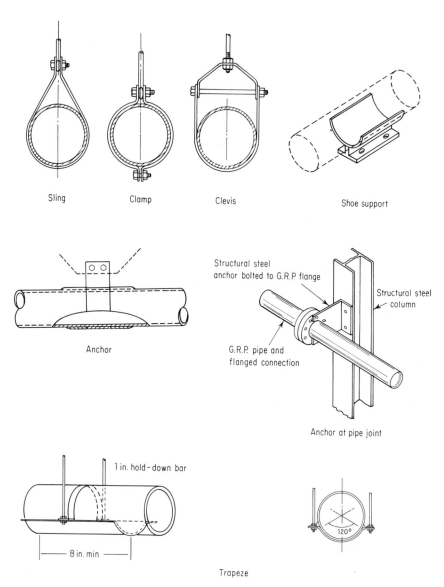

FIG. 8-1   Popular pipe supports and anchors.

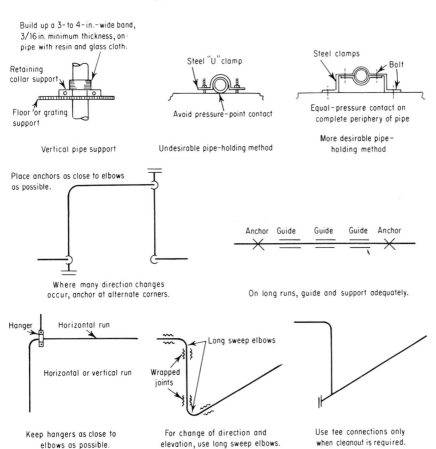

FIG. 8-2 Some suggested additional techniques for proper support of reinforced plastic pipe systems.[1]

8.2 RECOMMENDED INSTALLATION PRACTICE—POLYESTER PIPE[2]

The reader should refer to page 59, Table 6, which provides recommendations from the Commercial Standard covering the maximum spacing of pipe hangers for reinforced polyester pressure pipe. Remember that Table 6 is good for maximum spacing temperatures up to 180°F. There will be times, however, when service conditions at higher temperatures than this will be necessary. For this the reader is referred to the brief compilation below.

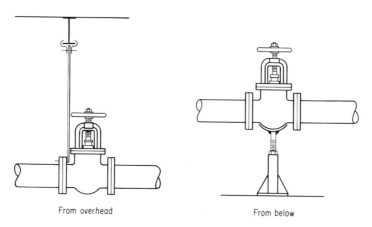

From overhead   From below

All valves should be supported
independently of the FRP line

FIG. 8-3   Typical valve support, independent of the line.

POLYESTER PIPE
Support spacing versus temperature

Temperature	*Percent reduction* (*from Table* 6, *page* 59)
Up to 180°F	0
200°F	3
220°F	6
240°F	9
260°F	12

In determining horizontal hanger spans, spacing is determined on the basis of three specific items:

1. Allowable deflection.
2. Stress occasioned by allowable deflection.
3. Stress in hanger area.

Trapeze-type hangers or shoe supports should have a very broad bearing area as shown in Fig. 8-1.   An 8-in. minimum support length is desirable and where we are dealing with larger diameter piping the support length should be approximately equal to piping diameter.

Note that the Standard recommends supporting a minimum of 180° of the pipe surface.   Certainly this is the safe and recommended approach, although some fabricators suggest 120° support is sufficient.   A 1-in. hold-down bar should be used with any 120° trapeze support.

The reader is referred to Chap. 9 for the special considerations necessary in installing pipe underground.

There may be occasion to design gravity feed distribution systems in which a series of long north–south runs are fed from an east–west header. If considerable temperature changes are involved, displacement of the east–west header will occur unless special precautions are taken to anchor the north–south runs immediately after leaving the east–west header. Further, make sure that the east–west header is anchored at several points to prevent an undesirable movement in the east–west direction.

There are several methods possible to provide the shortest installation time with the highest work index. These involve having made-on flanges in 40-ft pipe sections. To provide for the uncertainties, several sections should be furnished with flanges made-on one end only with a separate stub end. This permits cutting the unflanged end to the proper length, and wrapping the stub into position with the conventional butt joint. Where downtime is of critical importance these methods ensure the return of the system to active production in the shortest possible time.

8.3   SUPPORTING REINFORCED EPOXY PIPE

In glass-reinforced epoxy pipe there are slight variations in support spacing from vendor to vendor. These have been compiled from a large number of vendors for this type of pipe. Table 8-1 is an approximate guide in the spacing of pipe hangers in supporting epoxy pipe for most vendors providing a 150-psig pipe and with a liquid specific gravity of 1.25 and up to temperatures of 200°F.

LIQUIDS   HANGER SPACING, EPOXY PIPE
Maximum span-support spacing (Specific Gravity 1.25)

Table 8-1

	Pipe size, in.								
	1	1½	2	3	4	6	8	10	12
	Spacing, ft								
Temperature:									
100°F..........	7	8	9	9	10	11	13	15	16
200°F..........	6	7	7	8	8	10	12	13	14
Minimum band width per inch..	1½	1½	1½	2	2	2½	3	3	3½

NOTE: The above hanger spacing is conservative and is based on a maximum of ½-in. deflection of the pipe. The table may also be used to estimate the maximum span-support spacing when handling water or liquids of a specific gravity lower than 1.25.

GASES HANGER SPACING, EPOXY PIPE
Table 8-2

Pipe size, in.	Spacing, ft
1	7
1½	9
2	11
3	13
4	15
6	17
8	20
10	22
12	24

Where reinforced epoxy pipe is used for the handling of gases up to 200°F, Table 8-2 applies. The table should be used as a general guide only. Since design and construction of reinforced epoxy pipe may vary considerably from vendor to vendor, it is suggested that the engineer consult the vendor's catalogues for exact hanger spacing. Table 8-2 may be used for estimating purposes prior to the final design.

8.4 EXPANSION AND HOW TO DEAL WITH IT

Current practice in the installation of RP systems is almost completely without expansion joints. For example, a large chemical company installed 2 miles of RP piping without a single expansion joint. In another installation 8 miles of RP piping was installed, and not a single expansion joint was used.

Some reinforced plastic expansion joints of the slip type are manufactured to accompany specific systems. Where the engineer considers an expansion joint to be necessary, consideration should also be given to a bellows-type joint made of Teflon. A bellows-type expansion joint, such as molded TFE joint, Style E-1608, by John L. Dore Co., Houston, Tex., is recommended, since it operates with low axial force.

When we deal with the expansion characteristics of RP pipe, we enter a world completely different from metals. The comparative physical properties of metals and reinforced plastic are shown in Table 1-1. The main thing to note is that RP material has a very low Young's modulus of elasticity, which more than compensates for a higher coefficient of thermal expansion prevalent in the polyesters, and less in the epoxies.

Fiber stress in a pipe may be calculated by the formula $S = Ee$, where S, stress, is in pounds per square inch, E is Young's modulus, and e is the expansion per unit length. Thus the lower modulus of elasticity counteracts the higher expansion coefficients, so that most RP piping systems can be installed without expansion joints.   Bear in mind also that we are generally dealing with systems which may operate between 32°F and a probable maximum of 212°F.   The net result of these calculations is that the internal fiber stress in an RP system is generally only 5 to 10 percent of the allowable fiber stress.   This permits expansion to be taken up as an internal compressive stress.   The system, however, must be properly engineered with guides and anchors so that failure does not occur due to buckling.

Stresses by external constraint

Problem

Let us calculate the expansion of a 6-in. polyester line over a distance of 250 ft when operating between the temperatures of 60 and 160°F.   Assume a linear coefficient of expansion of 15×10^{-6}.

$$\text{Expansion in inches} = (12)(D)(K)(10^{-6})(\Delta t)$$
$$= \frac{(12)(250)(15)(160 - 60)}{1,000,000}$$
$$= \frac{4,500,000}{1,000,000}$$
$$= 4.5 \text{ in.}$$

where D = length, ft

K = coefficient of expansion

Δt = temperature difference, °F

Now suppose we anchored the lines at each end.   Calculate the maximum resultant stress in the pipe, with no provision being made for expansion:

$$S = Ee$$
$$= \frac{(800,000)(4.5)}{(12)(250)}$$
$$= \frac{3,600,000}{3,000} = 1,200 \text{ psi}$$

where S = stress, psi

E = modulus of elasticity (800,000 in this case)

e = strain, expansion per unit length

Since we have an allowable wall stress of at least 12,000 psi, the stress on this wall amounts to only 10 percent of the total.

If this has been a steel line, what would have been the resultant stress and expansion?   Assume a Young's modulus of 29 million and an expansion coefficient of $(6.5)(10^{-6})$.

$$\text{Expansion} = \frac{(12)(250)(6.5)(160 - 60)}{1,000,000}$$

$$= \frac{1,950,000}{1,000,000}$$

$$= 1.95 \text{ in.}$$

$$\text{Stress} = \frac{(29,000,000)(1.95)}{(12)(250)}$$

$$= \frac{55,000,000}{3,000} = 18,300 \text{ psi}$$

$$\frac{18,300}{1,200} = 15.31$$

The stress in the steel line would have been over fifteen times as much as in the RP line.

The low modulus of elasticity of RP material, as shown in the foregoing problem, permits construction simplifications through the practical elimination of expansion joints in nearly all work. Even on work running at the maximum temperature permissible with the material, a few changes in direction are all that is required.

Also observe that if we had not constrained the RP line, the line would have had a bow in it amounting to a line growth of 4.5 in.

The lateral buckling in such a line would amount to a distance of approximately 8 in. unless an external constraint were applied. If this is a free-swinging line hanging overhead, no particular problem exists. This stress in the pipe wall will have only a small effect on the pipe's original pressure performance, so that the axial stress and support spacing still hold. However, if we are dealing with a vacuum condition, it is conceivable that it would have an effect.

Compressive-stress limitations on hand-laid-up glass-reinforced polyester pipe are generally considered to lie in the 18,000- to 24,000-psi area.[3]

For sample expansion calculations the reader is referred to Fig. 8-4, which provides a handy rule-of-thumb approach for both hand-laid-up and filament-wound RP pipe. The graph was designed to provide average figures from a number of sources.

Conversely, if we experience a temperature drop on an anchored pipe, the pipe's pressure performance will be reduced, so that while, normally, an increase in temperature of 100°F would be no cause for alarm, a drop of more than 50°F can reduce the pipe's pressure performance. Let us reaffirm that this performance will be affected only if the pipe is securely anchored. Make sure that bending at elbows and tees is minimized by anchoring at or near the fittings. Never use an open-ended tee

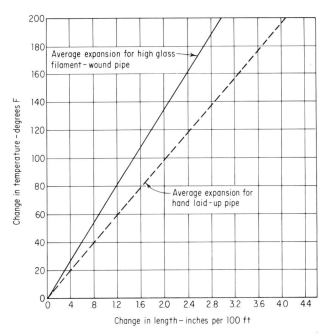

FIG. 8-4 Average expansion of filament-wound and hand-laid-up pipe.[4]

for a clean-out when making a 90° bend through the side outlet of the tee. While you may be able to get by with it, it is not recommended practice.

It is, of course, sometimes difficult to change accepted practice. To the engineer accustomed to dealing with steel lines, expansion loops and joints could no more be omitted than leaving out the drain and vent connections. There is certainly nothing wrong with providing a reinforced plastic piping system with suitable expansion take-up devices. Our point is that, due to the low Young's modulus, they are unnecessary. If, however, in the judgment of the engineer, they should be provided to minimize anchoring forces, then guidance in the use of loops and joints is appropriate. The following section covers some of the basic elements in both loop and joint design, such as:

- Primary guide spacing
- Secondary guide spacing
- Intermediate guide spacing as a function of design end load
- Suggested maximum end load with appropriate spacing
- Precompression

While expansion loops are not common, they have been employed successfully in the installation of long RP lines. Figure 8-5 indicates some of the design configurations which can be used in expansion-loop or -joint design.[5]   Additional factors are as follows:

$$PC = \frac{M(T_3 - T_1)}{T_2 - T_1}$$

where PC = precompression
M = rated movement of joint, in.
T_1 = minimum temperature, °F
T_2 = maximum temperature, °F
T_3 = installation temperature, °F

If, for example, we had a 4-in. polyester pipe 100 ft long where $T_1 = 50°F$, $T_2 = 200°F$, and $T_3 = 80°F$, then, from Fig. 8-4, the approximate expansion would be 3.1 in. per 100 ft.   The total expansion to be reckoned with would be 3.1 in.   Therefore

$$PC = \frac{(3.1)(80 - 50)}{200 - 50}$$
$$= \frac{(3.1)(30)}{150}$$
$$= 0.62 \text{ in.}$$

The expansion joint could therefore be precompressed approximately ⅝ in., leaving some 2½ in. to be taken up by the expansion joint itself.

Now let us suppose that in the above example we used an expansion loop instead of a joint.   Referring to Fig. 8-5,

$$A = 2B \qquad A = \frac{2.5}{0.02} = 125 \text{ in.}$$
$$B = 62½ \text{ in.}$$

Observe that each half of the loop is capable of absorbing the deflection shown.   A loop of this design would therefore be able to absorb satisfactorily the expansion of 100 ft on each side of the loop, or a total of 200 ft of 4-in. pipe.   Each end of the 200-ft run would be suitably anchored. This would be the length of each loop leg as determined from Fig. 8-5. A design such as this will absorb end loads and compressive strain in the system.   In addition, smaller end loads on the pipe due to internal pressure may approximate 10 to 15 percent of the maximum loading due to restrained thermal expansion.   Two other factors to be reckoned with are the expansion-joint load itself, which may be obtained from the supplier, and the load due to friction between pipe and pipe supports.

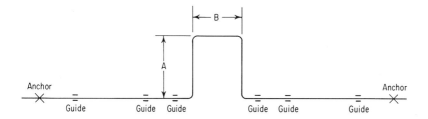

Allowable deflection of loop

Pipe sizes	Percent def.
2" & 3"	3%
4" & 6"	2%
8," 10," & 12"	1%

Expansion loop

FIG. 8-5 Expansion loop.

As a rule of thumb, primary and secondary guide spacing immediately adjacent to expansion joints or loops may be estimated from Table 8-3.

The purpose of the primary and secondary guides is to make sure that piping movement is guided in a truly axial direction and to eliminate any possibility of failure due to buckling. Intermediate guides are desirable at greater distances, generally, than the primary and secondary guides.

Intermediate-guide spacing is a function of the allowable end load. Table 8-4 suggests approximate intermediate-guide spacing in feet as a function of the allowable end load, rated by piping sizes.

DISTANCE FROM JOINT TO FIRST AND SECOND GUIDES[5]

Table 8-3

Pipe size, in.	Primary guide, ft	Secondary guide, ft
2	1	3
3	1	$3\frac{1}{2}$
4	$1\frac{1}{2}$	$4\frac{1}{2}$
6	2	7
8	$2\frac{1}{2}$	9
10	3	12
12	4	14

INTERMEDIATE–GUIDE SPACING AS A FUNCTION OF DESIGN END LOAD[6]

Table 8-4

Pipe size	End load, lb					Suggested design end load and spacing	Maximum end load at 1-ft, intermediate-guide spacing, lb
	100	500	1,000	5,000	10,000		
2″	15′	7′	...	...	...	800 lb at 5′	3,900
3″	28′	13′	9′	...	...	1,200 lb at 8′	7,000
4″	37′	16′	12′	...	...	1,550 lb at 9′	9,200
6″	75′	35′	24′	...	...	3,300 lb at 13′	18,000
8″	110′	50′	37′	16′	...	6,000 lb at 15′	25,000
10″	160′	75′	50′	23′	...	9,000 lb at 17′	31,000
12″	215′	100′	68′	30′	21′	13,000 lb at 18′	37,000

NOTE: This table applies to a heavy-walled epoxy pipe such as that made by Fibercast Company, specifically, grades OG-2025, BL-2025, and CL-2025. For a more comprehensive treatment covering the design of expansion joints and loops, the reader is referred to the Piping Design Manual published by Fibercast Company, Sand Springs, Okla.

MAXIMUM DESIGN END LOAD FOR 100–PSI HAND–LAID–UP POLYESTER PIPE

Table 8-5

Pipe size, in.	Cross-sectional wall area, in.²	Allowable stress, psi	Total allowable end load, lb*
2	1.29	900	580
3	1.85	900	830
4	3.35	1,200	2,000
6	4.8	1,200	2,875
8	8.1	1,350	5,475
10	12.25	1,500	9,150
12	17.0	1,500	12,750

* Assumes a force of this amount on each end of a run of pipe to produce the maximum allowable stress shown.

ANCHORING FORCE—EPOXY PIPE[7]

Table 8-6

Pipe size, in.	Force in pounds to anchor per 100°F
1	500
1½	750
2	820
3	1,200
4	1,550
6	3,300
8	6,000
10	9,000
12	13,000

8.5 ANCHORING FORCE—EPOXY PIPE

Table 8-6 is designed to aid in estimating anchoring force on reinforced epoxy pipe.

The pull-out strengths of good adhesive joints are at least sixteen times the anchoring force shown in the table on small sizes and about

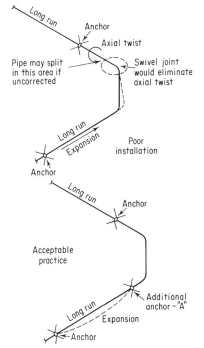

FIG. 8-6 Dealing with rotational moments in RP pipe.

eight times the anchoring force on the larger sizes. Coefficients of expansion on epoxy pipe run 9 to 13 times 10^{-6} in./($^{\circ}$F/in.).

8.6 ROTATIONAL TORQUE—RP PIPE

Some problems in RP piping have been caused by axial rotation of the pipe, in the manner shown in Fig. 8-6. The pipe literally split because it was being wrung out like a washcloth. Guard against any support and anchoring that would permit such a situation to occur.

Occasionally, in a system design, the epoxy or polyester pipe may be subjected to a twisting moment or a torsion effect. A typical example worked out below indicates the approach to such a problem.[8]

Problem

A 6-in.-dia epoxy pipe with an allowable shear stress of 5,000 psi is subjected to a torque of 10,000 in.-lb. Calculate the shearing stress in the pipe wall. Assume an ID of 6 in. and an OD of 6⅝ in. Also calculate the angle of twist, in degrees. The pipe is 50 ft long.

$$
\begin{aligned}
\text{Shearing stress} &= \frac{2TR}{\pi(R^4 - r^4)} \\
&= \frac{(2)(10,000)(3.31)}{\pi[(3.31)^4 - (3.0)^4]} \\
&= \frac{66,200}{125} \\
&= 530 \text{ psi shear stress}
\end{aligned}
$$

where T = torque pounds, in.
R = pipe OD, in.
r = pipe ID, in.

$$
\begin{aligned}
\theta &= \frac{2TL}{\pi(R^4 - r^4)G} \\
&= \frac{(2)(10,000)(50)(12)}{\pi[(3.31)^4 - (3.0)^4](2.3)(10^6)} \\
&= 0.0416 \text{ radian, or } 2.4^{\circ}
\end{aligned}
$$

where θ = twist angle, radians
L = pipe length
G = shear modulus, 2.3×10^6

In this problem the amount of torsion is relatively small.   If the pipe had been polyester with the same ID and OD dimensions, but with the modulus of $(1.0)(10^6)$, then our torsion would have been

$$\theta = \frac{(2)(10,000)(50)(12)}{\pi[(3.31)^4 - (3.0)^4](1.0)(10^6)}$$
$$= \frac{12,000,000}{(125)(10^6)}$$
$$= 0.096 \text{ radian, or } 5.4°$$

It will be observed from these calculations that we could have increased the shear stress in the epoxy pipe by over ninefold and still have been within the allowable limits.   Thus, at a shear stress of 5,000 psi, the angle of twist could have been moved up to 0.394 radian, which is equivalent to approximately 22°.

There are no swivel joints manufactured made of RP materials. Such a joint might be used to solve peculiar expansion problems associated with pumps and tanks, and could probably penetrate extensively into the swivel-joint market now dominated by the nickel alloys or other high-priced metals.   Such a joint would cost but a fraction of the price of high-nickel alloys and be competitive all the way to 316 stainless steel.

8.7 PRESSURE SURGES

Pressure-surge control and the elimination of hydraulic hammering are essential with reinforced polyester or epoxy systems.   It is good design practice to eliminate hammering and pulsating flows regardless of the system design, be it metallic or reinforced plastic.   This topic calls for discussion in some detail since the consequence of ignoring it can be serious system distress and, possibly, complete failure.   A reinforced plastic line that has been overstressed by surging or hammering will exhibit certain warning signs.   Hammering can be particularly intense and noisy while the system is still operating.   At this stage, examination of the internal walls on the pipe will reveal typical signs of the beginning of hoop-stress failure.   These are, invariably, longitudinal hairline cracks in the interior resin system.   The high chemical performance of the system has been lost, and the corrosive material will obtain rapid admission into the glass substrate.   Then, depending upon the rate of chemical attack, the wall of the pipe will become progressively weakened until, with repeated hammering, either weeping will occur, in the case of continuous-filament pipe, or a section of the pipe will blow out if it is a hand-laid-up pipe or a pipe made with sock-type construction.

A reinforced polyester pipe was judged largely unsuccessful when installed in a hammering condensate system.[9] The RP piping was, literally, beaten apart by heavy continuous hammering. When hammering is not present, the RP pipe appears to work very well in corrosive condensate service. And it should be added that the same hammering condensate service has split open heavy wrought-iron pipe, indicating the magnitude of the force that causes the problem.

Isolated sections of RP pipe have failed in hot, weak acid service. These are generally located on the discharge side of heaters, where continual pounding and hammering from condensate steam ultimately cause failures at the epoxy joints.[9] In most cases, however, they have outlasted their lead-lined steel predecessors by factors of 4:1.

Avoid RP polyester pipe (hand-laid-up) where intense hammering or pressure surges may create momentary pressures running to thousands of pounds. Such conditions can develop from hydraulic hammer caused by a check valve swinging shut or by any quick-closing valve. There are simple familiar ways to design around conditions of this kind, for example, equipping the system with periodic air domes that act as a cushion to sudden pressure changes of large magnitude. Also, the designer may elect to use a filament-wound epoxy or polyester, some designs of which show enormous hoop strength and are most resistant to failures of this type. The injection of air, where compatible with the process, may also prove beneficial.

In actual field service, over a long period, tremendous hammering on filament-wound or epoxy pipe has been reported. In such cases the filament-wound construction has withstood severe hammering for at least one year, and is still in operation.

Surge dome   The importance of shock control cannot be overemphasized. The air-chamber design should reduce the shock pressure so that it at least equals the rated pressure of the piping system. In this connection, it must be cautioned that inadequately sized air chambers do not possess sufficient capacity for shock control.

Two proposed designs have been developed for cushioning the shock of pulsating flows or hammering systems.[10] These are shown in Figs. 8-7 and 8-8.

Other methods of eliminating pressure surges in RP piping   Live steam fed into the circulating solution with condensation taking place a moment later is exactly the same phenomenon as that which occurs in a cavitating pump. The steam from an area of high pressure is introduced into an area of low pressure, and the steam balloon instantly goes into a state of collapse. In the vapor form at modest pressures, steam occupies a space

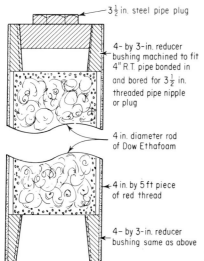

$3\frac{1}{2}$ in. steel pipe plug

4- by 3-in. reducer bushing machined to fit 4" R.T. pipe bonded in and bored for $3\frac{1}{2}$ in. threaded pipe nipple or plug

4 in. diameter rod of Dow Ethafoam

4 in. by 5 ft piece of red thread

4- by 3-in. reducer bushing same as above

FIG. 8-7 Foam-filled surge dome.[10]

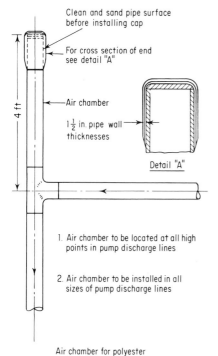

Clean and sand pipe surface before installing cap

For cross section of end see detail "A"

4 ft

Air chamber

$1\frac{1}{2}$ in. pipe wall thicknesses

Detail "A"

1. Air chamber to be located at all high points in pump discharge lines

2. Air chamber to be installed in all sizes of pump discharge lines

Air chamber for polyester pump discharge lines

FIG. 8-8 Air-chamber surge dome for polyester pump discharge lines.[10]

about 1,500 to 1,700 times its water equivalent. The implosion of the condensing vapor is translated into noisy operation, with considerable rattling and banging. If the bubble of steam collapses against the pipe walls, intense pressures over minute areas are created. If the collapse occurs in the mainstream, hydraulic hammer radiates to the walls. A hand laid on the pipe just below the heater reveals the intense disturbance being created within the heater and just downstream from it.

Actual tests have shown that in severely corrosive applications, heaters and downstream piping may last 6 weeks to 6 months when they are made from conventional corrosion-resistant metals. The same metals in any other part of the system would last 10 years. Corrosion products often protect the metal surface from further corrosion, but in an injection heater they are removed effectively and spontaneously by the succession of imploding steam bubbles. The metal literally wears out in short order.

There is a remarkably effective answer to the problem of the cavitating effect in steam-injection heaters. If, for example, each bubble of steam were furnished with its own built-in shock absorber (such as a bit of air), there would be a resilient cushion for the collapsing bubble to bounce against. This not only sounds good in theory, but works well in practice.[11]

A steam-injection heater equipped with a small air line, say, $\frac{1}{4}$ in., from a compressed-air source and tied into the steam line, can produce dramatic differences in the heater operation. The compressed-air source should, of course, be at a higher pressure than the steam. Compressed air should be fed in just downstream from the steam flow control valve. The air line should be equipped with a small pressure-regulating valve to ensure that the system will work well without constant attention. It takes only a few cubic feet of air per minute to do the job, even on relatively large heaters using 5,000 to 7,000 lb of steam per hour.

The demonstration of the effectiveness of this technique is best seen to be appreciated. The normal heater often operates with a considerable amount of banging and vibration. In the original heater design, anything done to reduce the size of the steam bubble will serve to produce a quieter unit operation. Basically, this is why many heaters, and especially large ones, are built with internal combining tubes. These combining tubes are generally lacking in smaller injection units. Unfortunately, the combining tube is often the first component to go as the heater wears out. Heater operation becomes progressively worse as steam enters the unit in successively larger bubbles. Rocking and shocking increase.

Large injection heaters in unstable operation have been known to shake big tanks on their foundations, causing pounding of heavy reinforced concrete floors to such an extent that safety of personnel in the immediate area became a matter of concern.

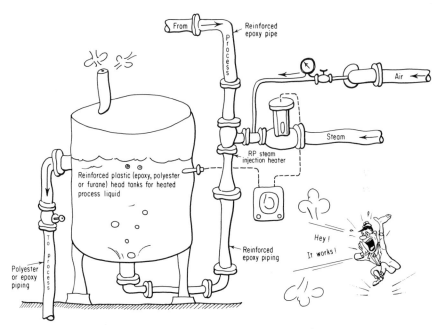

FIG. 8-9   Air injection into heated process circulating systems extends the life of piping and tanks.

Even the toughest heater responds to the compressed-air treatment. To test its effectiveness, the heater can be started in the normal manner. There will be rumbling, vibration, and noise.   As air is turned on slowly, the objectionable hammering will disappear.   Now if the air is abruptly turned off, the rattling is resumed.   A little adjustment of the air pressure, and it is safe to leave the unit.[12]

Reinforced plastic materials have outlasted lined metals four- to eightfold under severe erosive-corrosive conditions such as these.   For a flow diagram of such a system and its essential components the reader is referred to Fig. 8-9.

Safety pointers   There are three major safety considerations in using compressed air with steam-injection heaters and the plastic pipe which may lie beyond it:

1. Make sure that air and the process involved go together.   The addition of small quantities of air to the steam at the point of use will present no problem in most applications, but in one case in a hundred it may.

2. Quite often, heaters of the type described discharge into, or are located in, tanks. If it should happen that a particular process tank requires exhaust to maintain a negative pressure in the tank for the removal of toxic or obnoxious gases, make sure the air admitted does not adversely affect the exhaust balance. Normally, it will not, but it can.

3. Make the routine check of this equipment part of the operator's duties. It will pay handsomely in extending equipment life and providing quieter and smoother operation.

8.8   ELIMINATING PULSATING FLOWS ON RP SYSTEMS

Do not let your reinforced plastic piping system "crack the whip." There can be only one result, eventual failure. This applies to reinforced plastics or to any other material of construction. The destructive forces of pressure surging can break up the piping system, pull out anchors, or knock down the sides of buildings with equal facility.

Another way to eliminate pressure surges from a piping system is to eliminate pumping in "slugs." Circulating systems are installed that have, for example, a 1,000-gpm pump on an 800-gpm requirement. What this means is that liquid is pumped at a rate of 1,000 gpm followed by a 200-gpm slug of air, etc. This can only produce an intermittent hydraulic shock on the lines, so that at every change of direction we have a piston going back and forth, exerting tremendous destructive force on the line. This force can, ultimately, wreck any piping system, be it plastic or metallic, unless suitable anchoring is provided, or better yet, continuous pumping of the fluid stream through automatic control measures. Such automatic control measures may encompass anything from a solid-state pneumatic hydraulic valve to a recirculating level-control device in a tank activated by the most simple of level controls.

8.9   HYDRAULIC HAMMER WITH QUICK–CLOSING VALVES

The wider use of quick-closing valves, coupled with plastic piping, places new emphasis on the need for a better understanding of hydraulic hammer in liquid systems.

A moving column of liquid has a great deal of momentum, which is proportional to its weight and velocity. Stopping the flow by closing a butterfly valve quickly or by a check slamming shut converts this momentum into a high-pressure surge. The longer the line and the greater the liquid velocity, the higher the shock load will be. Repeatedly, in the past, it has been demonstrated that this will burst pipe, fittings, or valves.

PRESSURE GENERATED AS A FUNCTION OF VALVE CLOSING TIME

Table 8-7

Closing time of valve, sec	Water-hammer pressure, psig	Normal line pressure, psig	Total pressure, psig
5	50	70	120
3	80	70	150
2	125	70	195
1.5	175	70	245
1	250	70	320
0.5	550	70	620
0.3	950	70	1,020
0.1	3,000	70	3,070

For the sake of illustration we have chosen an acid return line with an overall length of approximately 400 ft and a liquid velocity of about 9 fps. Table 8-7 shows what happens when a butterfly valve near the end of this line is opened or closed quickly. The moral of the story is, take at least 3 to 5 sec to close any butterfly valve if you want to limit hammer and damage to the system. The table also shows what a quick-slamming check can do to a similar system.

When hammering occurs, the high-intensity pressure wave goes back through the piping system until it reaches the point of relief, or surge dome. The shock wave then ricochets back and forth between the points of relief and impact until the destructive forces are dissipated in the piping system. This surging wave accounts for the noise and vibration. At liquid velocities of 5 to 10 fps this wave may travel between the points of closure and relief at velocities of 4,000 to 4,500 fps. It is necessary to provide a means of absorbing and dissipating the energy causing the shock. Use of compressible gas is a most effective method.

The pressure in any system due to water hammer may be calculated from the formula

$$P = \frac{0.070VL}{T}$$

where P = pressure increase, psi, from the momentary surge
L = pipeline length, ft
V = liquid velocity, fps
T = valve closing time, sec

This formula may be applied to calculate the potential water hammer in any system, but several points should be borne in mind: (1) The damage may occur far from the source; (2) an air chamber full of liquid no longer acts as an air chamber; (3) the pressure surge is independent of line size; (4) if check valves are used with RP piping, their action must be almost instantaneous to prevent the moving column of liquid from attaining a high velocity.

8.10 HYDROSTATIC TESTING

RP piping systems should be pressure-tested prior to approval for service. The general recommendation is that they be tested at 150 percent anticipated working pressure.   It is also common practice to feed the testing liquid into the lower part of the system while bleeding the higher points to make sure that all the air has been removed from the system.

Although some manufacturers do not recommend testing with air, this advice is not universally heeded.   Other equally qualified manufacturers recommend it.   The author has practiced it, and by this method has found porous pipe and fittings which might have gone undetected in a water test.   Unless there is some specific reason why a system should not be air-tested, for example, some process incompatibility, the author recommends it.

8.11 DESIGN CHECK POINTS FOR ANCHORING, SUPPORTING, AND TRACING RP PIPE

1. In most plant RP piping installations, directional changes will be sufficient to provide for expansion and contraction due to temperature changes.   Only where this is not the case, and it can be clearly shown that overstressing may result, should expansion joints or loops be used.

2. Never use a metal expansion joint with RP pipe or, for that matter, wire-reinforced rubber expansion joints.   Both require action forces which are too high for this type of piping.   Stick to a Teflon bellows, if necessary, where action forces are low.

3. Teflon-bellows joints may serve not only as expansion joints, but also as vibration isolators, strain eliminators, and flexible connectors.

4. The basic disadvantages of loops are space requirements and a perpetual increase in pressure drop through the system.

5. Do not use a U bolt as an anchor.   Remember, it violates the principle of point loading.

6. Loose-fitting U bolts do well as guides, especially when used in pairs.

7. Guides and supports are not synonymous. Each one serves its own purpose.

8. It is sometimes necessary to steam-trace a line to prevent it from freezing. Spiral wrapping of the trace material is a necessity. Tracing of the line on one side will work a hardship on the pipe due to its low thermal conductivity, resulting in bowing. Wrap the combined tracer and pipe with some light insulation to contain the heat. Limit tracing temperatures of polyester pipe to 240°F (10 psig) and epoxy piping to 260°F (approximately 20 psig). It is tempting to simplify tracing installations by using polypropylene tubing, but this has not worked out well. Electrical tracing has been used sucessfully; but make sure the expansion of the RP pipe does not pull the electrical system apart. Allow generously for this expansion.

9. Operating bending moments of RP pipe should not exceed 25 percent of the allowable bending moment. Bending moments vary from vendor to vendor, depending on the type of piping and method of manufacture. For this reason the manufacturer's catalogue should be consulted.

10. Protect RP pipe as it passes through wall openings, or it will be damaged.

11. Make sure surge domes will continue to act as hydraulic-shock eliminators.

12. Watch out for twisting moments in RP pipe. Do not wring it out like a washcloth.

13. Support all valves independently.

14. Remember, generally, expansion joints or loops are not necessary, but there may be times when they are. If it makes you feel more comfortable about the installation, use them.

15. Think twice before you use a piece of rubber hose clamped on each end as an expansion joint.

Conclusion: Most of the RP piping system failures which occur are physical failures, due to violations of good supporting or design principles. This has been demonstrated repeatedly.

REFERENCES

¹ Mallinson, J. H.: Reinforced Plastic Pipe: Joining and Supporting, *Chemical Engineering*, Jan. 17, 1966.

² Recommended Product Standard for Custom Contact Molded Reinforced Polyester Chemical Resistant Process Equipment, TS-122C, Sept. 18, 1968.

³ Duracor Pipe Catalog 10-F, sec. 6.2, The Ceilcote Company, Berea, Ohio, March, 1966.

⁴ This chart was compiled from a number of sources, one of which was Duracor Pipe Catalog 10-F, published by The Ceilcote Company, Berea, Ohio, March, 1966. Other sources include various published data on expansion of filament-wound pipe. In some epoxy piping systems linear expansion is partly a function of piping diameter. Therefore the chart on filament-wound pipe can be only an approximation.

⁵ Piping Design Manual, Fibercast Company, Sand Springs, Okla.

⁶ *Ibid.*   This table is adapted from a graph in the reference publication.   The suggested maximum end load and spacing is the author's.

⁷ Kelly, Mark E., Jr., A. O. Smith Corp., Smith Plastics Div., Little Rock, Ark., personal communication, Feb. 19, 1966.

⁸ "Fibercast Technical Data and Specifications," Fibercast Company, Sand Springs, Okla., 1967.

⁹ Mallinson, J. H.: Plastic Pipe: Its Performance and Limitations, *Chemical Engineering*, Feb. 14, 1966.

¹⁰ Kelly, Mark E., Jr., A. O. Smith Corp., Smith Plastics Div., Little Rock, Ark., personal communication, Feb. 19, 1966.

¹¹ Mallinson, J. H.: How to Tame Injection Heaters, *Southern Engineering*, May, 1964.

¹² Mallinson, J. H.: Relieving Cavitation in Steam Injection Heaters, *Compressed Air*, May, 1965.

9

Reinforced Plastic
Industrial Sewer
and Drain Piping

9.1 INTRODUCTION

Glass-reinforced plastic industrial sewer and drain piping is furnished in a wide range of sizes up to 60 in. in diameter and in 10- and 20-ft lengths.

There are a number of advantages in RP sewer pipe, especially when handling corrosive liquids:

- Its light weight keeps down handling-labor costs.
- The pipe being nonmetallic, there is no electrolytic and galvanic corrosion.
- It is commonly completely resistant to attack by corrosive soils, bacteria, and groundwater.
- It will handle with ease dilute acid-bearing wastes and those containing hydrogen sulfide.
- All the RP pipe has an extremely smooth interior, permitting the use of lower slopes and sometimes smaller pipe. Flow coefficients of 150 (Williams' and Hazen's) are common for this type of pipe.
- Maintenance costs are low.
- It has a long life expectancy.

9.2 STANDARD JOINING METHODS

Numerous joining methods are employed in this type of construction, depending upon the manufacture. Some of the more common joints in use are as follows:

Standard butt joints   This is the conventional joint, made by wrapping layers of glass fiber impregnated with a catalyzed resin over the butted joint. For details on making a butt joint see Chap. 7. This joint is economical, permanent, and very satisfactory. It is especially useful if the drain or sewer line is expected to be under considerable pressure at various times.

Bell and spigot   There are various designs of a bell-and-spigot joint. Normally, it consists of a bell on one end of a pipe into which the next section of pipe is inserted. The seal is obtained by a round rubber gasket. Sometimes this is referred to as an "O-ring" joint. One of the advantages of this type of joint is that it allows some misalignment.

Redilok* coupling   This coupling is suitable for drainage work up to 5 lb and consists of a fiber-glass-reinforced polyester sleeve and a specially built internal rubber gasket. The coupling is self-centering, and joining by this method is rapid and easy. There are no metal parts subjected to external corrosion.

Flexible coupling   This is simply a rubber sleeve fitted over the butted end of two pipe lengths. Each end of the sleeve is then clamped with a stainless-steel band and tightened with a screw.

Flanged joint   This type of joint is commonly constructed with flanges made on at the factory. They are sometimes useful when it is necessary to connect with other types of pipe. They are not commonly used in buried-pipe construction.

9.3 EXCAVATION AND TRENCHING

It is generally good design practice to keep the width of the sewer trench to a minimum.[1] It is common practice to specify the minimum width of the trench at the top to be the OD of the pipe, plus 12 in. It may, however, be necessary to modify this somewhat if equipment used for densification of the backfill by compaction is required. Trench width will then be dictated by the necessary packing equipment and the number of people working it.

Ideally, the excavation of the trench should be to a depth which will provide a uniform bearing and support under the pipe. A uniform bedding equal to approximately one-third of the pipe circumference is desirable. If rock is encountered, the trench should be excavated at least 6 in. below the subgrade and then backfilled with a material which is thoroughly compacted. We have in mind, in each case, a flexible pipe

* Registered trademark of duVerre, a division of American Pipe Corp.

which requires care in installation if the job is to be satisfactory. It is good practice to make sure that all backfill is completely free of stones, boulders, hard clods of earth, and vegetation. Limit hydraulic compaction to free-draining soils to preclude the possibility of floating the pipe. Backfilling immediately adjacent to the pipe should be in well-compacted layers 6 in. at a time, including a 6-in. layer above the top of the pipe. It is important that the entire length of the pipe be properly bedded and supported, particularly if bell-and-spigot pipe is used. For a typical suggested detail on a sewer-pipe trench and backfill, see Fig. 9-1.

Reference is made to Standard Trench Loading Tables, which provide overburdened load forces of any type underground sewer pipe as a function of:

1. Pipe diameter
2. Trench width
3. Depth of overburden
4. Type of backfill

The reader is also referred to the recommendations of ASTM Committee D-20, covering the underground installation of flexible thermoplast sewer pipe, which are valid in this instance since the RP pipe should be considered a flexible conduit even though, strictly speaking, it is not a thermoplast.

Based on an earth fill of 120 lb/ft³, the maximum burial depth in feet may be approximated by the following formula:

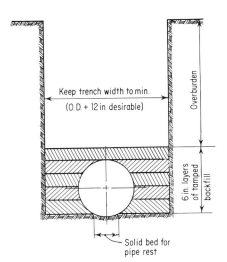

FIG. 9-1 Sewer-pipe trench and backfill.

$$BD = 0.8C$$

where BD = burial depth, ft

C = collapse pressure (ultimate), psi

For example, a reinforced polyester filament-wound pipe 4 in. in diameter may be designed for an ultimate collapse of 50 psi. Based on this equation our maximum burial depth would thus be approximately 40 ft.

9.4   TYPES OF SEWER AND DRAIN PIPING

Epoxy for sewer piping   Reinforced epoxy tubing for industrial and oil-field waste disposal, particularly in the smaller sizes such as 2 to 8 in., has been successfully used for the past twelve years.[2]   Collapse strengths in some types vary by sizes from 1,500 psi for the small sizes to 50 psi for the 12-in. size.   High tensile and joint strengths have permitted its use in depths up to 7,000 ft.   The oil industry in its oil-field operations has been a particularly large user of this type of tubing.

Since the production of reinforced epoxy pipe has not been standardized, collapse strength or crushing resistance varies from vendor to vendor, so that where the use of a particular reinforced epoxy pipe is intended, it is wise to contact the supplier.   As a guideline, one of the largest manufacturers[3] of filament-wound epoxy pipe suggests:

1. The 2- to 8-in. pipe may be buried to any practical depth encountered in normal chemical operations.

2. The maximum burial depth of 10- and 12-in. pipe is 12 ft.

3. Light traffic, such as autos and light trucks, may be allowed to run over RP epoxy pipe, provided the burial depth is not less than 2 ft.

4. When traversing under railbeds or roadways or in any areas where heavy-truck traffic may be encountered, the RP epoxy pipe should be encased in a conduit.

5. Good ditching practices should be followed, and pipe should be bedded top and bottom with sand and carefully backfilled.

Maximum bonding strength between RP epoxy piping and concrete may be achieved by adding microcrystalline silicates to the outer resin coat before it sets.[4]   The microcrystalline silicate increases the viscosity of the epoxy resin and, through introduction of thixotropic properties of the silicate, permits the applicator to apply the resin without difficulty. Microcrystalline silicates of this type have been developed by FMC Corporation, and are sold under the trade name Avibest-C.

These microcrystalline silicates also serve as a rheological control agent for polyester resins.[5]

Buried pipe.* The low modulus of elasticity of filament-wound structures, coupled with the relatively thin walls needed for most stress conditions, results in what is considered a flexible pipe. Dr. M. G. Spangler of Iowa State University did considerable work in deriving calculations for burying flexible conduit. This work is summarized in his book "Soil Engineering."[7] The formula he uses for the deflection of flexible pipe culvert is

$$\Delta x = D_1 \frac{KW_c r^3}{EI + 0.061Er^3}$$

where Δx = horizontal deflection of pipe, in.
D_1 = deflection lag factor
K = bedding constant, its value depending on the bedding angle
W_c = vertical load per unit length of pipe, lb/linear in.
r = mean radius of pipe, in.
E = modulus of elasticity, psi
I = moment of inertia per unit of cross section of pipe wall, in.^{4}/ in.

Chemically resistant filament-wound pipe is a composite structure of a liner and wall. The elastic properties of the two components are different. It was therefore deemed wise not to rely on a calculated modulus, but to use Dr. Spangler's suggested method of obtaining the stiffness factor by test, using the three-edge bearing test (ASTM C14-59) and the following formulas:

$$EI = 0.149 \frac{Wr^3}{y}$$
$$EI = 0.136 \frac{Wr^3}{x}$$

where E = modulus of elasticity, psi
I = moment of inertia per unit of cross section of pipe wall, in.^{4}/in.
W = vertical load per unit length of pipe, lb/linear in.
r = mean radius of pipe, in.
y = vertical deflection of pipe, in.
x = horizontal deflection of pipe, in.

---

* This section on the burying of filament-wound epoxy pipe is contributed by C. G. Munger, president of Amercoat Corporation, Brea, Calif.[6] The specific calculations apply to Amercoat's Bondstrand Series 4,000 or 5,000 filament-wound pipe. The same type of approach could be used to develop similar information for any other flexible conduit of which both the filament-wound and the hand-laid-up polyester pipe are a part.

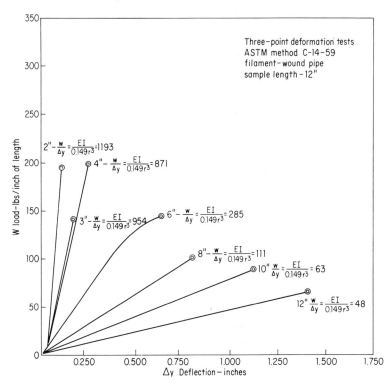

FIG. 9-2   Three-point deformation tests—filament-wound pipe.

Figure 9-2 shows the test results of typical filament-wound pipe. Dr. Spangler's suggested deflection limit of 5 percent of the diameter is well below the failure point and on the straight-line portion of the curves. Buried pipe using these limitations has been in service for many years. It is generally believed that this approach is conservative, but considerably more work will be required to change safely.

Furane sewer drain piping[8]   This piping is normally available in the smaller sizes only, that is, from 2 to 8 in.   Cement joints are commonly used with a furane cement and applied by means of a caulking gun into the deep recesses in each fitting.   Joints set, and may be handled, in about 30 min.   There are many chemical applications where furanes excel.   For a list of these the reader is referred to Chap. 5.   Normally, the furanes show excellent resistance to most acids, solvents, and alkalis, especially in the concentrations encountered in sewer and drain piping.

MAXIMUM DESIGN OVERBURDEN LOADS,[9] LB/LINEAR FT

Table 9-1

ID pipe, in.	Wall thickness, in.				
	³⁄₁₆	¼	⁵⁄₁₆	³⁄₈	½
4	2,000	2,800	3,400		
6	1,650	2,200	2,700		
8	1,400	1,900	2,300	2,900	
10	1,200	1,650	2,100	2,550	
12	1,000	1,450	1,900	2,400	3,300
14	.....	1,400	1,700	2,400	2,950
16	.....	1,150	1,550	2,000	2,750
20	.....	.....	1,250	1,650	2,350
24	.....	.....	1,100	1,450	2,200

Polyester hand-laid-up sewer and drain piping   Custom-contact-molded polyester pipe is the type commonly used for sewer work.   This is a hand-laid-up pipe with 25 to 30 percent glass and 70 to 75 percent resin.   This type of polyester pipe is available from many fabricators engaged in the manufacture of polyester pipe.

For the convenience of the engineer Table 9-1 gives the maximum design overburden loads in pounds per linear foot for various sizes and thicknesses of standard polyester pipe.

Reinforced polyester hand-laid-up sewer and drain piping should be designed with care to obtain the best results.   This material can be buried to almost any depth provided the piping is encased in a reinforced concrete shell.   This may be a good idea regardless of whether the drain pipe is shallow or deep, since the reinforced concrete will absorb both shock and pressing loads.   The practice of reinforcing pipe such as this is particularly prevalent in areas subjected to vehicular traffic.   To provide the designer with a good understanding of this type of construction, the cross section of a typical 36-in. acid-laden condenser water line is detailed with the typical pipe lay-up (Figs. 9-3 and 9-4).   Note, especially, the layer of 24-oz mat on the outer wall to provide a rough, uneven exterior surface next to the concrete.   With this type of design the RP pipe and reinforced-concrete shell have been literally locked together at thousands of points to provide a truly tough composite structure, designed to withstand shock loads from without and severe corrosive conditions from within.

Typical 36" RP polyester drain line specifications for
corrosive service at pressures not exceeding 50
PSIG $\frac{5}{8}$" wall minimum pipe constructed of the
following laminate:

Layer no.	Description
1	Gel coat with $\frac{1}{2}$ oz "C" mat
2 - 8	$1\frac{1}{2}$ oz mat.
9	24 oz woven roving
10	$1\frac{1}{2}$ oz mat
11	24 oz woven roving
12-15	$1\frac{1}{2}$ oz mat
16	24 oz woven roving
17	Hot coat

The above lay-up will actually
finish out at approximately .7 inches.

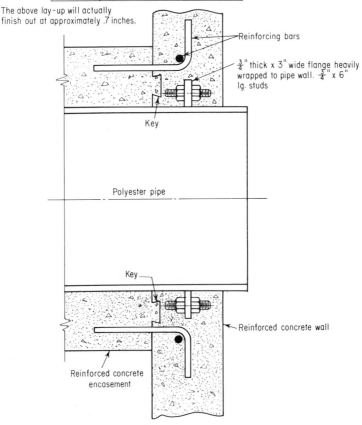

FIG. 9-3   Underground encased polyester pipe entering a seal well.

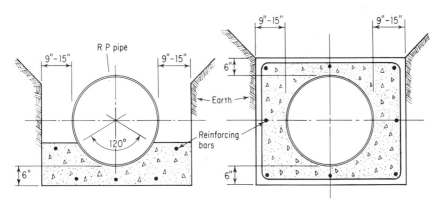

Reinforced concrete cradle   Concrete encasement detail of RP pipe

FIG. 9-4 Typical reinforced concrete cradle and encasement detail of RP pipe. (*Adapted from Ref.* 12.)

Reinforced plastic mortar pipe[10] This pipe, manufactured under the name of Techite, is a sand-filled polyester resin reinforced with continuous fiber-glass filaments constructed on a mandrel to provide a completely smooth resin-rich interior. It is available in two classes:

1. A low head, up to 25 ft (sewer and drainage)
2. A high head, up to 200 ft (pressure and irrigation)

A complete line of fittings is produced with bell-and-spigot joints of the "O-ring" type.

Prices are competitive with asbestos-cement and vitrified-clay pipe. It is available in sizes of 8 to 48 in. ID.

9.5 MANHOLES AND DRAIN-PIPING CONSIDERATIONS

No plan for the RP sewer piping would be complete without consideration of the manholes. Manholes are an integral part of any sewer system. They provide means for entering and maintaining the system. In industrial complexes they are quite often placed 300 to 500 ft apart. Where particulate matter obstructs the flow, cleaning through the manhole entrances may be accomplished by:

1. Hydraulic cleaning mechanisms
2. Drag buckets
3. Cables
4. Sewer-rodding methods

A typical manhole design is shown in Fig. 9-5.   There are, however, a number of accepted methods for tying the RP pipe into the manhole. These are shown in Fig. 9-6*A* to *D*.

The development of underground tank design would lead to the next step, and that is the development of a prefabricated all-glass-reinforced plastic manhole.   The conventional manhole consisting of structural walls of reinforced concrete, followed by glass-ply membrane, and finished off with acidproof brick laid in a resistant mortar, is expensive to

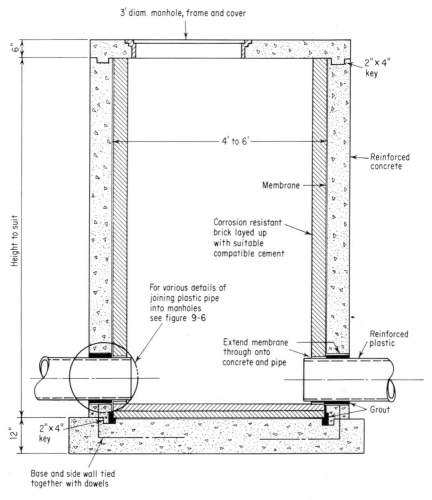

FIG. 9-5 Typical manhole details—brick, concrete, and reinforced plastic.

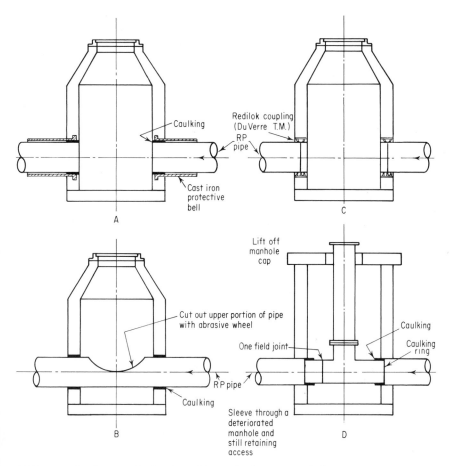

FIG. 9-6   Various methods of working RP pipe at manholes.

build.   A manhole of this type, 6 ft in diameter and 15 ft deep, normally costs $6,000.

9.6   HELPFUL HINTS IN INSTALLING REINFORCED PLASTIC PIPE

1. Be careful in handling pipe during transporting, storing, and installing.   Do not damage it.

2. Position the pipe along the right-of-way.

3. Excavate the narrowest practical trench to permit working, bedding, and compaction.   Pipe OD plus 12 in. is desirable at the top of the pipe.

4. Make sure the subgrade is solid. If on rock, excavate 6 in. below subgrade and backfill to subgrade elevation.

5. If bell-and-spigot construction is used, face the bell in the direction of laying.

6. Stabilize the pipe by providing backfill around and over it. Sand is an ideal backfilling material. Take care to bed equally up each side of the trench. Clay is also good backfill material. Keep stones and rocks away from the pipe.[11]

7. Reinforced polyester pipe is generally sufficiently flexible to handle minor changes in subgrade or alignment. Bell-and-spigot joints may handle a deflection of 2 to 3°.[11]

8. Although minimum radii of curvature vary with the piping used, curvature radii of 200 ft are common in sizes up to 36 in.[11]

9. Backfill above the pipe to the ground surface to obtain as nearly as possible a density equal to the original excavated earth.

10. For a number of different methods of treating piping at manholes, see Fig. 9-6. Study it carefully.

11. Have the factory make all the difficult connections. Design the system so that a minimum number of simple connections are left to the field forces. Branch connections and elbows should be put on by the factory.

12. A power saber saw, 22 to 28 teeth per inch, or a hand hacksaw, 22 to 28 teeth per inch, or an abrasive wheel can be used for cutting the pipe to fit.

13. A line may be tested with water and may be considered to have passed the test if the water loss does not exceed 100 gal/in. of diameter per mile per 24 hr when tested under 50-ft static head.[11] If wrapped joints have been used correctly, there will be absolutely no problem in meeting this test specification since the water loss should be, essentially, zero. The minimum test period should be 3 hr. A second test method which has been found satisfactory is to seal off each section of the sewer pipe and fill with water to a head of 4 ft. Under these conditions water exfiltration as measured by the loss of water in the head pipe should not exceed $\frac{1}{10}$ gal/in. of pipe diameter per linear foot of pipe per 24-hr day. Again, a 3-hr test is satisfactory.[12]

REFERENCES

[1] The Ceilcote Company, *Bulletin* C-88-4-R, p. 23, Berea, Ohio.

[2] Kelly, Mark E., Jr., and Richard Hof: Glass Fiber Reinforced Epoxy Piping Systems, *Materials Protection*, vol. 4, no. 10, pp. 50–53, October, 1965.

[3] Amercoat Corp. *Technical Data Sheet* B-102, R 11/63, Brea, Calif., November, 1963.

[4] *Modern Plastics,* vol. 45, no. 4, p. 114, December, 1967.

[5] Strunk, W. G.: Rheological Control Agent for Polyester Resins, *Modern Plastics,* October, 1967.

[6] Munger, C. G., Amercoat Corp., Brea, Calif., personal communication, 1967.

[7] Spangler, M. G.: "Soil Engineering," 2d ed., International Textbook Company, Inc., Scranton, Pa., 1960.

[8] Corrosion Resistant Piping, Beetle Plastics *Bulletin* P-100, Fall River, Mass.

[9] The Ceilcote Co. *Bulletin* 10-F, Berea, Ohio, March, 1966.

[10] "Techite Sewer Pipe," United Technology Center, Division of United Aircraft, TB-167, Sunnyvale, Calif., January, 1967.

[11] "Guide Specifications for Installation of Reinforced Plastic Mortar Sewer Pipe," United Technology Center, Sunnyvale, Calif., Specification No. 01018 (UTC-5052-9/66), 1966.

[12] Sewer Systems Standards, No. 1-01722-3, FMC Corporation, American Viscose Div., Marcus Hook, Pa., February, 1965.

10

Personnel—Labor Relations,
Training, and
Safety Precautions

10.1 LABOR RELATIONS

With the potential reduction of labor forces in the fields of pipe fitting, sheet metal, and lead burning, especially in those areas of construction where corrosion is a great problem, the use of reinforced plastics in the plant may be viewed with concern by labor, especially the craft unions. Although other unions, such as the painters' union, may be benefited by the use of these new materials, it is plausible to anticipate labor problems.

Unions earning high labor rates are naturally reluctant to see their work displaced by a lower-priced material and being taken over by another craft. Crafts within themselves can foresee this material outlasting previous materials of construction, with the resultant displacement of craft forces over the long run. With mutual good will on the part of labor and management, all these problems can be worked out. Reinforced plastics are simply one more link in the chain of technological progress, which can hardly be stopped or denied, any more than could the displacement of "buggies" be halted by the advent of the automobile.

Intercraft disputes may arise when reinforced plastic work is taken over by any individual craft. Such disputes have already arisen between pipe fitters and lead burners on piping and between sheet-metal workers and painters on ductwork. Most of these disputes can be settled with some good common sense, in give-and-take hard bargaining.

The craft unions may also use other reasons, in addition to technological change, for justifying rate increases within the craft while working on this material, such as an extrahazardous rating. An arbitrator's decision in this area is cited below in Sec. 10.2.

Although men are customarily employed in RP piping work, a number of places have employed females. Experience has shown that women will, in general, pay more attention to detail, and can turn out reinforced plastic work which is perfectly acceptable. They appear to excel in those areas where neatness is required and where the good appearance of a product is essential. In the employment of women in this area, however, labor laws promulgated for the protection of the woman worker must be followed. They can be used for making wrapped joints, small parts, or finishing up larger operations. Lifting heavy weights is prohibited; and they must receive "equal pay for equal work."

10.2   ARBITRATOR'S DECISION

As might have been expected, the use of RP piping came to an arbitration decision between a large union and an equally large corporation in the latter part of 1966.

In this grievance action, the union protested the company's using RP piping without negotiating with the union, on the basis that a change in method of piping had been instituted. They asked that the company negotiate a change in rate for two reasons:

1. The use of fiber-glass-reinforced epoxy or polyester pipe represented a technological breakthrough that would result in a decrease in employment in the pipe-fitting group.

2. In the union's opinion, the use of the cements and resins involved represented hazardous work and should provide extra compensation.

The request with regard to the first reason was denied on the basis that techniques of fabrication and repair have been nationally standardized and are commonplace and that the work involved is within the pipe fitter's typical job description. With regard to the second reason, the union's main thesis was that the use of the resin catalyst was sufficiently dangerous to health to justify a special rate, and the hazard rate in a craft group such as lead burners was cited as a precedent. It was brought out at the arbitration that a survey of 10 plants which used this material showed that they were not found to be a special hazard, nor was a premium rate paid for the job.

The arbitrator's conclusion, dated Nov. 8, 1966, denied the union's contention and found "that the elements which would support an increase in rate were substantially lacking. I am unable on what was presented to find any sufficient increase in hazard as would justify the requested rate increase." The request was therefore denied by the arbitrator. However, it may be anticipated that labor may attempt to achieve the same ends using different means or for different reasons.

10.3 TRAINING REQUIREMENTS

Training in the use of RP polyester or epoxy pipe involves a number of areas, as discussed below.

Joining   The butt-and-strap method of joining RP pipe is not difficult to master nor is the making of adhesive joints.   It would be advisable for any plant preparing to use this material to engage the services of the fabricator on an instructional basis.   The fabricator can readily provide training sessions by instructional demonstration and by having plant crews make experimental joints under his direction.   He also can advise on jigs and fixtures which will simplify making the joint.   Some particular kinds of RP pipe require tapering tools.   These are sold at a nominal charge by the fabricator, who is usually prepared to provide instruction also.

The advantages of this firsthand instruction cannot be overemphasized. The liability in not doing so can be measured in joints that come apart and a low order of reliability, with a great deal of disgruntlement on both sides of the fence.   Training involving several hours at a time should be extended to all groups of mechanics and craftsmen who may be working with the material.   These include pipe fitters, mechanics, painters (in some plants the painter's trade does more work in this area than any other), sheet-metal workers, etc.   This type of instruction should cover not only RP piping, but also ductwork and repairs to tanks or structures.

Man, being a creative creature, will soon forfeit the effects of training—almost before the instructor is out of the plant—in the hope of devising a better way to do a job.   This is only natural.   *However, history has shown that nine times out of ten the supposed advance is a step backward.*   To correct this, retraining becomes necessary.

At least for several years, even experienced crews should be subject to retraining every year or two by experienced personnel from the fabricator to make sure that bad habits are eliminated and that real improvements are adopted.   It is sometimes wise to switch training instructors, not because the first one might have been incompetent, but because different instructors have a slightly different technique and quite often add something to the craftsman's ability.   In a later section we shall take up the general subject of why joints fail, but here we may state that a great deal of it can be attributed to poor training and the inability to follow simple instructions.   The penalty for training failure can range from a leaky joint to the catastrophic failure of the entire system. *Training pays.*

In a reinforced plastics fabricating shop a laminate-lay-up man can be trained to produce acceptable work in approximately three weeks. Of course, he has not yet learned the skills of a craftsman, which would take much longer. Training in the use of spray lay-up may take 1 to 4 months. Then a spray-lay-up operator will continue to advance in skill, so that his work at the end of a year will be considerably better and more finished than it was at the end of 4 months. The aim is not, however, to produce a finished fabricator. Where only the making of wrapped or adhesive joints is involved, short periods of training followed by some practical experience under the watchful eye of a good supervisor will produce an acceptable level of competency.

Supporting   Training in supporting begins on the drawing board. Failures in supporting are generally due to "sins of omission" and designer's lack of knowledge of what is required. Training in this area can come only from a reference source or from the fabricator. The fabricator will tend to emphasize what *you should do* and not necessarily what *you should not do*. Reference should be made to the sections of Chap. 8 dealing with supporting, where there is considerable discussion of "sins of omission and commission."

The designer is all too prone to think of these materials in connection with his previous experience with metal piping, such as steel, copper, etc. He needs to reorient his thinking along a completely different line if the installation is to be successful. Two simple rules will serve in this area:

1. Never put an RP line in service until it has been adequately supported and anchored. Do not put it in "hanging on wire."

2. Obtain the services of the fabricator from whom you purchased the pipe, and have the chief design engineer sit down with him and review good supporting principles.

Storage and handling   The responsibility for the successful installation of an RP piping system rests not only with the original supplier, but also with the engineer who specifies it, the designer who performs the detail design work on the system, the stores department which receives it, and the field engineering group which installs it. Each group needs to achieve specialization in its area of responsibility.

RP pipe needs to be stored separately and to be handled with care. It should be protected from being hit by heavy objects. It is important that the stores group receive instructions of this kind. A surprising number of piping systems have known difficulty because the storage practices were less than desirable. RP piping which has been furnished with an ultraviolet inhibitor may be safely stored outdoors. Other piping must be stored indoors.

10.4   SAFETY INSTRUCTIONS—POLYESTERS

Purchasing and storing the resin   A wide variety of polyester resins are available.   In a chemical processing plant the general-purpose resins should usually be avoided, except for possible tote trucks or elevator buckets handling dry material.   Isophthalic resins have use in certain areas.   More often the engineer should consider a suitable bisphenol polyester or chlorinated polyester, to fit in with his particular requirements. Certainly, any good maintenance shop would want to stock both a bisphenol and a chlorinated polyester resin.   The resin itself, originally, is a solid material which is normally dissolved in 35 to 50 parts of styrene or vinyl toluene per 100 parts of solution.   The solid has an indefinite shelf life.

In many cases the engineer will find it to his advantage to maintain a stock of the polyester resin which has been preaccelerated through the addition of a small amount of cobalt naphthenate (generally, about 1 percent).   This procedure is recommended in preference to buying the liquid resin and adding the cobalt naphthenate in the plant.   The purpose of doing this is to make the procedure for use *ultrasafe* in dealing with the catalyst.   Quite commonly, methyl ethyl ketone peroxide (commonly abbreviated MEKPO) will generate a severe exothermic reaction when mixed with the cobalt naphthenate.   The cobalt-accelerated resin is no more hazardous to store than any other paint or vehicle.   This precludes the inadvertent mixing of MEKPO and cobalt naphthenate.

Normally, in a tightly closed container, the service life of the promoted resin will not be a problem (3 to 6 months).   This, however, does vary from resin to resin, and the purchaser should check with the vendor before laying in a stock of the material.   The quantity of stock to be maintained, of course, will vary from plant to plant.   Initially, the engineer must ask himself, what size container shall I stock?   The pint size is useful when dealing with 2- to 6-in. pipe.   The quart size is necessary in 8 to 16 in. and the gallon size above 18 in. in diameter.   Quite often the engineer, to suit his demands, will stock a mixture of pints, quarts, and a few gallons for larger jobs where big ducts and tanks are involved.   The gel time for a resin is usually short, on the order of 15 to 30 min at room temperatures of 75°F.   This gel time will be still further shortened by the addition of more catalyst or heat.   If the day is cool and the temperature in the 40s, the gel time with normal catalyst additions of 4 cm^3/pt is too long to be useful.   However, increasing the catalyst addition to 12 cm^3/pt will provide considerable acceleration of the gel time, as will raising the temperature.

In the throes of cost reduction, nearly every engineer wants to know the relative cost for larger quantities so that he may judge what economies can be achieved by purchasing larger amounts of polyester resin. Relative costs are, currently, approximately as follows:

Equivalent relative cost per pint

1 pt........................ 1.0
1 qt (2 pt)................. 0.83
1 gal (4 qt)............... 0.45

Unless the user is considering going into the reinforced plastic business himself, he will be hard put to justify anything beyond a 1-gal can. This is the experience gained from one of the largest individual users of reinforced plastics in the chemical-process industry. If, however, the stage has been reached where an RP pipe- and tank-fabricating shop has been established, the economics change; but this book is written for the user, not the manufacturer or fabricator.

The catalyst system. The engineer is offered the choice of several different catalyst-accelerator systems. One of these is a benzoyl peroxide catalyst–dimethyl aniline promoter. Another, and certainly more popular, system for use in the chemical industries is the methyl ethyl ketone peroxide (MEKPO) catalyst–cobalt naphthenate promoter. Our discussions here will be confined, principally, to the use of MEKPO catalyst with a cobalt naphthenate promoter-accelerator.

The use of a benzoyl peroxide catalyst system in process-plant work should generally be avoided because it normally comes in a paste and involves a weigh-out on a gram scale versus a simple volume measurement with MEKPO. In addition, benzoyl peroxide is a much greater fire hazard. Finally, benzoyl peroxide represents a much greater dispersal problem, so that if benzoyl peroxide is used, it will have to be thoroughly mixed in its own container to provide a homogeneous mass, weighed out on a gram balance, and dispersed in a pound of resin; then the pound of resin should be further mixed into a larger amount of uncatalyzed resin. Obviously, it is easier to use a simple volumetric measurement of MEKPO in a graduated cylinder.

MEKPO is normally supplied as a clear, colorless liquid containing 60 percent MEKPO in dimethyl phthalate. This is a great improvement over the handling problems associated with 100 percent MEKPO. The stability and safety of the 60 percent MEKPO is superior. The dimethyl phthalate's principal function is as an antidetonator compound. The purpose of the catalytic agent is to act upon the long-chain molecules and, through polymerization, provide a lattice structure. By this technique the liquid polyesters become a solid thermoset plastic. Actually,

MEKPO alone will cure polyesters at elevated temperatures of 100 to 200°F, but since most of the work done is at room temperature of 70 to 80°F, it is necessary to add an accelerator (cobalt naphthenate) to speed the polymerization reaction and to permit hardening to occur within a reasonable period of time. The purchase of the resin preaccelerated with cobalt naphthenate is to be recommended. Ideally, a warm room and low humidities provide the best background for good work. On the other hand, low temperatures and high humidity will inhibit the polymerization reaction. The amount of MEKPO to be used varies with (1) type of resin, (2) temperature, and (3) desired setting time.

The liquid MEKPO is easier to handle and volumetrically easier to add to the resin. The catalyst should be stirred in carefully and mixed thoroughly with the resin. Air bubbles must not be allowed to form, since these will have to be rolled out.

The 60 percent MEKPO solution is insoluble in water, slightly soluble in petroleum distillates, and completely soluble in liquid polyester resins. MEKPO is classified as an oxidizing material by the Interstate Commerce Commission. It may be shipped by freight or express, but must bear a yellow label, and it cannot be shipped parcel post. Carrier weight limits on unit packages must be observed. The recommended carrier is motor freight.

SPECIFICATIONS AND PROPERTIES—THE CATALYSTS[1]

Table 10-1

Methyl ethyl ketone peroxide	60%
Dimethyl phthalate	40%
Active oxygen	11.0% minimum
Appearance	Clear, colorless, nonviscous liquid
Specific gravity (77°F)	1.10
Freezing point	Below 22°F
Boiling point	Decomposes
Refractive index	1.45 (Nd^{25})
Flashpoint (open-cup)	125°F
Viscosity (77°F)	7.0 cps

Purchasing the catalyst   The following guidelines may be helpful in purchasing MEKPO:

1. Buy only MEKPO that has been diluted to a 60 percent concentration by the addition of 40 percent dimethyl phthalate (to improve catalyst stability and safety).

2. All MEKPO purchased should be supplied in 12-cm³ quantities, contained in individual ½-oz polyethylene bottles that are hermetically

sealed. The bottles should be plainly marked methyl ethyl ketone peroxide in dimethyl phthalate. Work out the details of purchasing with your supplier. And do not shop around—too much can go wrong, and the cost of this highly important item is small in relation to the entire system cost. MEKPO is always shipped in nonreturnable containers. Always use a graduated cylinder for measuring MEKPO.

3. All methyl ethyl ketone polyethylene bottles are shipped in vermiculite packing to soak up the material in case spillage occurs. This also provides some shock resistance for handling purposes.

4. The average plant should never consider buying the catalyst in large individual amounts, such as quarts or gallons. It is not convenient for the field personnel to handle it; dispensing problems are created; and field safety problems are magnified.

5. Absolutely no metal should be used as a storage container for an organic peroxide.

6. The MEKPO must not be contaminated with organic or inorganic materials. This contamination alone can cause rapid decomposition and resulting autoignition, especially at temperatures above 140°F. Although not easily ignited, MEKPO will burn. Since active oxygen is available from the material itself, the rate of combustion can increase rapidly and produce an intense fire which is difficult to extinguish. A dry chemical extinguisher is useful for this type of fire, or, in the early stages, water alone, if used in sufficient quantities, will do. Contamination is the greatest single hazard in the use of this material. Never pour unused MEKPO back into the supply container.

7. The reinforced plastics industry and their suppliers are continually on the lookout for safer catalysts. The introduction of MEKPO in a dimethyl phthalate carrier was one step forward which has been substantially standardized. This was a considerable advance over the original undiluted MEKPO. Another step in that direction has been the reportedly *self-extinguishing MEKPO,* obtained by adding water. While this might prove completely satisfactory for some polyesters, it should be viewed with reserve in chemical-resistance fabrication work: some studies have indicated that it does affect the final chemical resistance of the laminate; it also reacts with the thixotrope in the resin, increasing the viscosity of the resultant mixture, so that the system may be rendered unworkable. Improvements to curing systems will no doubt be introduced, but it is important to make sure that the chemical resistance of the laminate has not been adversely affected.

Storing the catalyst

1. All MEKPO should be stored at 35 to 40°F in refrigerators with magnetic door closure from which all sources of ignition have been

removed. (This is the temperature of the ordinary household refrigerator, which, however, must be stripped of all sparking sources.) Special refrigerators of this type are available to meet the highest specifications. Storage at 35 to 45°F provides maximum storage life of MEKPO without noticeable loss of active oxygen. When stored in the original containers at 35 to 45°F, the MEKPO will retain its activity for over a year. If, however, the MEKPO is used at a relatively rapid rate so that the stock turns over monthly, a storage of 65 to 70°F is sufficient. In general, the cooler any oxidizing material is kept, the less active it will become, although freezing (below $-22°F$) drops out ultrasensitive crystals.

2. Guard against exposing MEKPO to a high temperature (above 140°F) because it may cause rapid decomposition and subsequent autoignition. Sunlight and flame or sparks are also to be avoided.

3. It is wise to put an upper limit on the amount of catalyst to be stored in a refrigerator in any area at one time. Such a limit should not exceed 400 ½-oz 12-cm³ bottles of catalyst. In the event that any project requires larger quantities of MEKPO than can be adequately handled, as outlined above, the appropriately responsible personnel should have to approve such a shipment from the vendor. Large ductwork or piping installations will require considerable quantities of MEKPO, and special measures can be taken in such instances. It is recommended, however, that the purchaser not exceed a pint container.

4. Normally, the paint shop in any large chemical plant is separated from the main buildings because of the flammable nature of the products used. In some cases plants will erect so-called "red label" buildings. The purpose of the "red label" building (suitably equipped with high-capacity sprinklers) is to minimize damage in case of a resultant fire. All hazardous materials are stored in the "red label" building or in another separate building: this is just good business sense, to minimize potential loss. The storage of MEKPO should be treated in a like manner.

Catalyst control

1. Only one day's supply of MEKPO should be withdrawn at one time, and unused bottles returned immediately upon completion of the work.

2. To prevent half-used bottles from being thrown into trash barrels, all used MEKPO bottles should be returned to the source of origin for disposal, to be hauled to a dump and carefully burned.

3. A withdrawal sheet should be maintained at the point of storage, indicating the amount of catalyst withdrawn and the initials of the person withdrawing. This provides inventory control.

Other measures of catalyst and resin stocking and control   It will be recognized that the catalyst and resin purchasing, storage, and control measures outlined here have been essentially conservative. Much more liberal measures than those indicated have been adopted, with successful operation for years. One large plant purchases MEKPO in pint bottles, 24 to a case. This is stored in the same room in which the shop lay-up work is done. Cobalt naphthenate is also stored in the same shop, but in a separate cabinet. The resin is stored in 55-gal drums, not promoted. Refrigeration of MEKPO is totally absent. Five gallons of the resin is drawn from the drum at one time; a promoter is added to it; and a supply is then drawn from this 5-gal pail as needed. A mechanic going to the field has a polyethylene laboratory bottle which holds about 8 oz of MEKPO. This is similar to a laboratory wash bottle in which the MEKPO can be squeezed out easily into a graduated cylinder and then added to the resin as required.

It is obvious that the engineer must make a choice between an essentially conservative policy in handling catalyst and a more liberal one. While plants may report years of successful operation with either policy, the prudent engineer will tend to be conservative. The vendor fabricating this equipment on a large scale is poor counsel in this problem since he will tend to err to a point of recklessness.

10.5 INDUSTRIAL HYGIENE—POLYESTERS

The application of the elementary principles of industrial hygiene is all that is necessary to provide safe working conditions for plant personnel working in the polyester field.[2] (Many of these commonsense ideas are equally applicable to the epoxy field.)[3] For safe working conditions in this area five basic rules need to be followed:

1. Education of supervisors and workers
2. Good housekeeping
3. Adequate mechanical exhaust ventilation
4. Personal cleanliness
5. Use of protective creams, gloves, and safety clothing

A brief comment follows on each of the principal hazards and, in general, corrective actions that are necessary.

Sanding and buffing operations   Where this is done repetitively in a plant fabricating shop, a small exhauster is all that is needed. (A more elaborate installation may be one which includes a hood enclosure equipped with a mechanical exhauster.)

FIBROUS GLASS.   Fiber-glass particles in the air may produce, on some occasions, in some individuals, a dermatitis merely because of the mechanical friction of small broken fibers on the superficial layers of the skin.   It seems to be generally accepted in industry that persons with light skins are more susceptible to the so-called "glass itch" than those with darker complexions, although there is a lack of authoritative medical opinion on this aspect of the situation.   As a protection against this dermatitic irritation, workers should wear loose-fitting clothing, frequently laundered, and they should take a shower at the end of the day's work.

Although there has been a tendency in some quarters to minimize the effect of atmospheric dust from sanding and buffing operations of all sorts, good employee relations demand that these situations be kept under adequate control, and good working conditions generally result in higher labor productivity.

The tentatively suggested limit for fibrous glass is 5 mg/m³ of air.[3a]

ASBESTOS.   The Committee on Threshold Limits of the American Conference of Governmental Industrial Hygienists published, under a 1968 Notice of Intent, a limit of 5 mppcf (millions of particles per cubic foot of air) for asbestos in most forms, based on impinger samples counted by light field technicians.[3b]   One form of asbestos, however, known as crocidolite, is used as a corrosion-resistant inner liner in some epoxy piping.   It is also recommended for use in polyester lay-ups, with alternate layers of glass to produce a stiffer laminate.   There is evidence that crocidolite may produce, in addition to asbestosic inflammation, mesothelioma.   At the present time no safe limit has been established for this form of asbestos.   The recommendation has been made that workers exposed to particles of air-borne crocidolite be equipped with air-supplied helmets.   This would be particularly important in grinding operations in which crocidolite dust may be air-borne.

It is apparent that final recommendations relative to threshold limit values (TLV) on asbestos (all forms) have yet to be resolved.   The Threshold Limit Value Committee of the American Conference of Governmental Industrial Hygienists recommended in May, 1968, further modifications to the 1967 list.   Under this recommendation[3c] the provisional TLV for asbestos would become 12 fibers per milliliter greater than 5 microns in length, or 2 mppcf.   The limit of 12 fibers per milliliter is achieved by the membrane-filter method at 430X magnifications.   The 2-mppcf level is determined by the standard impinger light field counting technique.   The TLV for asbestos, however, remains a recommended value of 5 mppcf until at least 1970.   The user can only conclude that, as in the field of radiation exposure, asbestos exposure limits are being progressively tightened.   To the prudent consumer this should serve as a warning. It should not, however, preclude the use of purchased laminates containing asbestos. Only when the end user becomes involved in laminate modifications which may produce air-borne dust need special precautions be taken.

Polyester resins[2] The polymerized polyester resin is inert and non-toxic. A number of these resins have received the approval of the Federal Department of Agriculture and are used extensively in preparation of foods in tanks, pipelines, etc.

Styrene exposure in polyester lay-up The purpose of this section is to provide some basic useful information on styrene exposure in polyester lay-up work. The type of work covered here is that done in the normal application of polyester pipes, ducts, tanks, and structures, in routine plant fabrication work. Some of the factors related to styrene concentrations at the workman breathing level are:

1. Extent of fresh-air circulation
2. Rate and area of polyester application
3. Open or enclosed area of application
4. Spray- or hand-laid-up technique

There is little to be concerned about in the making of polyester wrapped joints and ordinary repair work in any normally ventilated room. In this case the exposure area of polyester is small. Repeated breathing-level samples in lay-ups of this type indicate styrene concentrations of 5 to 12 ppm at the nose level. Now, obviously, if a large vessel were being laid up and several quarts of resin at a time were being applied by a repetitive method on a rotating mandrel, the gross surface exposure to the atmosphere would be considerably higher. *It is apparent that gross surface exposure to the atmosphere, all other factors being equal, in general, determines the concentration of styrene in the air.*

As long as the craftsman is applying several square feet at a time, there is little to be concerned about, except possibly in an enclosed space. Under the conditions of use, the materials used and released—this includes the polyester, stryene, and MEKPO—are not toxic. Certainly, adequate ventilation or exhaust is the key to making working conditions as safe and comfortable as possible.

The decomposition products of MEKPO are methyl ethyl ketone and oxygen. Such small quantities of this compound are used in lay-up work that the release of the methyl ethyl ketone is virtually insignificant. The chemical reaction polymerization, or hardening of the polyester, is an autocatalytic reaction which proceeds with the generation of heat. As the reaction progresses and polymerization is complete, the joint hardens, or cures. As the cure progresses, small quantities of styrene are boiled off, and the MEKPO decomposes into methyl ethyl ketone and oxygen, one of the prime constituents of the atmosphere.

The styrene monomer in which the resin is dissolved is self-announcing,[4] with thresholds of perceptibility far below any toxic level. On the relative scale used by the American Industrial Hygiene Association for rating the severity of health hazards (nil, low, moderate, high, and extrahazardous), both the styrene monomer and the MEKPO are rated low. The irritating properties of both MEKPO and styrene serve as an effective warning because they occur at concentrations below that level at which symptoms become serious.

Human responses to styrene vapor   Nearly all the polyester resins use styrene as a solvent vehicle. The resin itself is not toxic, but all plants should be aware of the maximum allowable concentrations, or threshold limit values, of styrene.

The American Conference of Governmental Industrial Hygienists has established a threshold limit value of 100 ppm.[3a] This has been adopted by the following states and territories: California, Colorado, Connecticut, Florida, Kentucky, Maine, Maryland, Minnesota, New Hampshire, New Jersey, Ohio, Oregon, Pennsylvania, Puerto Rico, and Texas.

The human sensory responses to styrene vapor are shown in the following table.

Vapor concentration, *ppm*	*Human sensory responses*[4]
10–15	Odor noticeably sweet but not unpleasant
60	Odor easily detected
100	Odor strong and disagreeable to unacclimated
200–400	Odor strong and objectionable to most test subjects
600	Odor very strong and vapor very irritating to the eyes and nose

Air-sampling method and results for styrene   As a typical example, the following equipment may be purchased from Mine Safety Appliance Corp., for routine styrene checks in the working area:

MSA Universal tester, part No. 08-83500, at about $75
Styrene testing tube, part No. 08-93962, at about $5 per dozen

The sampling tester accurately controls the volume of air sampled and the rate of airflow in the test. With styrene, the detector chemical in the tube produces a yellowish stain. Length of stain in the tube converts to ppm concentration against a calibrated scale.

Other models on the market at higher cost give more precise readings, but for most plant work the MSA Universal testing kit (or its competitive equivalent) is completely adequate.

In a large fabricating plant doing all kinds of lay-up work, styrene concentrations were found to vary from 13 to 60 ppm. The 60 ppm occurred with an employee exposure laying up a large fiber-glass tank. Suffice it to say that, in a fabricating shop of a chemical plant, exposure will be but a fraction of that observed in large polyester-fabricating plants. The styrene content of polyester resin is approximately 50 percent. The monomer is, theoretically, completely reactive, and thus is not found upon completion of polymerization. The explosive range of styrene is 1.1 to 6.1 percent, and the autoignition temperature is reported as 914°F.[5]

Acetone (Threshold limit value 1,000 ppm) The practice of cleaning up equipment in acetone, styrene, or methyl ethyl ketone may be tolerated, but washing the hands in this type of material should be avoided. The effect is a defatting action on the skin. The employee should wear rubber gloves where possible, or at least wash hands frequently with liquid soaps and warm water. Fortunately, of all the solvents, acetone is about the least toxic and has the highest TLV.

Cobalt naphthenate The best recommendation regarding this compound is that the plant buy the resin preaccelerated with cobalt naphthenate so that there will be no contact with the latter material. However, occasionally, experimental work is required in which the plant must add the cobalt naphthenate. This is normally furnished in a solution containing 6 percent cobalt naphthenate by weight. A cobalt naphthenate solution tends to have a defatting action on the skin, so that, here again, rubber gloves are indicated when handling. There is no published information on the maximum allowable limit of exposure to this material.[2] (For additional information see Chap. 16.)

Methyl ethyl ketone The threshold limit value of methyl ethyl ketone recommended by the American Industrial Hygiene Association is 200 parts of vapor per 1 million parts of air by volume, based on an 8-hr exposure.

Methyl ethyl ketone peroxide Since this compound is a strong oxidizing agent, contact with the skin should be avoided. If contact occurs, the MEKPO should be removed immediately, and the skin washed with copious quantities of water and soap. The oxidation of methyl ethyl ketone peroxide forms methyl ethyl ketone (threshold limit value 200 ppm) plus the release of oxygen into the air. At this time no maximum allowable concentration has been established for MEKPO.

10.6 COLLISION HAZARDS—SAFETY

In the chemical industry today the use of fork trucks and other motor-powered material-handling devices is very common. Any piping system, be it steel, lead, or plastic, runs the hazard of ultimate collision with a fork truck or moving conveyance. In any plant where a large number of fork trucks are used, the amount of damage which can be done through the poor operation of these devices is something to be reckoned with. In addition, the destructive power of these trucks can be considerable. Piping nipples can be knocked from steel lines, electrical conduit severely bent, corridor doors torn down, and tile sections knocked out of wall facings.

How does reinforced plastic pipe fare under the menace of such battering rams?

In one plant in which nearly 50,000 ft of reinforced plastic pipe had been installed over a period of ten years, damage from fork-lift trucks occurred on two occasions, neither of which could be termed very serious. In one, the side of a 12-in. acid line was split for several inches, and a leak, perhaps equivalent to several gallons per minute, developed. With appropriate safety precautions the line was promptly repaired with a compression-band-rubber inserted sleeve. This repair is shown in Fig. 10-1. Several weeks later, during a plant break, a short section of 12-in. line was cut out, and a new section strapped in. This is shown in Fig. 10-2. The line was hit with sufficient force to dislodge it from the hangers.

Normally, in any installation, the line should be kept out of a traffic pattern. If it is necessary to route an RP pipe down a wall along which vehicular traffic commonly moves, the plastic pipe should be protected with a half-round metal shield anchored to the wall or a steel external pipe sleeve.

Because of the low modulus of elasticity, reinforced plastics may be subjected to considerable impact loads without suffering damage.[6] If a good lay-up job has been done, high-energy absorption can occur before the breaking down of the laminate. Damage is generally of a local nature and much easier to repair. The simplicity of repair mechanisms makes repair easy. Small wonder that reinforced polyesters are quite often used in a body shop for automobile repair. Composite lay-up

FIG. 10-1 Compression clamp makes temporary repairs to split reinforced polyester line after collision with a fork-truck boom. Example of use of a compression clamp for quick repair with no loss of production.

of reinforced plastics tends to limit crack propagation and subsequent damage.

The hazard, however, from this source on a statistical basis is so low that it would not be rated of any consequence. Strangely enough, nevertheless, this single reason is used by some companies as an excuse for not enjoying the economies of reinforced plastic piping.

FIG. 10-2 Permanent repair of a damaged pipe made on a normal maintenance break. Compression clamp is easily replaced with a new section of 10-in. pipe wrapped in place, with a typical butt joint showing at each end. The line is now as good as new.

10.7   SAFETY INSTRUCTIONS—EPOXY

The epoxy resins and hardeners generally carry a higher personnel hazard rating than the polyesters because of dermatitis potential.

Personnel protection   In the fabrication and assembly of RP epoxy systems, an epoxy adhesive is used to make the joint. This adhesive may be either a two- or three-component system, furnished in paste or liquid form by the vendor. In employing this material, the following safety instructions are generally applicable:

When mixing and applying the material, the employee should wear gloves, sleeves, and eye-protection equipment.[3] The material will stain clothing and discolor the skin; every effort should be made to avoid this.

If the material should come in contact with the skin, the area should be thoroughly flushed with large quantities of water, followed by a vigorous washing with a soapy solution. Solvents should not be used in attempting to remove the adhesion. Medical attention may be required.

When an allergic reaction to this material is anticipated, a water-removable protective cream may be used. It is good practice to wear disposable polyethylene gloves to avoid contact with the adhesive.

Contact with the eyes must be prevented. In case of any contact, the eyes should be flushed with water for at least 15 min, and medical attention procured promptly.

Some other simple safety rules to be followed in working with epoxy are:[3,7]

1. Educate the supervisor and worker in the importance of keeping the epoxy resin and hardener off the skin. Guard against dermatitis and dermatitis sensitization. Once an individual has become sensitized, smaller quantities of the materials will cause irritation, and curing is more difficult.

2. Do not permit prolonged exposure to vapors or dust from grinding epoxy resin.

3. Confine the epoxy materials to small work areas.

4. Keep the work area clean. Use layers of paper to cover workbenches and floors. Replace soiled paper frequently.

5. Use methyl ethyl ketone for cleaning apparatus and parts. It is a satisfactory solvent.

6. Mix the epoxy resins in disposable cardboard containers. Use disposable tongue blades for mixing and applying.

7. Make sure the area is well ventilated. A downdraft table and

local-exhaust ventilated hood are ideal. Suck the fumes away from the work area. Adequate general ventilation is also necessary.

8. Select workers who are least sensitive to the materials used.

9. Make sure that protective clothing is worn at all times.

Epoxy resins in liquid form for chemical application to laminate lay-up will be rarely used, representing probably less than 5 percent of the polyester work. They are much more difficult to work with than the polyesters. Purchased RP fabricated epoxy systems, however, have wide acceptance, principally in piping and to a lesser extent in tanks and other special structures. Normally, they will be met only as an adhesive component in the fabrication and assembly of RP epoxy systems. The adhesive generally consists of a filled epoxy plus a curing agent.

Dermatitis appears to be a function of the molecular weight of the epoxy resin.[7] Usually, these resins have molecular weights of 350 to 1,000. Resins used for laminating are, fortunately, in the higher-molecular-weight category and are less irritating to the skin than the lower-molecular-weight resins.

The curing agents, of which there are many, are strongly alkaline and in contact with the skin will produce a typical alkaline chemical burn.

Purchasing and storing the resin   For the process plant the epoxy resins may be purchased in suitable small containers, pints, quarts, or gallons, depending on the amount of work to be done.

The epoxy adhesive systems may be used with any of the polyester laminates and, where compatible with the chemical exposure, produce an excellent system. Low shrinkage and strong adhesion are characteristic of the epoxy adhesives. Maximum strengths of epoxy laminate systems are obtained through postcuring, which is best done in the fabricator's shop and generally is beyond the scope of a small or medium-sized process plant. However, where the capability for handling the epoxy resin exists, exceptionally strong laminates can be produced, probably due to the high glass-resin bond.

Purchasing and storing the adhesive   Adhesive kits are available from reinforced epoxy fabricators in varying sizes. In general, the choice depends on the size of the fittings and pipe to be assembled. Since the great bulk of piping work will probably lie in the 2- to 8-in. area, an adhesive-kit size of about 3 oz will be found to be most beneficial. These adhesive kits have a very limited pot life after mixing, and they must be used promptly. Where the engineer is dealing with larger piping sizes, such as 10 and 12 in., consideration should be given to stocking and storing larger kits, such as 8 or 12 oz.

Shelf life of epoxy resin and adhesive is, normally, not a problem, provided the container is kept tightly closed.

REFERENCES

[1] Reichold Chemicals, Inc., *Technical Bulletin* PR-21, White Plains, N.Y., January, 1960.

[2] O'Leesky, S., and J. Mohr: Handbook of Reinforced Plastics of the SPI, Reinhold Publishing Corporation, a subsidiary of Chapman-Reinhold, Inc., New York, 1964.

[3] Recommendations for Handling Epon Resins and Auxiliary Chemicals in Manufacturing Operations, Shell Chemical Company *Industrial Hygiene Bulletin*, New York, 1962.

[3a] "Threshold Limit Values for 1966," American Conference of Governmental Industrial Hygienists, May 16–17, 1966.

[3b] Committee of Threshold Limits of the American Conference of Governmental Industrial Hygienists, 1968 Notice of Intent, Nov. 22, 1967.

[3c] Committee of Threshold Limits of the American Conference of Governmental Hygienists, May, 1968, Suggested Modifications to the 1967 List.

[4] Safety Data Sheet—Styrene Monomer, p. 7, Shell Chemical Company, *Industrial Hygiene Bulletin*, New York, 1962.

[5] Sax, N. Irving: "Dangerous Properties of Industrial Materials," p. 1141, Reinhold Publishing Corporation, New York, 1957.

[6] Whitehouse, A. A. K.: Glass Fibre Reinforced Polyester Resins, paper 10, Symposium on Plastics and the Mechanical Engineer, London, Oct. 7–8, 1964.

[7] Shell Chemical Company, *Technical Bulletin* SC-60-39, rev., New York, 1960.

11
Duct Systems

11.1 AN INTRODUCTION TO DUCT SYSTEMS

Many chemical processes involve the handling of contaminated exhaust air or gases from process vats, tanks, or other equipment. In some processes the handling of exhaust fumes reaches fantastic proportions. Individual systems may range from a small unit handling perhaps 1,000 cfm to giant batteries of fans which may handle, in the aggregate, over a million cfm. In between are literally thousands of installations of 10,000 to 35,000 cfm. Quite often, for the sake of operating convenience and to keep duct and fan sizes in the manageable range, large duct systems may be broken down into a series of smaller duct systems and installed in multiples. One large exhaust installation, for example, designed to handle 300,000 cfm of acid-laden air from process machines, was broken down into ten smaller systems of five at 36,000 cfm and five at 24,000 cfm. Systems of this magnitude are ideally suited for the application of reinforced plastic ductwork, fans, and discharge stacks and dampers. Where chemical compatibility exists, excellent fire-retardant resins are now available. RP systems of this type have definite advantages over metal systems, which may corrode, or elastomeric-lined systems, which were the previous standard system of construction.

It is our purpose here to provide sufficient information based on field experience to provide the design engineer with the necessary reference material to permit him to adequately design, specify, estimate, and purchase complete duct systems constructed of RP material.

Pollution-control equipment The field of air-pollution control is growing at a rapid rate. In all probability local and federal legislation will set up regulatory bodies which will follow a path on air pollution similar to that previously followed on stream pollution. Since air pollution does cover such a wide sphere of activities, this section of the reinforced plastics business will probably enjoy the greatest rate of growth. Even today one fabricator reports making as many scrubbers as tanks. Another fabricator estimated that in 2 to 3 years air-pollution control would probably represent 50 percent of his total business. Today fans, ducts, and fume scrubbers can all be bought in reinforced plastics in a wide range of sizes.

A large-scale duct installation of ductwork, fans, butterfly dampers, etc., is shown in Fig. 11-1. Additional modern duct installations are shown in Figs. 11-2 and 11-3.

The reader is referred to Chap. 14 for a more detailed consideration of pollution-control equipment. In installation procedures the imagination of the engineer is severely challenged when installing large-diameter duct systems in high elevations and relatively inaccessible locations. This is particularly true where a duct system may be installed over plant

FIG. 11-1   Blower, ductwork, and butterfly valves are shown in hot wet chlorine service at Hooker's Niagara Falls, N.Y., plant.

FIG. 11-2   Intricate configurations of 20-in. RP ductwork.

roofs. Even after the supports are built, rigging long sections of duct-work can be a tedious, expensive job, notwithstanding the acknowledged light weight of the plastic duct material. For a fresh approach to problems of this type the reader is referred to Chap. 14 for studies of techniques of duct installation with the use of aircraft, which reduced project costs of installing by 80 percent in the rigging phase.

11.2 RECOMMENDED PRODUCT STANDARD FOR RP DUCTWORK WITH EXTENSIONS

Reference is again made to the Recommended Product Standard for Custom Contact Molded Reinforced Polyester Chemical Resistant Process Equipment, TS-122C, dated Sept. 18, 1968. This Standard is being adhered to by the majority of fabricators as a generally satisfactory standard which, with experience, will provide a good installation. Prior to the general adherence to this Standard there was considerable diversity

FIG. 11-3   A 42-in. reinforced polyester ductwork exhausts fumes from an acid process.   Stiffening wraps every 10 ft afford additional wall stiffening.   Supports are provided approximately every 10 ft, with a holddown strap completely encircling the duct, for protection in high winds.

in duct fabrication, design, and quoting procedures.   The industry has grown and matured.   By adherence to the Standard, the purchaser is reasonably sure of obtaining satisfactory performance for his purchasing dollar.   Occasionally, however, the purchaser's engineering group will be required to do their own duct design and to prepare engineering designs beyond the confines of the Standard.

In the following sections problems are developed to increase basic know-how in proceeding with the design for both vacuum and pressure service.[1]

AUTHOR'S NOTE: The following sections and table cover reinforced polyester round and rectangular ducts, and are quoted directly from:

TS 122C

September 18, 1968

RECOMMENDED PRODUCT STANDARD FOR
CUSTOM CONTACT-MOLDED REINFORCED POLYESTER
CHEMICAL-RESISTANT PROCESS EQUIPMENT

3.4   *Reinforced-polyester round and rectangular ducts**

3.4.1   *Duct size and tolerances*

3.4.1.1   *Round ducting*—The size of round ducting shall be determined by the inside diameter in inches. The standard sizes shall be 2, 3, 4, 6, 8, 10, 12, 14, 16, 18, 20, 24, 30, 36, 42, 48, 54, and 60 inches. Unless otherwise specified, the tolerance including out-of-roundness shall be $\pm\frac{1}{16}$ inch for duct up to and including 6 inch inside diameter, and $\pm\frac{1}{8}$ inch or ± 1 percent, whichever is greater, for ducting exceeding 6 inches in inside diameter.

3.4.1.2   *Rectangular ducting*—The sizes of rectangular ducting shall be determined by the inside dimensions. There are no standard sizes for rectangular ducting. Unless otherwise specified, the tolerance on ordered sizes shall be $\pm\frac{3}{16}$ inch for dimensions of 18 inches and under and ± 1 percent for dimensions of over 18 inches.

3.4.2   *Lengths*—Tolerances on overall lengths shall be $\pm\frac{1}{4}$ inch unless arrangements are made to allow for field trimming.

3.4.3   *Wall thickness*—The minimum nominal thickness of round ducts shall be in accordance with Table 2. For rectangular duct the minimum thickness shall be as in Table 2, substituting the longer side for the diameter. See also 3.3.6.

3.4.4   *Squareness of ends*—Ends shall be square within $\pm\frac{1}{8}$ inch for round duct through 24 inch diameter and rectangular duct through 72 inch perimeter; and $\pm\frac{3}{16}$ inch for larger sizes of both round and rectangular ducts.

3.4.5   *Fittings*—Tolerances on angles shall be $\pm 1°$ through 24 inches, $\pm\frac{7}{8}°$ for 30 inches, $\pm\frac{3}{4}°$ for 36 inches, $\pm\frac{5}{8}°$ for 42 inches, and $\pm\frac{1}{2}°$ for 48 inches and above. Wall thickness of fittings shall be at least that of duct of the same size.

3.4.5.1   *Ells*—Standard ells shall have a centerline radius of 1.5 times the duct diameter.

3.4.5.2   *Laterals*—Standard laterals shall be 45°.

3.4.5.3   *Reducers, concentric or eccentric*—Length of standard reducers shall be five times the difference in diameters $(D_1 - D_2)$. Minimum wall thickness shall be that required for the larger diameter duct as given in Table 2.

3.4.6   *Straight connections*

3.4.6.1   *Butt joint*—Strength of the butt joint shall be at least equal to that of the duct itself and shall be made in accordance with 3.3.5. Total minimum width of joint shall be 3 inches for $\frac{1}{8}$ inch thickness, 4 inches for $\frac{3}{16}$ inch thickness and 6 inches for $\frac{1}{4}$ inch thickness.

3.4.6.2   *Bell and spigot joint*—Straight duct shall be inserted into bell at least one sixth of duct perimeter or 4 inches, whichever is less, and over-wrapped in such a manner to provide strength at least equal to that of the duct. The opening between the bell and spigot shall be sealed with thixotropic resin paste.

3.4.7   *Flanges*

3.4.7.1   *Flange dimensions*—Dimensions of reinforced plastic flanges

* Rated at a minimum of 5 inch water vacuum and/or 50 inch water pressure. (See Table 2.)

REINFORCED–POLYESTER ROUND DUCT DIMENSIONS[1]

Table 2

ID inches	Wall thickness, min., inches	Allowable vacuum,[2] inches of water	Allowable pressure,[2] inches of water	Flange diameter, OD, inches	Flange thickness, inches	Bolt circle diameter, inches	Bolt hole diameter, inches	No. of bolt holes
2	0.125	405	750	$6\frac{3}{8}$	$\frac{1}{4}$	5	$\frac{7}{16}$	4
3	0.125	405	500	$7\frac{3}{8}$	$\frac{1}{4}$	6	$\frac{7}{16}$	4
4	0.125	210	410	$8\frac{3}{8}$	$\frac{1}{4}$	7	$\frac{7}{16}$	4
6	0.125	64	350	$10\frac{3}{8}$	$\frac{1}{4}$	9	$\frac{7}{16}$	8
8	0.125	30	180	$12\frac{3}{8}$	$\frac{1}{4}$	11	$\frac{7}{16}$	8
10	0.125	16	340	$14\frac{3}{8}$	$\frac{3}{8}$	13	$\frac{7}{16}$	12
12	0.125	9	280	$16\frac{3}{8}$	$\frac{3}{8}$	15	$\frac{7}{16}$	12
14	0.125	7	220	$18\frac{3}{8}$	$\frac{3}{8}$	17	$\frac{7}{16}$	12
16	0.125	6	290	$20\frac{3}{8}$	$\frac{1}{2}$	19	$\frac{7}{16}$	16
18	0.125	5	240	$22\frac{3}{8}$	$\frac{1}{2}$	21	$\frac{7}{16}$	16
20	0.125	5	190	$24\frac{3}{8}$	$\frac{1}{2}$	23	$\frac{7}{16}$	20
24	0.187	9	140	$28\frac{3}{8}$	$\frac{1}{2}$	27	$\frac{7}{16}$	20
30	0.187	7	100	$34\frac{3}{8}$	$\frac{1}{2}$	33	$\frac{7}{16}$	28
36	0.187	5	70	$40\frac{3}{8}$	$\frac{1}{2}$	39	$\frac{7}{16}$	32
42	0.250	10	120	$46\frac{3}{8}$	$\frac{5}{8}$	45	$\frac{7}{16}$	36
48	0.250	9	100	$54\frac{3}{8}$	$\frac{5}{8}$	52	$\frac{9}{16}$	44
54	0.250	7	80	$60\frac{3}{8}$	$\frac{5}{8}$	58	$\frac{9}{16}$	44
60	0.250	6	60	$66\frac{3}{8}$	$\frac{5}{8}$	64	$\frac{9}{16}$	52

[1] 5 to 1 design factor of safety on data in Table 1.   Also based on 10 foot lengths between stiffener rings for vacuum service.

[2] These ratings are suitable for use up to 180°F.   (82.2°C) in pressure service and ambient atmospheric temperatures on vacuum service.   For ratings at higher temperatures consult the manufacturer.   [Or refer to Fig. 11-5.   Also see p. 233.]

for round duct shall be in accordance with Table 2.   Flange thicknesses and width (O.D. − I.D.)/2 of flange faces for rectangular duct shall correspond to those for round duct having the same diameter as the longer side of rectangular duct.

    3.4.7.2   *Flange attachment*—Duct wall at hub of flange shall be at least 1.5 times normal thickness and taper to normal thickness over a distance of at least one flange width.   Fillet radius shall be at least $\frac{3}{8}$ inch at point where the hub meets the back of the flange.

    3.4.7.3   *Face of flange*—Face of flange shall have no projections or depressions greater than $\frac{1}{32}$ inch and shall be perpendicular to centerline of duct within $\frac{1}{2}°$. A camber of $\frac{1}{8}$ inch with respect to the centerline, measured at the O.D. of the flange, shall be allowable. The face of the flange shall have a chemically resistant surface as described in 3.2.4 and 3.3.1.

3.4.7.4 *Drilling*—Standard flanges shall be supplied undrilled.

3.4.7.5 *Flange bolting*—The bolt holes shall straddle centerline unless otherwise specified. Unless otherwise specified, the number of bolt holes and diameters of bolt holes and bolt circles shall be in accordance with Table 2. Rectangular flange width and bolt spacing shall be the same as that for diameters corresponding to the longer sides.

3.4.8 *Mechanical properties of ducts*

3.4.8.1 *Laminate*—The minimum mechanical properties shall be in accordance with Table 1.

3.4.8.2 *Deflection*—Maximum deflection of a side on rectangular duct shall not exceed 1 percent of the width of the side under operating conditions. Ribs or other special construction shall be used if required to meet the deflection requirement.

3.4.9 *Stacks*—Special engineering consideration is required for structural design of stacks, and the manufacturers should be consulted.

Suggested larger-duct-diameter specifications, going beyond the range of the published Standard, are also given below for the designer's guidance. A safety factor of 5 and stiffener rings on 10-ft centers are assumed.

LARGER–DUCT–DIAMETER SPECIFICATIONS

Table 11-1

ID, in.	Wall thickness, in., min.	Allowable vacuum, in. W. G.*	Allowable pressure, in. W. G.*	Flange diameter, in.	Bolt-circle diameter, in.	Bolt-hole diameter, in.	No. of bolt holes	Flange thickness, in.
72	$\frac{5}{16}$	12	43	$78\frac{3}{8}$	76	$\frac{9}{16}$	60	$\frac{3}{4}$
80	$\frac{5}{16}$	10	38	$86\frac{3}{8}$	84	$\frac{9}{16}$	66	$\frac{3}{4}$
84	$\frac{5}{16}$	10	33	$90\frac{3}{8}$	88	$\frac{9}{16}$	70	$\frac{3}{4}$
96	$\frac{3}{8}$	14	16	$102\frac{3}{8}$	100	$\frac{9}{16}$	78	$\frac{3}{4}$
108	$\frac{3}{8}$	10	12	$114\frac{3}{8}$	112	$\frac{9}{16}$	90	$\frac{3}{4}$

NOTE: With the vacuum calculations, duct thickness controls. With the pressure calculations, flange design controls.

* W. G. stands for inches water gauge.

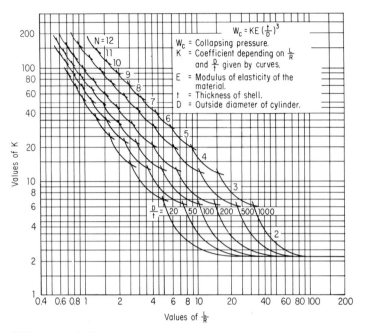

FIG. 11-4 Collapse coefficients; round cylinders with pressures on sides only, sides simply supported; $\mu = 0.30$.[2]

11.3 POLYESTER DUCT CALCULATIONS

Problem 1

Assume a 42-in.-dia duct to carry 28,000 cfm of acid-laden air. Maximum negative pressure is 7 in. H_2O. Wind load of 20 psf may be assumed. Wall thickness is $\frac{1}{4}$ in. (taken from the Recommended Product Standard). If the modulus of elasticity is 800,000, calculate the factor of safety. Assume stiffener rings on 10-ft centers. Design the stiffener ring in the form of a wrapped overlay.[3]

$$a. \ \ W_c = KE \left(\frac{t}{D}\right)^3 \quad \text{psi}$$

To solve for K, refer to Fig. 11-4.

$$\frac{D}{t} = \frac{42.50}{0.25} = 170$$

$$\frac{1}{r} = \frac{10 \times 12}{21.25} = 5.6$$

$E = 800,000$

$K = 12$

$$W_c = (12 \times 800,000) \left(\frac{0.25}{42.50} \right)^3 \text{ psi}$$

$= (9.6)(10)^6 \times (0.206)(10^{-6}) = 9.6 \times 0.206$

$= 1.99 \text{ psi} = \text{collapsing pressure}$

Wind load $= 20$ psf

7-in. H_2O vacuum $= 36.4$ psf

Total load $= 56$ psf, or 0.4 psi

Safety factor $= \dfrac{1.99}{0.4} = 5.0$, which is satisfactory   *Answer*

b. Assume a $\frac{3}{4}$-in. stiffener overlay on 10-ft centers. Calculate the width of the overlay.

$E = 1,000,000$

$$EI_c = W_s D^3 \frac{L_s}{24} \qquad \text{Ref. 3}$$

where W_s = pressure, psi

L_s = length between rings, in.

D = outside diameter, in.

I_c = moment of inertia, in.^{4} — combined moment of ring and that portion of the shell acting with the ring

$$= \frac{1.99 \times (42.50)^3 \times 120}{1,000,000 \times 24} = 0.77 \text{ in.}$$

$$I = \frac{bd^3}{12} \quad \text{or} \quad b = \frac{12 \times 0.77}{(0.75 + 0.25)^3} = 9.25$$

Overlay is $\frac{3}{4}$ in. thick by $9\frac{1}{4}$ in. wide on 10-ft centers.   *Answer*

Alternative stiffeners could be chosen which would serve equally well as the wrapped overlay as long as I_c was 0.77 or larger. If we desired to incorporate a safety factor of 5 on the stiffener, we should select an I_c of 3.85. For example, a stiffening ring 5 in. high $\times \frac{7}{16}$ in. thick would have an I_c of 4.5; and a half-round RP covered core $\frac{7}{16}$ in. thick wrapped over a core radius of 3 in. would have an I of 4.3. Either of these would be satisfactory.

Problem 2

Referring to Prob. 1, if we decreased stiffener spacing to 5 ft, could we get by with a $\frac{3}{16}$-in. duct wall and still have a factor of safety of 5?

$$W_c = KE \left(\frac{t}{D}\right)^3 \quad \text{psi} \quad \text{Ref. 2}$$

$$\frac{D}{t} = \frac{42.37}{187} = 226$$

$$\frac{1}{r} = \frac{5 \times 12}{21.187} = 2.82$$

$$E = 700,000$$

$$K = 38 \quad \text{from the graph}$$

$$W_c = 38 \times 700,000 \times \left(\frac{0.187}{42.37}\right)^3 \text{psi}$$

$$= 26.5 \times 10^6 \times 85 \times 10^{-9}$$

$$= 2,250 \times 10^{-3} = 2.25 \text{ psi}$$

Since our load factor was 0.4, 2.25 divided by 4 equals 5.6. We should therefore have a satisfactory design if stiffeners were used at 5 ft with a $\frac{3}{16}$-in. wall under the design conditions.   *Answer*

Problem 3 (Vacuum)

Duct design for interior service. No wind load is calculated. Assume we have an 8-in.-dia duct with a wall thickness of 0.125 and a modulus of 700,000 psi. Calculate the allowable vacuum in inches of water with a safety factor of 5. Assume stiffener rings on 10-ft centers.

$$\frac{D}{t} = \frac{8.25}{0.125} = 66$$

$$\frac{1}{r} = \frac{10 \times 12}{4.12} = 29$$

$$E = 700,000$$

$$K = 2.3$$

$$W_c = 2.3 \times 700,000 \times \left(\frac{0.125}{8.25}\right)^3$$

$$= 5.6$$

$$\frac{5.6}{5} \text{ psi} = 1.1 \text{ psi} = 30 \text{ in. of water}$$

This answer corresponds with the SPI table.

To be factually correct, the allowable vacuum figures given in Table 11-2 and suggested support spacing are based on laminate properties existing at 73°F. The average designer uses the table with some degree of latitude and actually, generally, finds them suitable for use up to 180°F, in pressure service. It should be stressed that these figures are applicable in vacuum service at ambient temperatures (presumably 73°F) or in that proximity. As the temperature increases, support spacing and collapse vacuum reduction factors come into play. The engineer may quickly estimate the effect of increased temperature by referring to Fig. 11-5 and applying these reduction factors to Table 11-2. In vacuum service, high temperatures have a pronounced effect on the laminate's ability to cope with collapsing vacuum, and this should be taken into account by the designer.

Now the engineer is presented with a number of alternatives for obtaining the desired stiffness. It is obvious that the effect of the stiffener is directly proportional to I. As we add breadth to the section, additional stiffening is directly proportional to the width, but the effect of adding depth to the section is proportional to the depth cubed. Theoretically, a depth/width ratio of about 12:1 for a rib section approaches a design optimum, so that it would appear that the ideal stiffening device is a standing ring of whatever depth necessary to handle the particular conditions. Suffice it to say that the ring needs to remain rigid. Some near-optimum designs, with their calculated I, are shown in Table 11-3.

Quite often, if we built a core material of wood or foam or cardboard, we could overlay some reinforced plastic material to provide sufficient rigidity. On small-diameter round ductwork, a solid RP standing ring or flange or stiffening overlay may be the most economical, but as we reach larger diameters, the core material combined with the best production techniques may provide the least costly method of producing a stiffener. Generally, but not always, the most desirable

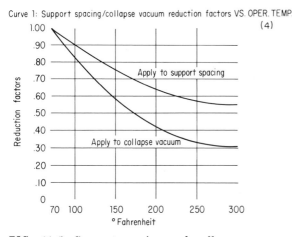

Curve 1: Support spacing/collapse vacuum reduction factors VS. OPER. TEMP.

FIG. 11-5 Support spacing and collapse vacuum reduction factors versus operating temperatures.[4]

EFFECT OF STIFFENER SPACING ON ALLOWABLE VACUUM IN AN RP DUCT SYSTEM WITH SUGGESTED MAXIMUM SUPPORT SPACING[4]
Allowable Vacuum in Inches of Water

Table 11-2

Duct ID, in.	Minimum wall thickness, in.	No stiffener rings	Stiffener rings on 10-ft centers	Stiffener rings on 5-ft centers	Suggested maximum support spacing, ft
2	0.125	405	405	405	8
3	0.125	405	405	405	8½
4	0.125	210	210	260	9
6	0.125	64	64	78	10
8	0.125	30	30	40	10
10	0.125	16	16	26	10
12	0.125	9	9	20	9½
14	0.125	6	7	16	9
16	0.125	4	6	13	8½
18	0.125	3	5	11	8
20	0.125	2	5	9	7½
24	0.187	4	9	21	10
30	0.187	2	7	15	9
36	0.187	1	5	12	8
42	0.250	2	10	24	11
48	0.250	1½	9	19	10
54	0.250	1	7	17	9½
60	0.250	¾	6	15	9

method of stiffening is the one that provides the greatest depth of section common to the neutral axis.

Some additional configurations usable as stiffeners in ductwork design are shown in Tables 11-4 and 11-5.

The upper range of these stiffeners is very heavy for ductwork stiffening, but might be applicable in stiffening designs for vacuum vessels. The smaller sizes of these stiffeners would be applicable to ductwork problems.

Another type of stiffener which can be used is the half-round RP covered core.

Now stiffener and cost design vary from vendor to vendor, and depend upon the engineer's ability to combine the design necessities, such as flanged and wrapped joints, with the functional requirements,

I VALUES FOR RIB–TYPE STIFFENER RINGS $\quad I = \dfrac{bd^3}{12}$

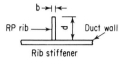

Rib stiffener

Table 11-3

d, in.	b, in.	I
3	¼	0.56
4	⅜	2.0
5	⁷⁄₁₆	4.5
6	½	9.0

I VALUES FOR HOLLOW–CORE HAT STIFFENER
I INCLUDES TOP FLANGE AND BOTH WEBS ONLY

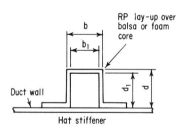

Hat stiffener

Table 11-4

b	d	t	I
2	2	¼	1.0
3	3	¼	2.0
4	4	⅜	6.9
5	5	½	17.7
6	6	½	31.4
8	8	½	77.1

I VALUES FOR HALF–ROUND STIFFENER

$$I = (0.1098)\,(R_1{}^4 - R_2{}^4) - \frac{0.283(R_1{}^2)(R_2)^2(R_1 - R_2)}{R_1 + R_2}$$

Half–round–RP–covered core

Table 11-5

R_1	R_2	t	I
$1\frac{1}{4}$	1	$\frac{1}{4}$	0.11
$2\frac{1}{4}$	2	$\frac{1}{4}$	0.76
$2\frac{1}{2}$	2	$\frac{1}{2}$	1.75
$3\frac{1}{4}$	3	$\frac{1}{4}$	2.3
$3\frac{1}{2}$	3	$\frac{1}{2}$	5.1

such as stiffeners. *Stiffening costs, it thus can be concluded, are almost directly a function of the ease of application.*

Obviously, the greatest ease of application is in the vendor's shop. Stiffening that can be applied in the vendor's shop will cost the minimum amount. Where we are dealing with modest vacuums, so that our Ic requirement is not above 1.0, serious consideration should be given to combining the wrapped joint with an overlay to produce the desired thickness. Quite often this combination will provide the minimum functional cost. Precalculated stiffener-overlay designs are furnished in Tables 11-3 through 11-6. From these tables the designer may make the proper selection of stiffener overlay.

This typical example shows Ic^4 for a $\frac{3}{4}$-in. overlay on a $\frac{1}{4}$-in. duct and represents the combined Ic of duct plus overlay. The sample calculation is for an 8-in.-wide overlay.

Supporting calculations for Table 2—reinforced polyester round ducts[5]
This section on supporting calculations for Table 2 of the Product Standard was prepared by Mr. Eugene W. Hanszen, President of Hanszen Plastics Corp., Dallas, Tex. The section is part of a paper presented by him at the 23d Annual Technical and Management Conference, February, 1968, Society of the Plastics Industry, Reinforced Plastics Division, at Washington, D.C. The title of the paper was "A Discussion of Those

Portions of the Standard Relating to Reinforced Polyester Round and Rectangular Duct Design." Only that portion of the paper relating to Table 2 of the Standard is used here. Since the mechanical properties for various thicknesses in Table 1 were used to calculate Table 2, Table 1 is repeated in this context.

Basis on which Table 2 was compiled   (table of ductwork specifications in Product Standard TS-122C)

1. Mechanical properties of duct and flanges will be based on those in Table 1 for appropriate thickness.
2. Standard duct diameters 2 to 60 in. as listed in sec. 3.4.1.1 will be tabulated.
3. Duct will be suitable for ambient operation under at least 5 in. water vacuum, and a theoretical overdesign of 5:1 will be applied, so that actual calculations will be based on 25 in. of water additional external

STIFFENER–OVERLAY TABLE—I[4]

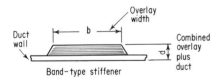

Band–type stiffener

Table 11-6

Duct thickness	Overlay thickness	Combined duct + overlay, d	Overlay width, in., b					
			4	6	8	10	12	14
0.125	0.625	0.75	0.14	0.21	0.28	0.35	0.42	0.49
0.125	0.750	0.875	0.22	0.34	0.45	0.56	0.67	0.78
0.187	0.50	0.687	0.11	0.16	0.22	0.27	0.32	0.38
0.187	0.625	0.812	0.18	0.27	0.36	0.45	0.54	0.63
0.187	0.750	0.937	0.27	0.41	0.55	0.68	0.82	0.95
0.250	0.50	0.75	0.14	0.21	0.28	0.35	0.42	0.49
0.250	0.625	0.875	0.22	0.34	0.45	0.56	0.67	0.78
0.250	0.750	1.0	0.33	0.50	0.66	0.83	0.99	1.17

$$Ic = \frac{bd^3}{12} = \frac{(8)(1^3)}{12} = 0.66$$

SUGGESTED FLANGE–ATTACHMENT DESIGN FOR DUCTS

Table 11-7

				Stiffening action of flanges—I^4		
Duct diameter, ID, in.	*Wrap thickness, in.*	*Suggested min. shear length, in.*	*Flange, I^4*	*Wrap + duct wall, $I_w{}^4$*	*Individual stiffener, $I_f{}^4 + I_w{}^4$*	*Total I_T flange joint, $I_T{}^4$*
2	0.125	2	0.22	. . . . . .	0.22	0.44
3	0.125	2	0.22	. . . . . .	0.22	0.44
4	0.125	2	0.22	. . . . . .	0.22	0.44
6	0.125	2	0.22	. . . . . .	0.22	0.44
8	0.125	2	0.22	. . . . . .	0.22	0.44
10	0.125	2	0.32	. . . . . .	0.32	0.64
12	0.125	2	0.32	. . . . . .	0.32	0.64
14	0.125	2	0.32	. . . . . .	0.32	0.64
16	0.125	2	0.43	0.0026	0.43	0.86
18	0.125	2	0.43	0.0026	0.43	0.86
20	0.125	2	0.43	0.0026	0.43	0.86
24	0.125	2	0.43	0.0045	0.43	0.86
30	0.125	2	0.43	0.0045	0.43	0.86
36	0.125	2	0.43	0.0045	0.43	0.86
42	0.125	2	0.54	0.01	0.55	1.10
48	0.125	3	1.66	0.02	1.68	3.36
54	0.125	3	1.66	0.02	1.68	3.36
60	0.125	3	1.66	0.02	1.68	3.36
72	0.187	4	2.00	0.04	2.04	4.08
80	0.187	4	2.00	0.04	2.04	4.08
84	0.187	4	2.00	0.04	2.04	4.08
96	0.187	6	2.00	0.06	2.06	4.12
108	0.187	6	2.00	0.06	2.06	4.12

pressure over internal pressure. The length between pairs of flanges or other stiffener rings was set at 10 ft for all diameters of pipe.

4. A minimum wall thickness of $\frac{1}{8}$ in. is required for integrity in a corrosive condition, and wall thickness will be increased only in multiples of $\frac{1}{16}$ in. as required to meet vacuum requirements.

5. Flange OD and drilling will be similar to dimensions originated by du Pont. Information from du Pont and other industry data were used to determine flange thicknesses for various diameters.

6. A minimum flange thickness of $\frac{1}{4}$ in. is required for proper gasket

sealing with bolting as specified in point 5. Flange thickness will be increased only in multiples of $\frac{1}{8}$ in. as required to provide vacuum sealing and stiffener-ring requirements.

7. Pipe and flange dimensions were set by vacuum considerations. These thickness values were then used to compute allowable pressure ratings, with a 5:1 theoretical factor of overdesign, based on Table 1 mechanical properties. For use with pressure, the flange alternatives discussed for sec. 3.4.7.1 may be desirable, and will increase the allowable pressure ratings and/or factor of safety on the pressure rating.

8. Since tensile strength of the laminate does not decrease with increasing temperature as rapidly as does flexural modulus of elasticity, the pressure ratings in the table are suggested to 180°F, while the vacuum values are for ambient temperature values only.

The method of par. UG-28 of sec. VIII of the ASME Pressure Vessel Code has been adopted for use with reinforced plastics by creating a theoretical "Chart for Determining Collapsing Pressure of Cylindrical and Spherical Vessels under External Pressure When Constructed of Glass Reinforced Plastic."[6] A copy of this chart is included as Fig. 11-5a. *No safety factor* has been included in the chart. Values of P obtained from this chart are the theoretical collapsing pressures for the pipe or vessel geometry, based on room-temperature values of E for the laminate used. Appropriate safety factors can be applied separately.

REQUIREMENTS FOR PROPERTIES OF REINFORCED-POLYESTER LAMINATES

Table 1

	Thickness, inches			
Property at 23°C. (73°F)	$\frac{1}{8}$ to $\frac{3}{16}$, psi	$\frac{1}{4}$ psi	$\frac{5}{16}$ psi	$\frac{3}{8}$ and up, psi
Ultimate tensile strength, minimum*................	9,000	12,000	13,500	15,000
Flexural strength, minimum†................	16,000	19,000	20,000	22,000
Flexural modulus of elasticity (tangent), minimum‡........	700,000	800,000	900,000	1,000,000

* See 4.3.2.
† See 4.3.3.
‡ See 4.3.4.

AUTHOR'S NOTE: The chart of Fig. 11-5a is based on the (D_o/t) lines, scales, and grid used in app. V of sec. VIII of the ASME Pressure Vessel Code. Superimposed are the second set of lines, which depict a series of idealized stress-strain curves for FRP laminates. These sloped lines are straight, on the assumption that the tensile modulus of elasticity does not vary as the stress level is increased. The FRP chart does not include a safety factor, which must be applied after determination of the theoretical collapsing pressure for a given cylinder or head. At elevated temperatures the safety factor should be increased to include the effective temperature on the modulus of elasticity. This FRP chart assumes Poisson's ratio is 0.3. Assuming that the actual value of Poisson's ratio lies between 0.25 and 0.35, then FRP chart values might be in error by as much as 3 percent. The composite chart is under study by the ASME and has not been adopted by them at this time.

Directions for use of the chart for external pressure on FRP pipe or vessels are as follows [taken from par. UG-28(c) of sec. VIII of the ASME Code]:

1. Assume a value for wall thickness, t. Determine the ratios of length to outer diameter (L/D_o) and outer diameter to wall thickness (D_o/t).
2. Enter left-hand side of chart at value of (L/D_o) from (1).
3. Move horizontally to (D_o/t) line from (1).
4. From this intersection, move vertically (up or down) to intersect the materials line for the appropriate value of E.
5. From this intersection, move horizontally to the right and read the value of factor B.
6. Compute the collapsing pressure by the following formula:

$$P = \frac{B}{D_o/t}$$

Using the chart in this way, collapsing pressures have been calculated for FRP pipe.

A number of pieces of ductwork with 10 ft between stiffeners have been taken to failure with vacuum in our laboratory to study how and where failure occurs. In all cases failure has occurred at external pressures somewhat higher than the values calculated from Fig. 11-5a, and we therefore feel that, for duct made with good control of wall thickness, glass content, and cure, this curve will give conservative values for failure of fiber-glass duct if good values of flexural modulus are used in the calculation.

Method of determining wall thickness required for 5-in. water vacuum
Using Fig. 11-5a, allowable vacuum was calculated for each diameter

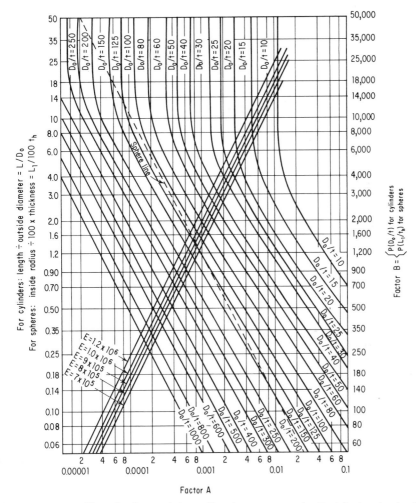

FIG. 11-5a Chart for determining collapsing pressure of cylindrical and spherical vessels under external pressure when constructed of glass-reinforced plastic.[6]

and wall, and when allowable vacuum fell below 5 in. of water, the wall thickness was increased.

Problem 1

2-in. ID × 0.125 wall $E = 700,000$

$$\frac{L}{D_o} = \frac{120}{2.25} = 53.4 \text{ in.} \qquad \frac{D_o}{t} = \frac{2.25}{0.125} = 18$$

From the chart,

$$B = 4{,}300 = P\left(\frac{D_o}{t}\right) = P(18)$$

$$P = \frac{4{,}300}{18} = 238 \text{ psi}$$

$$= \frac{238 \times 27.5 \text{ in./psi}}{5 \text{ factor}} = 1{,}315 \text{ in. water}$$

Full vacuum = 405 in.; so 2 in. ϕ is good for full vacuum.

Problem 2

24 in. ID $\times$ 0.125 wall $E = 700{,}000$

$$\frac{L}{D_o} = \frac{120}{24.25} = 4.94 \text{ in.} \qquad \frac{D_o}{t} = \frac{24.25}{0.125} = 194$$

From the chart,

$$B = 130 = P(194)$$
$$P = 130/194 = 0.67 \text{ psi}$$
$$= \frac{0.67 \times 27.5 \text{ in./psi}}{5 \text{ factor}} = 3.7 \text{ in. allowable} \qquad \textit{not enough}$$

Try 0.187-in. wall. $E = 700{,}000$

$$\frac{L}{D_o} = \frac{120}{24.38} = 4.93 \text{ in.} \qquad \frac{D_o}{t} = \frac{24.38}{0.187} = 130$$
$$B = 225 = P(130)$$
$$P = 225/130 = 1.73 \text{ psi}$$
$$= \frac{1.73 \times 27.5}{5 \text{ factor}} = 9.5 \text{ in.} \qquad \textit{adequate}$$

Thus were the other vacuum walls and ratings filled in.

Method of determining flange thickness required as vacuum stiffener
The required moment of inertia of stiffener per ASME Unfired Pressure Vessel Code, par. UG-29, is

$$I_s = \frac{D_o{}^2 L(t + A_s/L)A}{14}$$

where I_s = moment of inertia, in.[4]
D_o = OD, in.
L = center-to-center distance between rings, in.
t = wall thickness, in.
A_s = cross-sectional area of ring, in.[2]
A = S/E (factor A on Fig. 11-5a)
P = external pressure, psi

Problem 1

For 8-in. pipe with flanges at $\frac{1}{4}$ in. thickness and wall of 0.125 in.

Vacuum = 30 in. at 5:1

$$= \frac{30 \text{ in.} \times 5 \text{ factor}}{27.5} = 5.5 \text{ psi}$$

$$I = \frac{bd^3}{12} = \frac{(0.25)\left(\dfrac{12.375 - 8.0}{2}\right)^3}{12}$$

$$= \frac{(0.25)(2.187)^3}{12} = 0.215 \text{ in.}^4$$

$$A_s = (0.25)(2.187) = 0.54 \text{ in.}^2$$

$$\frac{A_s}{L} = \frac{0.54}{120} = 0.0045$$

Factor B from chart $= \dfrac{PD_o}{t + A_s/L} = \dfrac{(5.5)(8.25)}{0.125 + 0.0045} = 349$

Thus, at $E = 800{,}000$, from chart, $A = 0.00022$.

$$I_s = \frac{(8.25)^2(120 \text{ in.})(0.125 + 0.0045)(0.00022)}{14}$$

$$= \frac{(8{,}180)(0.130)(0.00022)}{14} = 0.017 \text{ in.}^4 \quad required$$

Since 0.215 in.[4] is available, flange is adequate.

Problem 2

For 60 in. $\phi \times$ 0.250-in. wall at 6 in. water with $\frac{5}{8}$-in. flange at $66\frac{3}{8}$ in.
OD, $E = 10^6$.

$$\text{Vacuum} = \frac{6 \text{ in.} \times 5 \text{ factor}}{27.5} = 1.09 \text{ psi}$$

$$I = \frac{(0.625)\left(\dfrac{66.375 - 60}{2}\right)^3}{12} = \frac{(0.625)(3.187)^3}{12} = 1.69 \text{ in.}^4$$

$$A_s = (0.625)(3.187) = 2.0 \text{ in.}^2$$

$$\frac{A_s}{L} = \frac{2.0}{120} = 0.0167$$

$$\text{Factor } B \text{ from chart} = \frac{(1.09)(60.50)}{0.25 + 0.0167} = \frac{66}{0.267} = 247$$

Factor A from chart $= 0.00012$

$$I_s = \frac{[(60.50)^2 \times 120](0.25 + 0.0167)(0.00012)}{14}$$

$$= \frac{(4.4 \times 10^5)(0.267)(0.00012)}{14} = 1.0 \text{ in.}^4; \text{ so } I \text{ of } 1.69 \text{ is adequate.}$$

Method of determining allowable pressure rating on duct with wall fixed by vacuum considerations   The following formula was used to compute the allowable pressure:

$$f \text{ allowable} = \frac{Pd}{2t}$$

where f = stress, psi
   P = pressure, psi
   d = inside diameter of pipe, in.
   t = thickness of pipe wall, in.

From Parker, "Simplified Mechanics and Strength of Materials," p. 229, f was reduced to 1/10 of values in Table 1 for that wall thickness to provide a 10:1 factor of overdesign for pressure.

Problem 1

2-in. pipe at 0.125-in. wall

$$f = \frac{9,000}{10} = 900 \text{ psi}$$

$$= 900 = \frac{P(2)}{(2)(0.125)}$$

$$P = (900)(0.1250) = 112 \text{ psi}$$

Flanges control here.

Problem 2

60-in. pipe at 0.250-in. wall

$$f = \frac{12,000}{10} = 1,200$$

$$= 1,200 = \frac{P(60)}{(2)(0.250)}$$

$$P = \frac{1,200}{120} = 10 \text{ psi}$$

Flanges control here also.

Method of determining allowable pressure rating of duct flanges already fixed by vacuum considerations   Use Taylor forge system as outlined by Brownell and Young, "Process Equipment Design," Eq. 12,109, p. 245, 1959.

$$t = 0.72 \sqrt{(M_oY)/(B)}(f \text{ allowable in flexure})$$

where t = flange thickness
$\quad M_o$ = total moment acting on the flange, in pounds
$$K = \frac{A}{B} = \frac{\text{flange OD}}{\text{flange ID}}$$
$\quad Y$ = a function of K, and can be read from fig. 12.22, if K is known

Also

$$M_o = M_d + M_g + M_t = (H_d \times h_d) + (H_g \times h_g) + (H_t \times h_t)$$

where $H_d = 0.785 \ B^2p$
$\quad h_d$ = radial distance from bolt circle to circle where H_d acts (near pipe OD)
$\quad H_g = W_{m1} - H = (H + H_p) - H = H_p$
$\qquad = H_p = 2b\pi Gmp$
$\quad h_g$ = radial distance from gasket load reaction to bolt circle, in.
$\quad H_t$ = difference between hydrostatic end force and hydrostatic end force on the area inside of flange = $H - H_d$
$$h_t = \frac{R + g_1 + h_g}{2}$$

Problem 1

For 36-in.-dia duct with 0.187-in. wall, a 40⅜-in. OD flange at ½ in. thick, a 39-in.-dia bolt circle with 32 bolts at ½ in. diameter.

$$f \text{ allowable in flexure} = \frac{22,000}{5 \text{ factor}} = 4,400 \text{ psi}$$

$$t^2 = \frac{(0.72)^2(M_oY)}{(\text{flange ID})(f_{\text{allowable}})}$$

$$M_o = \frac{(t^2)(\text{flange ID})(f_{\text{allowable}})}{(0.72)^2Y}$$

$$K = \frac{\text{flange OD } 40.375}{\text{flange ID } 36.375} = 1.11$$

and from fig. 12.22, in Brownell and Young,

$$Y = 19$$

Then

$$M_o = \frac{(0.50)^2(36.375)(4,400)}{(0.72)^2(19)}$$

$$= \frac{(0.25)(36.375)(4,400)}{(0.52)(19)} = 4,050 \text{ in. lb}$$

$$M_d = H_d \times h_d$$

$$= (0.785)(36.375)^2P \times \frac{(39 \text{ in.} - 36.375)}{2}$$

$$= 1,040p \times \frac{2.625}{2}$$

$$= 1,370_p$$

$$M_g = H_g \times h_g$$

$$= 2b\pi Gmp \times \frac{39 - 36.375}{4}$$

$$= (2)(1)\pi(37.68)(0.5)p(0.67)$$

$$= \pi(37.68)(0.67)p = 78p$$

$$M_t = H_t \times h_t$$

where $h_t = \dfrac{R + g_1 + h_g}{2}$

$$M_t = (0.785)p(G^2 - B^2) \times \frac{1.5 + 0.8}{2}$$
$$= (0.785)p[(37.688)^2 - (36.375^2)](1.15)$$
$$= (1,420 - 1,320)(0.89p)$$
$$= (100)(0.89p) = 89p$$

Then

$$M_o = M_d + M_g + M_t = 1,370p + 78p + 89p$$
$$M_o = 1,537p = 4,050 \text{ in. lb}$$

Thus

$$p = \frac{4,050}{1,537} = 2.63 \text{ psi} = 73 \text{ in. water, rounded off to 70 in.}$$

Problem 2

For a 14-in.-dia duct with 0.125-in. wall, an 18⅜-in.-OD flange at ⅜ in. thick, a 17-in.-dia bolt circle with 12 bolts at ⅜ in. diameter.

$$f \text{ allowable in flexure in flange} = \frac{22,000}{5 \text{ factor}} = 4,400 \text{ psi}$$

$$K = \frac{18.375}{14.250} = 1.29$$

Y from fig. 12.22 = 7.8 (Brownell and Young)

Then

$$M_o = \frac{(t^2)(\text{pipe OD})(f_{\text{allowable}})}{(0.72)^2 Y}$$
$$= \frac{(0.375)^2(14.250)(4,400)}{(0.52)(7.8)}$$
$$= \frac{(0.141)(14.250)(4,400)}{(0.52)(7.8)} = 2,180 \text{ in. lb}$$
$$M_d = H_d \times h_d$$
$$= (0.785)(14.25)^2 p \times \frac{17 \text{ in.} - 14.25}{2}$$
$$= 159p \times 1.375 = 218p$$

$$M_g = H_g \times h_g$$
$$= 2b\pi Gmp \times \frac{17 - 14.250}{4}$$
$$= (2)(1)\pi(15.625)(0.5)p \times 0.688$$
$$= 34p$$
$$M_t = H_t \times h_t$$
$$= (0.785)(G^2 - B^2) \times \frac{3.0/2 + 0.688}{2}$$
$$= (0.785)[(15.625)^2 - (14.250)^2](1.1)$$
$$= (0.785)(244 - 203)(1.1)$$
$$= (0.785)(41)(1.1) = 36p$$
$$M_o = 2,180 = 218p + 34p + 36p = 288p$$
$$P = \frac{2,180}{288} = 7.6 \text{ psi} = 208 \text{ in. water, rounded off as 210 in.}$$

11.4 DESIGN AND PURCHASING INFORMATION FOR DUCT SYSTEMS

To obtain minimum cost on a duct system the design engineer needs to be cognizant of certain facts and practices in the industry. The following hints are intended to aid the designer in this effort.

1. Most of the larger vendors normally have standard mandrels, in accordance with the standard round-duct dimensions as published in the proposed Commercial Standard table. Ductwork in the smaller sizes will generally be furnished in 20-ft lengths, wherever 20-ft mandrels are available. Some vendors' mandrels are in other lengths, such as 7 and 12 ft. The vendor is, of course, completely capable of factory-joining these lengths in his shop.

2. Making factory joints in a 10-ft length combines the joint with the stiffener overlay and thus can materially reduce the cost of assembling a duct system. This should be done if possible.

3. The least expensive place to make any duct joint is in the factory. Have the vendor make every butt joint, stiffener overlay, or branch joint possible. Strive to make the field joints an absolute minimum. Make sure that all the field joints are as simple as possible. Confine the field joints to straight butt joints or flange bolt-ups.

4. If the vendor is furnishing a complete duct system, it is altogether satisfactory to have the vendor do the flange drilling. If, however, it is necessary to mate to another duct system, the flanges should be furnished undrilled. See Table 11-11 for approximate flange-drilling costs.

5. It is common practice to have the vendor furnish field joining kits as part of the purchase order for the duct. Field joining kits include glass mat, catalyst, resin, and brushes. Cost of the field-joining-kit material varies widely from vendor to vendor and with the duct size. In sizes up to 24 in. the field joint-material kit cost may be approximated as $0.50/in. of duct diameter. In sizes above 24 in. field joining kits run about $1/in. of duct diameter.

6. The problem of shipping the ductwork from the vendor to the job site finally limits the subassembly that can be done at the shop. Quite often an elbow can be strapped onto a straight piece of ducting but the joint on the opposite side of the elbow will become a field joint.

7. When a system requires more than about six field joints, the field joining material is generally shipped in bulk form. Make sure that catalyst-control procedures are observed. You may reserve the right to furnish all catalyst on the job and so instruct the vendor to ship kits less catalyst. This will permit continuous catalyst control. In wrapping large duct systems the amount of catalyst required can be considerable.

8. Generally, duct shipments are limited to the following approximate dimensions: 92 in. wide by 96 in. high by 39 ft long. If you are shipping by truck, ship shop-assembled all ductwork that will fit inside the carrier's vehicle. Rail shipments may permit somewhat longer assemblies.

9. The vendor will quite often require an additional cutting charge if the lengths in question are other than those of his standard mandrels. Typical cutting charges for other than standard lengths vary in the industry but may be approximated from the following table:

Size, in.	Cost per cut
2–12	$2–$3
14–24	$4–$5
26–48	$6–$7

10. Quite often the most economical duct design will be other than that covered in the proposed Commercial Standard. This will be especially true when the designer requires large duct sizes, designed for specific pressure losses. The designer, of course, has the option of using other than a standard size, especially if the duct system is of considerable magnitude. For example, the standard duct jumps from 36 to 42 in., with nothing in between. Now suppose that the designer's requirements can be met with a 40-in. duct. Many vendors have low-cost expandable-mold techniques in which the next-lowest-size mold is expanded to meet the new design requirements. In this way substantial economies can be obtained by adhering to the 40-in. size, rather than assuming that the

42-in. size is the nearest standard size obtainable. The same can be done with fittings.

11. Some vendors charge nearly as much for a 45° elbow as they do for a 90° elbow. Considerable economies can be realized by ordering a 90° elbow and then simply cutting it in half to make two 45° elbows.

12. Bear in mind, in your duct design, that stiffening the duct every 10 ft is a necessary basis for the use of the proposed Commercial Standard table. Slight adjustments to the stiffeners will permit extending this to 12 ft where such mandrels are available. Your entire system should be designed with the utmost care to ensure that stiffeners are properly located, that stiffeners and wrapped joints serve a dual function, and that flanges are strategically located to serve in a stiffening capacity also. Attention to these details is of great importance in maintaining minimum system costs.

13. The use of reinforced plastics offers an ideal combination of corrosion resistance and high strength/weight ratio. Duct systems are light and, in the smaller sizes, are easily handled.

14. Although the butt joint is the standard recommended joint for use in ductwork fabrications, other types of joints have specific advantages in areas where easy disassembly is required. Obviously, the only way to disassemble a system made of butt joints is, literally, to saw it in two. Three easily disassembled joints are given below (excluding the butt-and-flanged joint):

a. Flanged joints, while expensive, serve a number of purposes:
   (1) They permit easy access to the system and rapid reassembly.
   (2) They serve a second function as a stiffening member.
   (3) They permit tie-ins to existing flanged-duct systems.
   (4) They are completely compatible over the full design range of vacuum and pressure conditions for which the duct system itself is suitable.
   These joints are illustrated in Fig. 11-6.

b. A Redilok* coupling will serve as an effective field assembly method in place of butt joints and permits ease of disassembly. By simply inserting the duct into the Redilok coupling and pushing together, assembly is complete. The Redilok gasket has a stop ring which allows the duct to enter the coupling only to the required amount. Multiple elastomer annular rings provide the seal. A coupling of this type, however, should be used for low-pressure or vacuum work only.

c. Morris spanner coupling† (IP coupling) is a composition-type joint of light stainless steel or galvanized sleeve in which has been

* Registered trademark of duVerre, a division of American Pipe Corp.
† Registered trademark of Morse Coupling & Clamp Co., Erie, Pa.

inserted a soft Neoprene or elastomer sleeve. By pulling up gently on several bolts, an effective seal is obtained in a duct system. It may be used in place of a butt joint and should be used where ease of disassembly is necessary. Large quantities have been used for this purpose on ductwork-system installations and have performed satisfactorily. This coupling is good only for very light pressure or vacuum conditions. See Fig. 11-7 for a typical installation.

11.5 EXPANSION JOINTS IN DUCTWORK

Many engineers will choose to employ expansion joints on long duct runs, especially if a considerable swing in temperature may result. In a

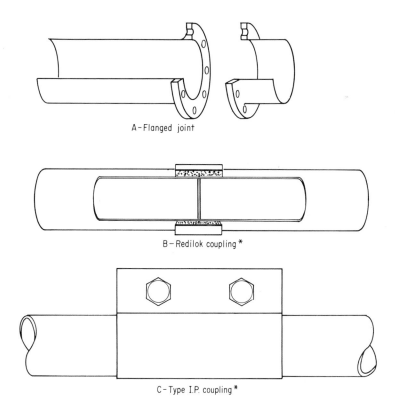

A – Flanged joint

B – Redilok coupling*

C – Type I.P. coupling*

*Du Verre trademark

FIG. 11-6 Three easily disassembled joints used in ductwork.

normal duct, laminate consisting of 25 percent glass and 75 percent resin, each 50°F of change will result in 1-in. expansion per 100 ft of pipe, according to Table 11-8.

DUCTWORK EXPANSION

Table 11-8

Temperature change, deg	Expansion, in. per 100 ft
25	0.5
50	1.0
75	1.5
100	2.0
125	2.6
150	3.1
175	3.6
200	4.1

Again, as in piping design, reinforced plastic ductwork has a relatively low modulus of elasticity, so that the formula is

$$S = Ee \qquad S = \frac{E \times \text{change in length}}{\text{original length}}$$

FIG. 11-7 Morris Spanner coupling installation.

where S = stress, psi

E = flexural modulus (see p. 82)

e = expansion from Table 11-8

A sample problem, using this formula and Table 11-8, indicates how a duct system properly guided and anchored can be protected against failures from buckling.

Problem

Calculate the stress in a duct operating over a temperature range of 100°F. Assume 40 in. diameter, ¼-in. wall, and 300 ft length.

$$S = \frac{800,000 \times 2 \times 3}{300 \times 12}$$

$$= \frac{4,800,000}{3,600}$$

$$= 1,330 \text{ psi}$$

Since the duct flexural-strength minimum is 19,000 psi, we should have a factor of safety of over 14:1.

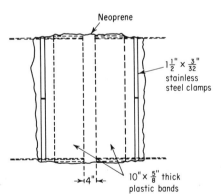

Neoprene

$1\frac{1}{2}" \times \frac{3}{32}"$ stainless steel clamps

←|4"|← 10"x $\frac{5}{8}"$ thick plastic bands

FIG. 11-8 Expansion-joint detail.

The designer, however, may conclude that it is much easier to install simple expansion joints periodically in the line than attempt to anchor and guide. For a simple expansion joint of this type see Fig. 11-8.

This permits the designer to proceed with simple band-type duct hangers. Possible alternatives should be assessed at each installation.

11.6 SUPPORTING DUCTWORK

Supporting of ducts will be most generally satisfactorily accomplished by using a band-type duct hanger covering either 120 or 180° of support contact.   See Fig. 11-9.

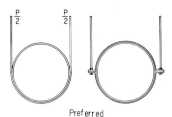

Preferred

FIG. 11-9 Band-type duct hanger.

Wall stresses produced by supporting bands at least 4 in. wide and covering 120 to 180° of arc are generally so low as to be negligible.   It will be found that wall stresses are almost independent of the width of the band, within reasonable limits.   However, the designer should stay away from point loading, which can be generated by a trapeze type of duct hanger, shown in Fig. 11-10.   This method is not desirable because the point loading will often produce a wall stress above the maximum allowable, unless the duct wall is specifically strengthened to take care of the point loading.

Another argument against the trapeze hanger and point loading is the undesirable effect of expansion and vibration across the point, which may literally abrade it so that in time a hole appears in the duct.

Figures 11-11 to 11-15 may aid the designer in preparing different types of supports.

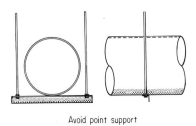

Avoid point support

FIG.   11-10 Trapeze-type duct   hanger   (avoid   point support).

The suggested maximum support spacing shown in Table 11-2 is, essentially, a conservative design.   It can be used safely under almost any conditions for duct design and eliminates tedious calculations to determine true allowable support distances.   A rigorous investigation of support distances for ductwork and correlated deflection is, however, a

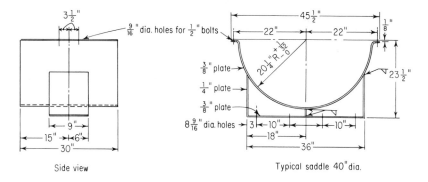

Side view   Typical saddle 40″dia.

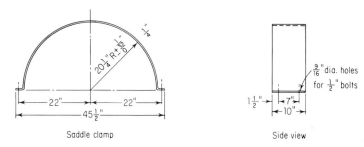

Saddle clamp   Side view

FIG. 11-11   Typical saddle-type support.

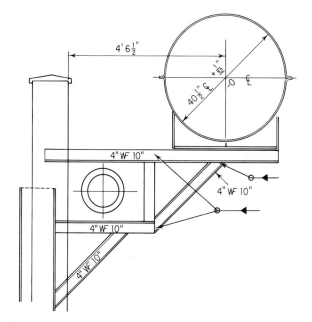

FIG. 11-12   Adding to an existing pipe rack for duct support.

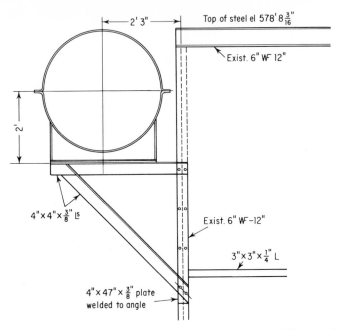

FIG. 11-13 Supporting ductwork from existing building steel.

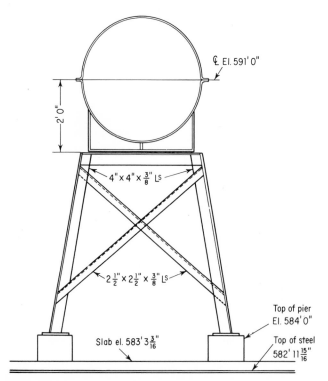

FIG. 11-14 Duct support in open area.

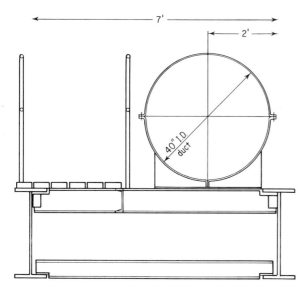

FIG. 11-15 Heavy support for roadway combining walkway, ductwork, and utility carrier.

necessary development from time to time. In crossing roads and railroad tracks and taking advantage of existing building supports and roof beams, the engineer can often develop a much more economical supporting design than by simply referring to the table. If the duct itself is considered as a simply supported member between supports (L = the distance apart), then the conventional formula used in stress analysis will apply. The following problem is representative.

Problem

Assume a 40-in.-dia duct with ¼-in. walls to span a roadway. Supports are 30 ft apart. Assume an evenly distributed wind load of 20 psf. The duct weighs 750 lb. Assume additional stiffeners at 250 lb, making the total duct weight between supports 1,000 lb.

$$\text{Total uniform load} = \underset{\text{wind load}}{2{,}000\text{ lb}} + \underset{\text{dead weight}}{1{,}000\text{ lb}} = \underset{\text{total}}{3{,}000\text{ lb}}$$

$$I = \frac{\pi(D^4 - d^4)}{64} = \frac{(3.14)[(40.5)^4 - (40.0)^4]}{64} = 6{,}337$$

$$\Delta M = \frac{5Wl^3}{384EI} = \frac{(5)(3{,}000)(360)^3}{(384)(800{,}000)(6{,}337)}$$

$$= 0.36\text{ in.}$$

This indicates that, deflection being nominal, stresses in the duct are probably also conservative.

To calculate the stress in the duct:

Maximum moment at center,

$$M_{max} = \frac{WL}{8} = \frac{(3,000)(360)}{8} = \frac{1,080,000}{8} = 135,000 \text{ in. lb}$$

$$\text{Section modulus} = \frac{\pi(R_1{}^4 - R_2{}^4)}{4R}$$

$$= (0.785)\frac{(20.25)^4 - (20)^4}{20.25} = 315$$

$$\text{Maximum unit stress} = \frac{M}{S} = \frac{135,000}{315} = 428 \text{ psi}$$

which is satisfactory.

The end beam reaction at the support points would be 3,000 lb.

$$\text{Shear stress} = \frac{\text{reaction}}{\text{duct-wall area}} = \frac{R}{A} = \frac{3,000}{31} = 96 \text{ psi}$$

This, however, is an average stress. Local stresses in the duct wall are independent of supporting bandwidth.

Bearing load = 3,000 lb at each support

Unit stress at which buckling will occur with a safety factor of 6 is 360 psf. The area required is

$$\frac{3,000}{360} = 8.3 \text{ ft}^2$$

The supporting-band design, therefore, with a long span with such a high reaction would be designed to contain 8.3 ft^2 over a 120° arc. To accomplish this we should require a bandwidth of about 28½ in. The band would be made 30 in. wide.

11.7 HELPFUL HINTS IN REINFORCED FABRICATION OF DUCTS, HOODS, AND STACKS

1. Normally, the reinforced plastic material will be a reinforced polyester and the material fabricated by the contact-molding process.

2. In the specification of reinforced plastic ductwork and other air-handling equipment, it is wise to specify the maximum in fire retardancy compatible with the chemical-process requirements.

3. Generally, no more than 50 percent styrene by weight should be used in the liquid polyester resin.

4. A thixotropic agent such as CAB-O-SIL may be added to the resin.

5. Occasionally, pastes or polyester putty will be required to smooth out joints. To formulate this a silica flour or clay of 100 to 300 mesh may be used as a filler. A very short chopped-strand filler may also be helpful.

6. Pigmentation is an essential part of a duct system, especially for exterior ductwork. It will add greatly in preventing ultraviolet degradation. Choose a color scheme and stick to it. A deep green is very pleasing and will hold up well in many years of service.

7. Commonly, ductwork is not subjected to high stress and is generally built of chopped-strand type E glass mat in sizes $3/16$ in. and under. Above that, glass cloth or roving may be used. However, if woven roving or cloth is used, it should be covered on both sides with a chopped-strand mat. Both interior and exterior surfaces should be finished with a C mat. A typical duct lay-up might be as follows, from the inside out:

a. 10- to 20-mil type C glass embedded in a resin-rich surface

b. Two $1\frac{1}{2}$-oz layers of chopped-strand mat

c. Mat or roving as necessary to build up to the desired thickness

d. Final layer of $1\frac{1}{2}$-oz mat suitably hot-coated

e. At the designer's option, a layer of C glass embedded in the hot coat to provide the best protection where exterior corrosion is a factor

Commonly, ductwork will run 25 to 30 percent glass content and 70 to 75 percent resin content. It is generally conceded that stack design, hoods, and rectangular ducts have so many combinations and special problems that no standardization exists. In the case of stacks, each one must be considered on an individual engineering basis. Hoods and rectangular ducts are in the same category.

8. To simplify duct construction a single-line drawing is completely acceptable.

9. It is common practice for the vendor to furnish the flanges undrilled, the drilling to be done in the field.

10. No glass fibers are to be exposed. Small spots in the duct may be 80 percent of the design thickness without being cause for rejection.

11. Butt joints and branch-connection lay-ups should be at least as thick as the heaviest section being joined, so that the joint is as strong as the duct itself.

12. All raw or mating edges should be sealed with a coat of filled polyester resin to present a smooth interior or exterior.

13. Wrapped overlay requires a roughened surface which is wider than the overlay. In the finished joint the roughened surface and the overlay are completely covered with a pigmented thixotropic resin.

11.8   REPAIR OF EXISTING DUCT SYSTEMS

Plant maintenance staff is often confronted with deterioration of metal or lined-metal duct systems. Here is another case where glass-reinforced polyester lay-ups can be attached most profitably to repair and completely rehabilitate the existing ductwork. Ductwork that has deteriorated or that requires replacement can be effectively replaced *in situ* by using the existing duct for a form and literally encapsulating it with a wraparound system of glass-reinforced polyester. Here the mode of construction should be to use a $\frac{1}{2}$-oz C glass on the interior, followed by the appropriate lay-up of $1\frac{1}{2}$-oz E grade mat saturated with a duct-type resin containing a fire-retardant material. As a guide to the thickness of the wrap covering, refer to the ductwork table found in the Recommended Product Standard. If the ductwork is finished with a heavy pigmented resin, it will be sufficient protection against ultraviolet degradation. The joints of the duct should receive careful attention to make sure they are adequate. In the end the old duct may corrode away, leaving in its place a glass-reinforced polyester duct which will provide service for many years. Wooden ducts, tanks, and flumes have also been repaired in this manner. Round surfaces are best since the polyester contracts on curing. An excellent bond is generally obtained provided reasonable precautions have been taken with temperature and surface preparation.

11.9   ALTERNATIVE DUCT CONSTRUCTION—
THE DUCTWORK OF THE FUTURE?

The combination of hand-laid-up and filament-wound construction was applied initially in RP tanks and structures. Following this, it was extended to piping, and then to ductwork. In combination the best qualities of each technique produced a standard constructional procedure with economic advantages meriting the consideration of every engineer engaged in specifying RP ductwork. The new technique offers the advantages of the high-corrosion-resistant qualities of hand-laid-up construction and the ultrahigh strength available in the filament-wound approach.

A typical construction of a duct from the inside out using both techniques is as follows:

1. An internal gel coat reinforced with a 10-mil surfacing mat representing perhaps 5 percent glass and 95 percent resin. This is the all-important corrosion inner barrier so necessary for the best performance.

2. Two layers of 1½-oz mat giving about 25 percent glass and 75 percent resin. This second layer represents the second line of attack. It still possesses an extremely high resistance to corrosion since it is made of a random mat.

3. A suitable thickness of filament-wound material running perhaps 60 to 70 percent glass and 30 to 40 percent resin. This filament winding at the proper helix angle provides immense tensile strength in the duct wall. The thickness may be varied to suit the design conditions dictated by the problem to be solved.

4. Following the filament winding, an outer layer of C glass is applied. The resin used to saturate the C glass may be pigmented if desired (it is probably a good idea to do so).

5. Finally, over all this is applied a pigmented "hot" coat. An aquamarine green is very pleasing, but the engineer can choose the color coding. The pigmented hot coat and the undercoat of the C glass provide excellent resistance to weathering and ultraviolet degradation. Some duct systems are pigmented throughout, at the option of the engineer. Pigmenting covers up imperfections in fabrication and reduces the chemical resistance of the ducting. The latter, however, is not generally of concern, unless such ductwork is installed outdoors. The provision for resistance to ultraviolet degradation is obligatory, where outdoor installation is contemplated.

11.10 PREVENTION, CAUSES, AND PROPAGATION OF DUCTWORK FIRES

A study of fires which have occurred in reinforced plastic ductwork permits some useful generalizations.

Prevention There is a great deal that the purchaser can do to reduce the possibility of fire occurring in a ductwork system or, if such a fire should occur, to limit the damage and, in many cases, literally, save the plant. Some suggestions based on experience in this area are as follows:

1. By all means specify and use a resin which has a fire-spread rating of 25 or less and to which a "snuffing" compound has been added, such as antimony trioxide. A chlorinated polyester with 5 percent antimony trioxide added is an excellent material of construction for a duct system.

This resin-glass system has a fire-spread rating of 25 or less, which is rated self-extinguishing; that is, the fire will go out when the source of ignition is removed.

2. A snuffing compound may be added to many resins to provide fire retardancy; however, the chlorinated polyester with the snuffing compound will provide the fire retardancy to the maximum degree. This is accomplished through synergistic action.

3. Moving-picture studies of simulated duct fires show conclusively the advantages of the snuffing-type compound with the chlorinated polyester. When the source of heat is shut off, combustion ceases almost instantaneously. A straight bisphenol resin with high gas velocities will burn with considerable rapidity, as evidenced by its flame-spread rating of 150 to 200. For this reason bisphenol resins should not be used in duct construction unless the corrosion problem cannot be solved with a chlorinated polyester resin (as in the case of high caustic content).

4. Sprinkler protection should be installed commensurate with the value of the duct and property involved.

5. Buildup of combustible residues in a duct system is always potentially dangerous. For this reason all duct or exhaust systems should be cleaned out on a reasonable schedule to keep this buildup to a minimum. Some buildups, especially where sulfur compounds are involved, complicate the task of fighting a duct fire because of the quantities of sulfur dioxide which may be liberated.

Causes of duct fires   Most duct fires which have been reported have been caused by an external source. Some of the causes of ignition have been:

1. Electrical short circuits
2. Welding or burning
3. Overheated laboratory or process equipment
4. Poor housekeeping and carelessness
5. Static electricity
6. Lightning

Propagation   A ductwork fire will propagate itself with astonishing speed, only a little slower than the gas velocities in the duct. Heavy smoke generally accompanies the fire, and will be black and voluminous while the fire is in progress. However, where a chlorinated resin has been used with 5 percent antimony trioxide as an additive, the clouds of smoke will turn white almost instantly (in 9 to 15 sec) when the source of ignition has been removed.

Ignition temperatures of non-fire-retardant polyester resins generally run 750 to 800°F.[7]

On the positive side, there is a case history of an intense fire in a polyester duct which remained confined to the duct. The duct materials of construction were credited with saving the plant itself from severe damage or complete destruction. *No duct system yet built can be classed as fireproof, especially where we are dealing with ductwork construction in corrosive areas.* To maintain reasonable costs, ducts are often built of rubber-lined steel, elastomeric-lined steel, PVC, elastomeric-lined plywood, or concrete. All too often duct systems are then subjected to further deposition of organic materials, sulfur, dirt, grease, etc., so that even if the duct were constructed solely of reinforced concrete, a fire hazard of combustible products would still be considerable.

In addition, the environment of an organic material fed by high-velocity air streams of 1,000 to 3,000 fpm provides ideal conditions for rapid flame propagation and extension along the interior of the duct surface. Duct fires are generally of brief duration, during which time considerable damage may ensue. Early and aggressive action is thus one of the paramount steps in successfully combating a duct fire, once it has begun. *For this reason sprinkler systems provide a constant defense and, when combined with a good self-extinguishing resin system, constitute probably as good fire protection as present-day engineering can devise.*

In addition to the usual fire-protection devices in any large duct installation, consider providing a quick-access arrangement of about 18 to 24 in. in diameter every 100 ft (see Fig. 11-16). Do not make this a normal bolt-up manhole, or its accessibility purpose will be defeated. Should a fire occur, hose streams with fog nozzles may be a good method of combating the fire. Carbon dioxide is also a very effective extinguishing agent for small duct fires.

Special considerations are necessary to combat static-electricity problems. A full discussion of the methods and practices involved in duct systems subjected to static charges will be found in Chap. 13.

Duct fires—experimental studies   Extensive studies of simulated duct fires have been made by the Ceilcote Company, Berea, Ohio. Some of the observations resulting from these tests should be understood by the design engineer.[8]

Briefly, this instrumented test setup provided a measured means of studying duct fires in terms of methods of extinguishing, fire-retardant compounds, length of time to snuff out, and duct rigidity under actual burning conditions.

The tests involved used a propane torch to set a duct section 10 ft long on fire with an air velocity of approximately 200 fpm. When the duct caught fire, during a preset ignition period the propane torch was extinguished after 90 sec and air velocity increased to approximately 2,000 fpm, which is normal duct design. The time required for complete snuff-out to occur was measured, starting with the propane fuel shut off at zero. Some of the important conclusions observed from these tests were as follows:

1. With the duct on fire, internal duct temperatures of 1700 to 2500° were achieved. Remember, autoignition temperature for polyester resins is in the order of 800°F.

2. A good fire-retardant resin will produce consistent flameout in 9 to 16 sec from the time the source of the fire is removed.

3. CO_2, used as a fire-extinguishing medium on systems which continue to burn, promptly acts to extinguish the fire.

4. Tests on general-purpose resins showed the danger of using them for ductwork construction. Complete burn-through of the duct resulted. With a good fire-retardant resin, the fire was contained within the duct, and at the end of the test the duct was still structurally sound.

5. The use of at least 5 percent antimony trioxide (Sb_2O_3) as a snuffing compound is essential for maximum fire retardancy. The antimony trioxide, however, must be coupled with a chlorine or bromine

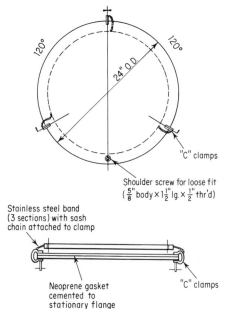

FIG. 11-16 Detail of quick-release access door in large-duct systems.

atom from the polyester resin. This reacts to produce antimony oxychloride (SbOCl) in the case of a chlorinated polyester. The visible evidence of this is that the combustion products from the burning duct change in color from black to white. This snuffing compound is its own built-in fire extinguisher. The engineer should be warned that the general-purpose and isophthalic resins, in addition to many of the high-performance chemical-resistant bisphenols and hydrogenated bisphenol resins, do not contain this all-important chlorine or bromine atom. Fire-spread ratings of polyester resins lacking the chlorine or brominated addition are relatively high, being higher than red oak in combustibility (flame-spread rating of 100). The addition of antimony trioxide to any of these resins lacking the chlorine or bromine atom is not a means of providing fire retardancy.

6. Because a duct material is rated as nonburning, such as PVC, the engineer should not delude himself into thinking that the possibilities of duct fires have been eliminated. Many duct fires are fed by ignited deposits in the duct. The deposit itself may be extremely combustible. In the case of PVC duct systems, with their low melting point, collapse of the duct systems occurs early in the fire, and the melted material, along with the ignited deposits, may drop into the plant or room and become the source of a general conflagration. Plastisol duct systems may spew flaming jelly into an area. The containment of duct fires within the duct is thus of paramount importance.

7. Some of the polyester duct fires have resulted in extensive amounts of smoke damage since the compound burns with a voluminous amount of dense black smoke. When used in clean-room areas, smoke damage alone is something to be reckoned with.

While the tunnel test* is the standard reference used for determining fire-spread ratings of materials in ductwork construction, a word of advice is in order. The tunnel needs to be calibrated frequently to make sure that the results achieved reflect a true rating. For example, on large orders of ductwork material purchasers have insisted on tunnel tests being performed on samples of the materials to be furnished. If a tunnel test is desired, make sure that the tunnel has been recently calibrated. For example, material fire-spread ratings of 30 to 40 have been obtained solely because the tunnel was out of calibration. The same materials with a calibrated tunnel tested 17 to 25. Off-standard results may mean material manufactured to an improper specification, but it may also mean a tunnel in need of calibration.

---

* ASTM E84-61, Standard Method of Test for Surface Burning Characteristics of Building Materials.

FLAME–SPREAD CLASSIFICATION[*,9] (To be Used in Conjunction with Table 11-10)

Table 11-9

Flame spread	Classification
0– 25	Noncombustible
25– 50	Fire-retardant
50– 75	Slow-burning
75–200	Combustible
Over 200	Highly combustible

* Based upon common tests as outlined by the Underwriters' Laboratories.

FLAME–SPREAD RATING TABLE—REINFORCED POLYESTER RESINS[10] (With Other Reference Materials)

Table 11-10

Item no.	Trade name	Generic type	Flame-spread rating*	Hardness
1	....	Asbestos	0	
2	Hetron 72, 92, or 92TG with 5% Sb₂O₃ added..........	Chlorinated polyester	20–25	50
3	Furane....................	Furane	20–35	
4	Atlac 382 FRB with 5% Sb₂O₃ added.............	....	25–35	
5	Hetron 72, 92, or 92TG......	Chlorinated polyester	50–75	50
6	Red oak.....................	....	100	
7	Masonite....................	....	130	
8	General-purpose, isophthalic, bisphenol, or hydrogenated bisphenol with 5% Sb₂O₃ added.....................	....	150–200	38
9	Plywood.....................	....	200	

NOTES: In general, the chlorinated polyesters with 5% antimony trioxide added are excellent fire-retardant resins for ductwork specifications.

Furane has good solvent resistance and may be used as a liner with a reinforced polyester overlay. Furane liners usually add about 10 percent to the cost of a duct system.

* ASTM E84-61 tunnel test.

Recently one of the major resin suppliers has introduced a new fire-retardant polyester resin Atlac 711-05A (a halogenated polyester) which reportedly possesses a flame-spread rating of 15 when compounded with 5 percent Sb₂O₃. Initial reports on this resin indicate a high degree of corrosion resistance comparable to the bisphenols. Major field testing of this resin is in progress.

11.11 FLAME AND BURNING THEORY — CHLORINATED POLYESTERS

The four primary methods for fighting any fire are:

1. Removal of fuel
2. Reduction of heat
3. Reduction of air
4. Inhibition of flame chain reactions

The addition of chlorine or bromine to the polyester molecule has the important advantage of lowering the combustion enthalpy.[11]   This lowers flame temperatures and markedly slows down flame speeds.   Heat losses from the flame are much higher than usual, so that rapid cooling of the combustion products follows as they leave the reaction zone.

In order to maintain long-term fire and chemical resistance, the chlorine must be chemically bonded into the resin molecule.   Some claims indicate that at least 24 percent chlorine is required for adequate fire-resistant performance.[12]

Various snuffing compounds have been tried to lower fire-spread susceptibility.   Many of them, however, tend to be lost from the resin with aging, and the fire resistance drops with time.   The addition, however, of 3 to 5 percent (preferably 5 percent) antimony trioxide (Sb_2O_3) reacts with the chlorine in the resin to form an antimony oxychloride (SbOCl), a most effective snuffing compound, providing excellent long-term fire resistance.

The evaluation of flame-retardant synergists in a laminate shows the antimony trioxide to be superior to most of the other snuffing compounds tested.   In general, flame-retardancy synergists are made from phosphorous or antimony compounds.   Most of the data published by researchers in this area indicate the advantage of the antimony trioxide, provided there is no objection to the opacity produced.   In the event there is, then Phosgard* C-22-R may be used.   Exposure to corrosive solutions generally results in an increase in the self-extinguishing time.[13]

Again, however, the antimony trioxide resin laminates appear to lose less of the self-extinguishing properties than the phosphate compounds.

The use of snuffing compounds in the resin lowers the chemical resistance of the resin, and this must be considered.   However, in most cases (but not all) the corrosion problems in ductwork are of less intensity than in the parent solution.   For this reason, ductwork may not require

* Registered trademark of Monsanto Company.

the ultimate in corrosion resistance.   Generally, it can be stated that the chemical-grade chlorinated polyesters with 5 percent antimony trioxide added will meet the corrosion-resistant standards typical for the polyester field.

Flame temperatures found are generally less than those which have been calculated, probably due to heat losses through radiation.[14]

The engineer may wish to conduct his own quantitative burning tests for various types of duct materials.   The advantage of the chlorinated polyester molecule to which 5 percent antimony trioxide has been added can be remarkably demonstrated.

11.12   FIRE PROTECTION IN RP
VAPOR–HANDLING STRUCTURES

It is good engineering practice to protect all fiber-glass-reinforced fire-retardant-type polyester vapor-handling structures by sprinklers, either in the structures or overhead in the area.   This is commonly worked out with the insurance company, and the benefit of their advice obtained. It is a good practice to submit drawings of the proposed sprinkler system to the insurance carrier when the engineering is being done.   In each case the dollar value of the duct system and related damage which might be caused needs to be weighed against the cost of the sprinkler system and the protection provided.

It is extremely important to protect exhaust-fan inlets and major duct systems.

A suggested sprinkler spacing would be as follows:

Duct size, in.	Sprinkler spacing, ft
Up to 36	25
42 to 60	20
Above 60	15

Fires in reinforced plastic ductwork have generally been attributed to:

1. A combustible buildup in the duct
2. Ignition from an outside source
3. Improper ductwork specifications

Refrain from specifying non-fire-retardant resins for ductwork, and always specify a resin to which the snuffing compound Sb_2O_3 has been added. Do not be tempted by the false economies of a general-purpose polyester duct system.

The great attraction of reinforced polyester duct systems is their resistance to corrosion and substantial freedom of maintenance over long periods of time. The sprinkler head, of course, must also be protected, and it is common to provide wax-coated heads, stainless-steel heads, or heads made of an alloy resistant to the chemicals. In addition, a unique design of a sprinkler for polyester ductwork has been developed in which a conventional sprinkler head is completely protected from a corrosive atmosphere by the use of a pyroxylin cover.[15] Figure 11-17 shows a typical installation for such a cover. This type of protection has been approved by some fire-insurance companies.

11.13 BUTTERFLY VALVES AND DAMPERS

In the balancing of complicated duct systems, the flow of air to a specific area becomes a must. One of the most convenient ways of controlling this flow is by means of dampers or butterfly valves. Table 11-11 shows the approximate price range of duct dampers versus duct size. Figure 11-18 shows a large butterfly valve constructed almost entirely of reinforced polyester.

More simple designs are also possible, with the damper built into a short section of straight duct, which is then butt-strapped into the system. A nominal clearance of $\frac{1}{16}$ in. all around the butterfly is maintained, so that these units normally do not provide tight shutoff.

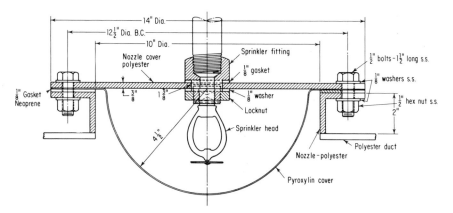

FIG. 11-17 Sprinkler assembly (full-size) with protective covering.

11.14 DUCTWORK COST ESTIMATING

Table 11-11 COST ESTIMATING PRICE RANGES—STANDARD DUCTWORK AND FITTINGS

Size, in.	Duct-wall thickness, in.	Approximate round-duct cost per foot	Butt-joint, factory	Butt-joint, field	90° elbow	Inter-section joints, factory	Single flange, undrilled	Single-flange drilling	Duct dampers	Duct blast gates	Duct stub ends
2	0.125	$ 2.00–$ 3.00	$ 5.00–$ 8.00		$ 10–$ 15	$ 9.00	$11–$16	$ 3			
3	0.125	2.25– 3.25	5.25– 8.00			9.35	12– 18	3			
4	0.125	2.50– 3.50	5.50– 8.00		10– 17	9.75	15– 20	3			$15–$20
6	0.125	3.25– 5.00	6.00– 9.00	$ 6.50–$ 9.50	15– 25	10.50	17– 22	6	$ 25–$ 35	$ 55–$ 67	17– 22
8	0.125	3.50– 5.50	6.50– 10.00	7.00– 10.50	17.50– 30	12.00	19– 24	6	28– 38	57– 70	19– 24
10	0.125	4.25– 6.25	7.50– 11.50	8.00– 12.00	22– 37	13.00	20– 26	8	32– 42	60– 75	21– 26
12	0.125	4.75– 7.00	8.25– 13.00	9.50– 13.50	28– 45	15.50	22– 27	8	33– 43	65– 80	23– 27
14	0.125	5.50– 8.25	9.25– 14.50	11.50– 14.00	37– 57	17.50	23– 29	8	37– 47	70– 85	26– 32
16	0.125	6.00– 9.50	10.00– 16.00	13.00– 16.00	48– 60	20.00	24– 30	12	40– 50	75– 90	28– 34
18	0.125	7.00– 10.50	12.00– 17.00	14.50– 18.00	55– 75	21.50	25– 32	12	43– 55	80– 100	30– 36
20	0.125	7.75– 14.00	12.50– 19.00	16.00– 20.00	75– 100	24.00	26– 34	15	47– 60	90– 110	32– 38
24	0.187	9.75– 15.00	15.50– 23.00	19.50– 24.00	90– 135	29.00	28– 37	15	55– 70	100– 130	
30	0.187	15.00– 22.00	19.00– 27.00	23.50– 29.00	130– 190	34.00	30– 41	22	68– 85	130– 160	
36	0.187	17.50– 27.25	23.00– 31.00	29.00– 36.00	175– 290	38.00	32– 46	24	83– 104	155– 200	
42	0.250	27.50– 36.00	26.00– 36.00	33.50– 41.00		42.00	36– 50	27	95– 120	175– 220	
48	0.250	32.00– 45.00	30.00– 40.00	38.50– 48.00		46.00	40– 56	33	110– 130	190– 240	
54	0.250	......	......	......		......	......	33	......		
60	0.250	......	......	......		......	......	39			

NOTE 1. The above approximate costs are based on reinforced polyester ductwork made up in accordance with the current Commercial Standard (TS-122C, dated Sept. 18, 1968). Costs have been obtained from a variety of sources and a good cross section of vendors.

2. Butt joints are assumed to be of the same thickness as the duct wall.

3. 90° elbows are plain-end.

4. Flange drilling would be in accordance with the Commercial Standard.

5. A duct stub end is an undrilled flange integral with a short section of duct.

6. Duct laminate is assumed to be made up of a high-grade chemical-resistant resin with 5 percent antimony trioxide added for maximum fire retardancy and a fire-spread rating of 25 or less.

7. Some comments are applicable to these data to explain why costs may vary over a considerable range in the industry. Normally these are (a) area labor costs, (b) price factors relating to yearly volume purchases, (c) size of a purchase order, (d) competitive bidding position of the vendor, and (e) normal supply-and-demand influences.

FIG. 11-18 A large butterfly valve constructed almost entirely of reinforced polyester, used to control the flow of air to a specific area in a complicated duct system.   (Photo, courtesy of Leep Zelone Photographers, of an AnCor Plastics installation at Hooker Chemical Corp., Durez Div., Niagara Falls, N.Y.)

REFERENCES

[1] Recommended Product Standard for Custom Contact Molded Reinforced Polyester Chemical Resistant Process Equipment, TS-122C, Sept. 18, 1968.

[2] Sturm, R. G.: A Study of the Collapsing Pressure of Thin Walled Cylinders, *Engineering Experimental Station Bulletin*, ser. 329, Nov. 11, 1941, vol. 39, no. 12, p. 24, University of Illinois, Urbana, Ill.

[3] *Ibid.*, p. 70.

[4] Duct Systems, Amercoat-duVerre Products, *Bulletin* R-6/66, Amercoat Corp., Brea, Calif., 1966.

[5] Hanszen, E. W.: A Discussion of Those Portions of the Standard Relating to Reinforced Polyester Round and Rectangular Duct Design, presented at 23d Annual Technical and Management Conference, February, 1968, Society of the Plastics Industry, Reinforced Plastics Div., Washington, D.C.   (Reprinted with permission.)

[6] Private Correspondence, A. K. Shadduck, E. I. du Pont de Nemours & Company, Engineering Department, Wilmington, Del., 1968.

[7] Silva, P. A., Beetle Plastics Corp., Fall River, Mass., personal communication, Jan. 23, 1964.

[8] McMahon, R.: Paper presented at NACE Conference on Reinforced Plastics, Niagara Falls, N.Y., Aug. 8, 1967.

[9] Tunnel Test, ASTM E84-61.

[10] Flame Spread Rating Table drawn from a variety of sources.

[11] Krischer, B.: The Rate of Flame Propagation and the Composition of Mechanism of Chlorinated Hydrocarbons, *Chemic-Ingenieur Technik*, vol. 35, pp. 856–886, 1963.

[12] O'Leesky, Sanmel: Handbook of Reinforced Plastics, p. 460, Reinhold Publishing Corporation, New York, 1964.

[13] Chae, Y. C., et al.: Chemical Resistance and Flame Retardant Polyesters, SPI 22d Annual Technical Conference, Washington, D.C., 1967.

[14] Little, J. P., Factory Mutual Insurance Co., Providence, R.I., personal communication, Dec. 15, 1966.

[15] Approved Design for Sprinkler Installation in Polyester Ductwork Conveying Corrosive Gases, FMC Corporation, American Viscose Div., Philadelphia, Pa., 1964.

12

Storage Tanks

12.1 RP TANK DESIGN—GENERAL CONSIDERATIONS

Reinforced plastic tanks have been made in a great many different shapes and sizes. The geometry of construction and the manufacturing methods currently prevalent in the industry dictate certain basic parameters for the designer's consideration. Tank construction, in order of popularity, is as follows:

Cylindrical tanks   The shape and configuration of these tanks is covered by the SPI table on tanks (Table 7 and Table 8 of the TS-122C), for normal construction up to 20,000 gal. Such a tank may be furnished in either hand-laid-up contact-molded RP material or one of the numerous filament-wound methods. Either fabricating process uses a rotating steel mandrel on which has been wrapped a thin Mylar film. The tank is laid up on top of the Mylar film. As the large steel mold rotates past the applicator, glass and resin are applied until the product is finished. Normally, filament-wound RP tanks cost less than hand-laid-up RP tanks. The reason for this is that labor costs on a filament-wound RP tank are considerably less than on hand-laid-up tanks. Higher mechanical properties are developed with the filament-wound RP tank, so that in filament winding we may obtain conservative-design hoop tensiles of 40,000 psi (and in many cases much higher figures than this). Compare this with hand-laid-up tensiles in the order of 15,000 psi. The end product is a thinner wall with the filament-wound RP material.

The total consideration, however, is not covered so simply and easily. Better corrosion resistance in the hand-laid-up area more than offsets the physical property differences. Some filament winders use a loose filament wind to obtain the maximum combination of strength and chemical resistance. In addition, each manufacturer furnishes the inside of the tank with a heavy gel coat in which is embedded a C glass for corrosion protection. An excellent lay-up for a tank in chemical service consists of an inner surfacing veil, followed by one or more layers of 1½-oz chopped-strand glass mat, with a filament wind to provide the necessary structural strength. On the outer surface, again, a layer of 1½-oz chopped-strand mat, followed by a chemical-glass overlay with a hot coat, provides excellent wall construction.

Due to its ease of construction, the cylindrical tank per se is by far the most preferred tank in use for holding process vessels. In the storage of process liquids more tanks of this type are built than all others combined. The cylindrical tank will generally provide the greatest amount of storage per dollar spent.

Minimum cylindrical tank costs are achieved by using a flat bottom, graduated wall heights where necessary, and some form of a dished or lidded top. Tanks of this type generally require a continuous bottom support, although this can be modified by increasing the thickness of the tank bottom so that a grid support on multiple beams can be used.

Generally, tanks of this type are used for the static storage of liquids, although with extra consideration they are finding an increasing use in the field of process vessels, where liquids may be heated, crystallized, agitated, or filtered. The designer who is considering RP tanks for the static storage of process liquids is advised to give the cylindrical tank his first consideration.

Figure 12-1 shows 11 different shapes of tanks currently being built of reinforced plastics.

As the engineer's knowledge of cylindrical-tank construction grows, certain aspects of the economy of tank construction will engage his interest. One of the inescapable conclusions is tank standardization. This is then followed by the corollary of building tanks by automated procedures. Nearly every vendor subscribes to standardization, but modifies it to suit the demands of the customer. In hand-laid-up tank fabrication each tank is thus custom-made on standard-sized mandrels. Even the so-called standard-sized mandrels can be interpolated through slick techniques of mandrel expansion, so that for a very low cost, the custom fabricator can provide RP tanks to suit the preferences of the vast majority of process designers, locating nozzles, vents, overflows, and drains to suit each particular design with very little increase in cost. About the only limit on shop-fabricated construction is the size which can be shipped, that is, some 12 ft in diameter by 24 ft high for a single tank. By special techniques one vendor can do 16 ft diameter by 30 ft high for a single tank when built in sections. This height can be exceeded by wrapping on additional sections in the field for greater storage as desired, or at his option, the engineer may choose winding a tank in place, a technique discussed elsewhere in this chapter. Winding tanks in place permits the possibility of advancing to gigantic-sized vessels.

Automation of tank manufacture, while somewhat limited in the contact-molded field, finds much easier application in the filament-wound line. Here programmed systems automatically apply glass and resin in properly premixed quantities to the face of a rotating mandrel to provide a truly standardized tank. Stiffeners are cleverly designed, of wood and cardboard, which wrap onto the tank in specific intervals and serve as a form for holding additional filament-wound material, to provide the true stiffening necessary.

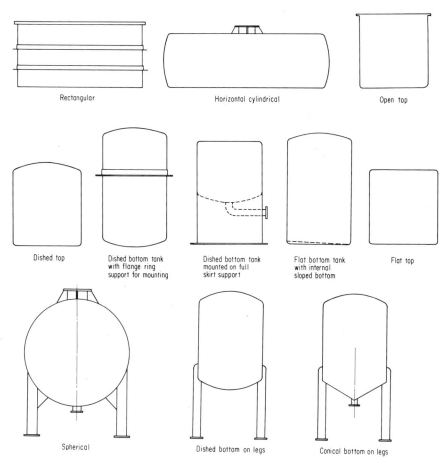

Rectangular — Horizontal cylindrical — Open top

Dished top — Dished bottom tank with flange ring support for mounting — Dished bottom tank mounted on full skirt support — Flat bottom tank with internal sloped bottom — Flat top

Spherical — Dished bottom on legs — Conical bottom on legs

FIG. 12-1 Eleven shapes of tanks constructed of reinforced plastics.

Flat bottoms or domed heads can be built rapidly by two men, one of whom sprays a chopped-glass–resin mix onto a rotating cap, while the second man rolls it out. These caps are subsequently wrapped into the tank sidewalls.

There are many modifications to this procedure and as many different lay-ups as there are vendors. The custom molder will generally build an extremely stout dome or bottom, interlaying chopped-strand mat and woven roving to provide maximum physical and chemical resistance. Great care must be taken in the design of a tank bottom if the vessel is alternately filling and emptying. Liquid fed into the top of a high tank onto an empty floor can result in considerable buffeting and the containment of a great deal of energy. It is sometimes difficult to provide absolutely flat bottoms in large design tanks. If possible, lead the liquid gently into the tank, unless the tank is being operated with several feet of liquid in it at all times.

The same principles that assure cost economy in the manufacture of automobiles, radio sets, or salt shakers govern the economics of RP tanks. Long production runs, one-shot engineering, maximum use of productive facilities, lowest labor cost per tank built, portion control of resin and glass, and standardized handling facilities, all point the way to the lowest-cost tank construction. When these automated facilities are operated at their maximum capacity, the lowest unit costs are achieved. Today a 6,000-gal tank can be built by automated procedures, using an isophthalic resin for $1,000 to $1,200. The secret of lowest-cost tank procurement is standardization and the ability to use those tanks built by highly automated procedures. Unfortunately, at the present time, the selection of tanks available through these highly automated methods is relatively small, covering nominal capacities of 1,000 to 12,000 gal in about nine sizes. Tanks of 6,000-gal capacity and high-chemical-performance resin, using a modified filament-wound process, can be purchased in the $1,800-to-$2,200 range, while custom-fabricated tanks built to a detailed design cost about $2,700. Bidding from vendor to vendor varies considerably, depending upon many of the other common economic factors that produce variations in bidding. For detailed suggestions on how the designer may obtain the most process vessel for the money, consult Chap. 15.

Rectangular tanks   While the designer will probably purchase and use 10 cylindrical tanks for each rectangular tank, there are certain cases where the problem parameters can be solved only by the use of a rectangular tank. Normally, this arises where maximum volume must be obtained.

The design of a rectangular tank is much more difficult than the design of a cylindrical tank. The engineer must check his design for wall stress, deflection in the wall, and possible failure at the corners, and he must provide any but the smallest rectangular tanks with both horizontal and vertical stiffening. Tank-top problems are also more difficult with rectangular tanks. All these items add up to the conclusion that a rectangular tank is a premium tank which costs more to do the job because the job is more difficult to do. For a typical rectangular-tank design showing stiffener spacing, both horizontal and vertical, see Figs. 12-2 and 12-3.

Again, rectangular tanks must be provided with continuous bottom support. They may be built in sections and assembled on the site, where space requirements make such an assembly necessary.

Sometimes tanks are subjected to physical abuse little dreamed of by the original designer. One rectangular reinforced polyester tank stood the buffeting of continually emptying and filling from a dilute hot sulfuric acid solution for 6 years, without any sign of trouble—then one morning a gas explosion in the tank ruptured several of the upright and bottom seams where the sides joined each other and the base joined the sides. The tank did not burn. Its repair was accomplished by temporarily bypassing the liquid and laying in new interior and exterior straps. It is a matter of conjecture, but had this been a round tank (space did not permit this), it is considered engineering judgment that the tank would have withstood the explosion with little damage.

Spherical tanks   Geometrical considerations indicate that the largest volume in relation to surface area can be contained in a shape whose surface is a sphere. With this fascinating bit of geometry, and with the knowledge that the material ingredients of RP tanks represent approximately 50 percent of the total tank costs, it could be reasoned that the sphere would be a most attractive shape from a cost standpoint. In addition, since a ball can be filament-wound with relative ease, we should also be achieving minimum labor costs, which normally represent some 20 percent of the cost of the tank. (Overhead is the remaining 30 percent.) All these statements are true, and for a time, in the early 1960s, process storage vessels were being constructed in the shape of a sphere. Figure 12-4 is a picture of one of these which has been in operation for some five years, in hot dilute sulfuric acid service.

Now, unfortunately, other factors than pure geometry or process economics seem to govern in the marketplace. The sphere was not a popular shape for a tank. Engineers preferred the cylindrical tank, even though it cost slightly more for the same service conditions. The ball as a tank shape in RP thus died out. It is still of interest, how-

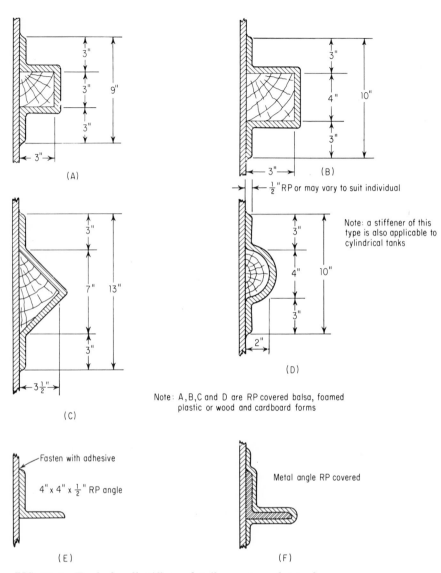

FIG. 12-2 Typical wall-stiffener details—rectangular tanks.

ever, to examine the theoretical aspects of using a ball shape for process pressure vessels. Here the design puts all the tank fibers in tension. It also has the distinct advantage of minimizing evaporation losses where such a feature is necessary. Where pressure storage is required, the engineer would do well to consider the ball shape, which, coupled with

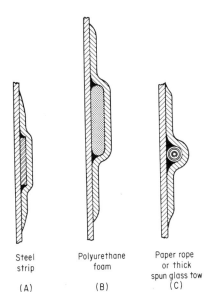

Steel
strip

(A)

Polyurethane
foam

(B)

Paper rope
or thick
spun glass tow
(C)

FIG. 12-3 Typical wall-stiffener details—cylindrical tanks.

FIG. 12-4 Acid wash tank, 2,500-gal capacity.

filament winding, might provide a useful solution to some difficult problems.

At the present time standards have not been established covering filament-wound fiber-glass-reinforced tanks. A tentative proposed standard has been prepared by Mr. H. D. Boggs of the Amercoat Corporation, Brea, Calif. This is reproduced in Appendix C. The reader may find it useful for guidelines in an area that is not too well defined at this writing.

12.2 INTRODUCTION TO PRODUCT STANDARD TS–122C[1]

Any textbook dealing with reinforced plastic tanks would be remiss if it did not include the appropriate section on Tanks—Stationary Non-pressure Vessels, as set forth in TS-122C, Recommended Product Standard for Custom Contact Molded Reinforced Polyester Chemical Resistant Process Equipment, as prepared by the Society of the Plastics Industry. The section dealing with tank standards consists of some four pages. Intimate and detailed knowledge of this section of the Standard is extremely important to the design engineer because today this is an industry authority. Often the designer will simply specify his tank to be built in accordance with this Standard. As the designer gains knowledge in the design and construction of RP tanks, however, he soon recognizes the limitations of the Standard. Only by the broadest interpretation can it be applied to filament-wound tanks with an advanced design beyond simple hand lay-up. A designer will also learn that the Standard is rather loosely and widely interpreted by the fabricators, so that, for example, three fabricators, equally competent, quoting on a tank to SPI specifications, will quote widely divergent prices. Notwithstanding these drawbacks, however, the Standard is a good beginning for a young industry, and will doubtless be amplified in future years.

Following the Product Standard, a large part of this chapter is intended to provide the engineer and designer with concrete suggestions for handling each phase of the design of RP tanks and structures and to furnish him with information on cost, items that influence cost, design techniques for long life, supporting, and nozzles, and on all the practical details that go into making an installation which will operate satisfactorily over a long period of time. It is important that the designer be familiar with things to avoid. While the Standard is positive in defining in a general manner procedures to be followed, it does not define what not to do. It is easy for the engineer to stumble into gray areas. This text seeks to reduce these pitfalls and provide a practical guide for successful installations.

AUTHOR'S NOTE: The following section, covering both cylindrical and rectangular tanks, is an excerpt from:

RECOMMENDED PRODUCT STANDARD FOR

CUSTOM CONTACT-MOLDED REINFORCED POLYESTER

CHEMICAL-RESISTANT PROCESS EQUIPMENT

TS 122C

September 18, 1968

3.6 *Reinforced-polyester tanks (stationary nonpressure vessels)*

3.6.1 *Cylindrical flat bottom vertical tanks*

3.6.1.1 *Sizes*—Standard tank sizes are 2, 2½, 3, 3½, 4, 4½, 5, 5½, 6, 7, 8, 9, 10, 11, and 12 feet in inside diameter.

3.6.1.2 *Dimensions and tolerances*—The tank diameter shall be measured internally. Tolerance on nominal diameter, including out-of-roundness, shall be ±1 percent. Measurement shall be taken with tank in vertical position. Taper, if any, shall be increasing and shall be added to the nominal diameter. Taper shall not exceed ½° per side. Tolerance on overall height shall be ±½ percent, but shall not exceed ±½ inch. The radius at bottom to wall shall be a minimum of 1½ inch.

3.6.1.3 *Wall thickness*—The minimum wall thickness shall be in accordance with Table 7. See also 3.3.6.

3.6.2 *Horizontal cylindrical tanks*

3.6.2.1 *Sizes, dimensions, and tolerances*—These shall be the same as for vertical cylindrical tanks (see 3.6.1). Standard end closures shall be standard convexed, domed heads with a maximum radius of curvature equal to the tank diameter. The knuckle radius shall be a minimum of 1½ inch.*

3.6.2.2 *Support cradle*—Two support cradles shall be provided. The cradles shall be at least 6 inches wide supporting at least 120° of the tank circumference. Wear plates (reinforced areas) 12 inches wide covering 180° of support surface shall be provided when required: laminate construction and minimum thickness shall be as agreed upon between fabricator and purchaser. Tanks longer than 24 feet require special design and support consideration.

3.6.2.3 *Wall thickness*—The minimum wall thickness shall be in accordance with Table 8. See also 3.3.6.

* Larger knuckle radii are commonly used, such as for ASME torispherical heads.

MINIMUM WALL AND BOTTOM THICKNESS OF VERTICAL TANKS RELATIVE TO DIAMETER AND DISTANCE FROM TOP[1]

Table 7

Minimum wall and bottom thickness for tanks of diameter:

Distance from top, feet	2 ft	2½ ft	3 ft	3½ ft	4 ft	4½ ft	5 ft	5½ ft	6 ft	7 ft	8 ft	9 ft	10 ft	11 ft	12 ft
2	3/16	3/16	3/16	3/16	3/16	3/16	3/16	3/16	3/16	3/16	3/16	3/16	3/16	3/16	3/16
4	3/16	3/16	3/16	3/16	3/16	3/16	3/16	3/16	3/16	3/16	3/16	3/16	3/16	3/16	3/16
6	3/16	3/16	3/16	3/16	3/16	3/16	3/16	3/16	3/16	3/16	3/16	3/16	1/4	1/4	1/4
8	3/16	3/16	3/16	3/16	3/16	3/16	3/16	3/16	3/16	1/4	1/4	1/4	1/4	1/4	5/16
10	3/16	3/16	3/16	3/16	3/16	3/16	3/16	1/4	1/4	1/4	1/4	1/4	5/16	5/16	5/16
12	3/16	3/16	3/16	3/16	3/16	1/4	1/4	1/4	1/4	1/4	5/16	5/16	5/16	5/16	5/16
14	3/16	3/16	3/16	1/4	1/4	1/4	1/4	1/4	1/4	5/16	5/16	5/16	5/16	3/8	3/8
16	3/16	3/16	3/16	1/4	1/4	1/4	1/4	1/4	1/4	5/16	5/16	3/8	3/8	3/8	3/8
18	3/16	3/16	3/16	1/4	1/4	1/4	1/4	5/16	5/16	3/8	3/8	3/8	3/8	7/16	7/16
20	3/16	3/16	1/4	1/4	1/4	1/4	5/16	5/16	5/16	3/8	3/8	7/16	7/16	1/2	1/2
22	3/16	1/4	1/4	1/4	1/4	5/16	5/16	5/16	5/16	3/8	3/8	1/2	1/2	1/2	1/2
24	3/16	1/4	1/4	1/4	1/4	5/16	5/16	5/16	3/8	3/8	7/16	1/2	1/2	9/16	9/16

[1]Based on a safety factor of 10 to 1, using the mechanical property data in Table 1 and a liquid specific gravity of 1.2. For tanks intended for service above 180°F (82.2°C) consideration in design should be given to the physical properties of the material at the operating temperature. Tanks with physical loadings, such as agitation, should be given special design consideration.

MINIMUM WALL AND HEAD THICKNESSES FOR REINFORCED–POLYESTER HORIZONTAL CYLINDRICAL TANKS USING TWO SUPPORT CRADLES[1]

Table 8

	Minimum wall and head thickness for tanks of diameter:[2]							
Tank length, ft	2 ft, in.	3 ft, in.	4 ft, in.	5 ft,[3] in.	6 ft,[4] in.	8 ft,[5] in.	10 ft,[6] in.	12 ft,[7] in.
8	$3/16$	$3/16$	$1/4$	$1/4$	$5/16$	$5/16$	$7/16$	$9/16$
10	$3/16$	$1/4$	$1/4$	$5/16$	$5/16$	$3/8$	$7/16$	$9/16$
12	$3/16$	$1/4$	$1/4$	$5/16$	$5/16$	$7/16$	$1/2$	$5/8$
14	$1/4$	$1/4$	$5/16$	$5/16$	$3/8$	$1/2$	$9/16$	$3/4$
16	$1/4$	$5/16$	$5/16$	$3/8$	$3/8$	$9/16$	$11/16$	$13/16$
18	$1/4$	$5/16$	$3/8$	$7/16$	$7/16$	$5/8$	$13/16$	$15/16$
20	$5/16$	$5/16$	$3/8$	$7/16$	$1/2$	$11/16$	$7/8$	$1\,1/16$
22	$5/16$	$3/8$	$3/8$	$1/2$	$9/16$	$3/4$	$15/16$	$1\,3/16$
24	$5/16$	$3/8$	$7/16$	$1/2$	$5/8$	$1\,3/16$	1	$1\,1/4$

[1] Based on 5 to 1 safety factor, using the mechanical property data in Table 1, a liquid specific gravity of 1.2, and support cradles located $1/2$ of tank length from each end. For tanks intended for service above 180°F (82.2°C) consideration in design should be given to the physical properties of the material at the operating temperature. Tanks with physical loadings (such as agitation), other support designs, stiffening rings or for use in situations requiring higher safety factors, should be given special design consideration. In the use of more than two support cradles, maintenance of uniform support of the tank at all points of support is essential.

[2] For intermediate standard tank inside diameters given in 3.6.1.1, the minimum wall and head thickness shall be that given in this table for the next higher diameter.

[3] Wear plates required for 8 foot tank length.

[4] Wear plates required for 8, 10, and 12 foot tank lengths.

[5] Wear plates required for tanks 8 to 18 feet long, inclusive.

[6] Wear plates required for tanks 8 to 20 feet long, inclusive.

[7] Wear plates required for all tank lengths.

3.6.3   *Rectangular tanks*

3.6.3.1   *Sizes*—There are no standard sizes for rectangular tanks.

3.6.3.2   *Dimensions and tolerances*—The length and width shall be measured internally. Tolerances on nominal dimensions of length and width shall be $\pm 1/4$ in. or $\pm 1/4$ percent, whichever is greater. Overall height tolerance shall be $\pm 3/8$ inch. Taper is increasing and added to the nominal dimensions. Taper should not exceed $1/2°$ per side.

3.6.3.3 *Side wall*—Deflection shall not exceed ½ percent of span at any location when tested by filling with water.

3.6.3.4 *Wall thickness*—Since the design of rectangular tanks is considerably more complex than that of cylindrical tanks, no simple chart of wall thickness can be given. However, the minimum wall should be similar to that for cylindrical tanks with consideration given to the height of the tank relative to loadings and the largest span relative to deflection. External ribs shall be used to prevent side wall deflection exceeding the tolerance in 3.6.3.3. See also 3.3.6.

3.6.4 *Mechanical property requirements for tanks*—The minimum mechanical properties shall be as specified in Table 1.

3.6.5 *Shell joints*—Where tanks are manufactured in sections and joined by use of a laminate bond, the joint shall be glass fiber reinforced resin at least the thickness of the heaviest section being joined. The reinforcement shall extend on each side of the joint a sufficient distance to make the joint at least as strong as the tank wall and shall be not less than the minimum joint widths specified in Table 9. The reinforcement shall be applied both inside and out with the inner reinforcement considered as a corrosion resistant barrier only and not structural material. The inner reinforcement shall consist of a minimum of three ounces of glass per square foot, followed by a 0.010 inch to 0.020 inch of surfacing material (see 3.3.5).

3.6.6 *Flanges*

3.6.6.1 *Flanged nozzles*—Flanges for liquid inlets and outlets shall meet the same requirements as for pipe (see 3.5.7 to 3.5.7.3 inclusive). At assembly there shall be a minimum dimension of 4 inches from the flange face to the tank. Where angular loadings are anticipated, the flange nozzle shall be supported by a minimum of three gussets or by other suitable means of structural support.

3.6.6.2 *Assembly of flanges*—Standard orientation will have bolt holes straddling principal centerline of vessel unless otherwise specified.

3.6.6.3 *Tolerances*—Tolerances on flange construction shall be the

MINIMUM TOTAL WIDTHS OF OVERLAYS FOR REINFORCED-POLYESTER TANK SHELL JOINTS

Table 9

Tank wall thickness, inches	$\frac{3}{16}$	$\frac{1}{4}$	$\frac{5}{16}$	$\frac{3}{8}$	$\frac{7}{16}$	$\frac{1}{2}$	$\frac{9}{16}$	$\frac{5}{8}$	$1\frac{1}{16}$	$\frac{3}{4}$
Minimum of outside overlay width, inches..	4	4	5	6	7	8	9	10	11	12
Minimum of inside overlay width, inches..	4	4	5	5	6	6	6	6	6	6

same as for pipe flanges (see 3.5.7 and Table 5). Location of nozzles on the vessel shall be held to $\pm \frac{1}{8}$ inch.

    3.6.7   *Recommended installation practice*

    3.6.7.1   Flat bottom tanks should be supported on a flat surface or on properly-spaced dunnage. It is recommended, where possible, that a flat surface, preferably a reasonably-soft surface (confined sand or cinder-filled pad, plywood-surfaced concrete or a concrete grout) be used. Where full bottom support is not possible, special bottom design is required.

    3.6.7.2   Closed tanks should have a properly sized vent.

12.3   WALL THICKNESS OF RP TANKS

The author would suggest that in any engineering design of an RP tank the designer give second thought to the use of tanks constructed of RP material less than $\frac{1}{4}$ in. in thickness. A large section of Table 7, showing the minimum wall and bottom thickness of vertical tanks relative to the distance from the top, specifies as a suggested minimum design a wall thickness of $\frac{3}{16}$ inches. While there is no question that theoretical design relative to the structural strength of RP material at that thickness will provide suitable basic strength, experience has indicated the desirability of providing additional ruggedness. This can be achieved at the minimum cost by keeping minimum wall thickness $\frac{1}{4}$ in. or above. Going from a $\frac{3}{16}$- to a $\frac{1}{4}$-in. minimum wall will hardly exceed 10 to 15 percent of the cost of the vessel—a small price to pay for satisfactory performance over the long term. Improper hanging of outlet valves, weak nozzle gusseting, buffeting from internal heaters, all require the extra safety provided by the minimum $\frac{1}{4}$-in. thickness. This precept has been followed hundreds of times in the design and fabrication of RP tanks and has well withstood the test of time.

12.4   GRADUATED WALL HEIGHTS FOR ECONOMIC CONSTRUCTION[2]

One of the secrets of providing maximum economies in tank construction is to apply the principle of graduated wall heights. This applies primarily to hand-laid-up tanks, although it may also apply in the larger sizes of filament-wound construction. Except in extremely heavy walled construction—and by this we mean above $\frac{5}{8}$ in. wall—segments of different wall thickness may continue to provide additional economies in wall depths of not less than 36 in. A typical example of a tank 12 ft in diameter by 24 ft high with graduated-wall construction is shown in Fig. 12-5. Such a tank is perfectly capable of storing liquids with temperatures up to

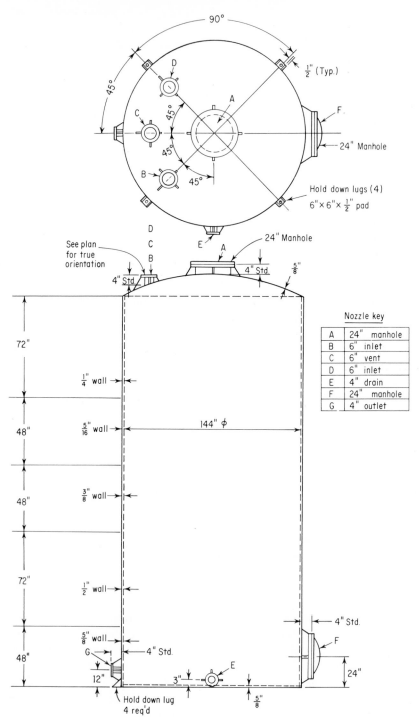

FIG. 12-5 RP tank with a graduated wall, showing important construction details.

200°F and may have a life expectancy of 12 to 15 years. It can be built with savings of at least 30 percent over lead-lined steel and for less than rubber-lined steel. Now it is very easy for the neophyte designer to simply specify a ⅝-in. bottom, a ⅝-in. wall, and a ⅝-in. top. Compare this with the graduated design that has been prepared. The premium the designer pays for not using a graduated wall is in the order of 6 to 35 percent of the total purchase price of the vessel. Whether it is advantageous to provide a graduated wall thickness can speedily be determined by looking at the basic SPI table on vertical tanks. Where only one graduation exists, that is, ³⁄₁₆ and ¼ in., the maximum savings would be in the order of 6 to 8 percent, even on relatively big tanks. This in essence comprises the left one-third of the table. In the center of the table may be found savings ranging from 15 to 25 percent by following the graduated-wall method. On the extreme right of the table, on big tanks, savings range from 18 to 35 percent by going to a graduated-wall design.

These factors are extremely important for the cost-conscious engineer who is seeking to minimize capital costs. As a rule of thumb, the tank top and the bottom should be constructed of material of the same specified thickness, regardless of the graduated wall used. Any tank whose capacity is less than 2,200 gal will gain only marginal savings, if any, in the graduated-wall-thickness method. Normally, the greater the tank depth or tank diameter, the greater the possible savings. In larger tanks of 15,000 to 20,000 gal the possible dollar savings in tank construction, using the graduated-wall method, can be very substantial.

Some concrete examples of actual tank prices for graduated versus nongraduated walls in varying sizes are given in Table 12-1.

12.5 FILAMENT-WOUND TANKS— ON-SITE CONSTRUCTION*,[3]

Generally, the basic maximum size of a tank which can be constructed in a fabricator's shop and shipped to the purchaser as a completely fabricated unit is about 12 ft in diameter by 36 ft high, containing approximately 30,000 gal. Beyond this size the fabricator develops shipping problems and must resort to other methods. This is true regardless of whether we are dealing with reinforced plastic or any other standard material of construction. At the present time a 30,000-gal shop-fabricated tank would be somewhat less expensive than a field-fabricated tank, and even two 30,000-gal tanks would still provide a small economy over a 60,000-gal field fabrication. As we approach the 100,000-gal size, field fabrication

---

* The information in this section was supplied by Mr. Norbert J. Kraus of Justin Enterprises, Inc., Fairfield, Ohio.

SAVINGS WITH A GRADUATED–WALL DESIGN[4]

Table 12-1

Tank capacity, gal	2,500	3,400	3,500	5,000	6,800	10,000	14,000	20,000
Tank size (diameter × height), ft	6 × 12	6 × 16	5 × 24	6 × 24	7 × 24	12 × 12	12 × 17	12 × 24
Cost of graduated-wall construction	$1,078	$1,490	$1,712	$2,160	$2,615	$3,400	$4,883	$6,230
Cost of ungraduated walls (bottom, walls, and top same thickness)	$1,166	$1,578	$1,964	$2,702	$3,094	$3,778	$6,582	$8,238
Extra cost of ungraduated walls	$88	$88	$252	$542	$479	$378	$1,699	$2,008
Percent premium	8%	6%	15%	25%	18%	11%	35%	32%
Cost per gallon	$0.43	$0.44	$0.49	$0.43	$0.39	$0.34	$0.35	$0.31

NOTE: Tank costs vary with tank configuration. The table assumes a round tank constructed, basically, in accordance with SPI specifications, and applies to hand-laid-up cylindrical tanks only. Except in the case of the 2,500-gal tank, all the other tanks used in the table include in their cost a $400 allowance for nozzles and manholes. The 2,500-gal tank has a $200 allowance. The tank prices are complete with tops.

288

then becomes more economical than clustering an equivalent volume of smaller tanks. Undoubtedly, as this field of construction develops further, the economics pointed out here will change and field fabrications of tankage above 30,000 gal will become more attractive.

Today, reinforced plastic tanks as large as 250,000 gal have been constructed for chemical-plant use. One of these, installed in West Virginia, is a 46-ft-dia by 20-ft straight shell. Designs and quotations have been prepared to push this on-site field-fabrication capability to 500,000 gal capacity. Even a 100-ft-dia tank with a capability of storing over 1 million gal is theoretically feasible. The practical limitations of this fascinating end of the business are yet to be seen.

To bridge the gap in sizes and in economics between the completely shop-fabricated tanks and the completely field-fabricated tanks, Justin Enterprises is now shop-fabricating tanks in the 14- to 16-ft-dia range. These sizes permit shipment to most locations by means of depressed-bed flatcars. Fabricated cylindrical sections of about 10 ft of straight-shell height are placed lengthwise on the car and shored to maintain a good rigid cylinder.

Field assembly is still required, but is a much simpler operation than complete field fabrication. The field assembly is facilitated by the bell-and-spigot joint design. The bell portion is filament-wound into the straight-shell portion of the tank wall in the process of fabricating the cylinder. This provides for ease in field assembly and lends greater confidence to this concept, since the field joint is required merely to effect a seal against the liquid head, but is not truly a load-bearing joint, at least in the hoop direction.

Tank fabrication by means of field winding methods can be accomplished even in restricted areas as long as a minimum clearance of 24 in. around the OD of the tank is available. And even this limit can sometimes be circumvented since it is possible to rotate the tank in front of a fixed winding station or to rotate the winding station around the tank. As the tank is rotated, it is possible to work in areas of even smaller clearance. The tanks themselves may be projected upward through floor openings.

The general fabrication procedure is to work with a mold some 4 ft in height with the shell portion of the tank. Both the top and bottom must be prefabricated and set on the bottom or top of the mold. The first shell segment is wrapped to the top or upper portion of the mold, thus freeing the greater mold portion of the surface to fabricate the second shell surface, which is then tied in by overwinding the raised segment. Because overwinding is used, ribs are automatically built into the tank shell. The tank wall in the shell sections may be tapered in accordance with stress requirements.

The use of filament winding provides a much greater strength than is obtainable by hand-lay-up procedures. It also provides good uniformity of laminate structure and is a most economical fabricating method. Some judgment must be used to obtain the maximum degree of corrosion resistance commensurate with the desirable strength. By winding at a lower tension, wall tensile strengths of 60,000 psi are common, although 40,000 psi is generally used for design. An excellent blend of stress and corrosion resistance occurs with 40 to 45 percent resin content in the filament-wound portion of the wall.

In this filament-winding method a good lay-up might be as follows:

1. A flaked-glass or glass-surfacing-mat inner liner on the tank interior surface. Dynel or other surfacing mats may also be used.

2. A chopped-strand mat is suggested between each cycle of filament winding to provide rigidity to the tank wall.

3. Depending upon the service conditions, the tank may be finished with a C mat or provided with an ultraviolet inhibitor. If the final coat is a pigmented polyester (white makes a very pleasing appearance), the tank will be adequately protected against ultraviolet degradation.

It is also possible to lower construction costs of a tank of this type by building a composite tank using a high-chemical-resistant resin for the interior, or perhaps the initial 125 mils, and then finishing the tank with an isophthalic resin. This will allow some economies.

In extra-large sizes on-site filament-wound tanks may be constructed on previously prepared foundations for as low as 20 cents per gallon.

A large on-site filament-wound 250,000-gal tank is shown in Fig. 12-6. It is used in hydrochloric acid service. On-site filament-wound fabricated tanks are in service storing such chemicals as

25° Bé sulfuric acid at 170°F
22° Bé hydrochloric acid
Pigment slurry

The use of on-site filament-wound tanks has another attractive aspect. Not only does it permit supplying tank sizes which are far too large to ship, but the winding equipment, resin, and structural glass permit the development of a market for tanks in areas which are completely inaccessible otherwise. All the simple components are easily transportable by air.

The on-site filament-wound tank is completely free of any structural seams or joints and offers the higher-tensile-strength properties of the filament-wound structures.

FIG. 12-6 On-site filament-wound tank, 250,000-gal capacity, used for storage of hydrochloric acid.

12.6 KABE–O–RAP TANKS*,5

Another unique method in the field of reinforced plastic tank construction is that called the Kabe-o-Rap method, which is patented in the United States and many foreign countries. In this method the tank is built in molded sections either at the plant or in the fabricator's shop. Simple wooden and Formica sections are used for the segmented mold. As assembly proceeds, the flanges on the tank sections are mated, buttered with polyester resin putty, and then overlaid on the interior seam with a conventional lay-up of polyester resin and fiber-glass mat.

After the tank shell has been assembled, the flanges are slotted, and the outside of the tank is helically wound with a steel cable. The slots in the flanges are utilized to serve as guides and properly space the cable on the reinforced plastic column. As the hydrostatic head in the tank increases from top to bottom, it follows that the cable spacing is closer together at the bottom of the plastic column than at the top.

As the plastic laminate comes under load with the tank being filled, the reinforced plastic sidewall moves outward fractionally and transmits the

* This section was contributed by Metal Cladding, Inc., North Tonawanda, N.Y.

load to the cable. The cable is hand-stressed on the column. Since the cable is in one piece, the load is transmitted through its entire length, thus picking up the hoop-stress load from the reinforced plastic column. It will be observed that the principle of this design is based upon utilizing the difference in Young's modulus and the difference of expansion between the two materials. A principle similar to that used in suspension-bridge construction.

The main claims made for this process are that they develop the highest safety factors in the industry. The plastic laminate is being used to contain the liquid, while the steel cable actually carries the load. Common safety factors in this design are usually 40:1.

Since this process is proprietary with Metal Cladding, Inc., or their licensees, the tanks may be purchased in shop-assembled units up to 12 ft in diameter and field-erected beyond 12 ft in diameter. This is a transportation limitation which is common to all tanks, regardless of manufacturing type. Where field-assembly problems exist, the tank can be literally built in place regardless of size, being put together from molded parts which can easily pass through most doorways. Tanks of this type have had a wide distribution and application, particularly in paint manufacturing, chemical industry, food processing, and pulp and paper. The tank may be purchased in all the high-performance chemical-resistant resins. Tanks are fabricated on a customized basis to suit the need of the individual, so that a price list is not published. However, rough costs for guideline purposes indicate that a 5,000-gal tank may be purchased for approximately $2,400 and a 10,000-gal tank for $3,400. Figure 12-7 indicates the method of construction of Kabe-o-Rap tanks.

Although we have constantly referred to a steel cable, actually, the cable may be furnished in other materials of construction, such as double-galvanized steel, galvanized steel with extruded vinyl covering, Type 316 stainless steel, or any special alloy which is found to be necessary.

12.7 TANK NOZZLES

Flanged nozzles   When constructed according to the SPI, the general specifications call for nozzle inlets and outlets to be the same as pipe. This is a good specification. The author would suggest that the designer adhere to the 100-psi specification to obtain the necessary ruggedness. A minimum dimension of 4 in. from the flange face to the tank has been suggested. This, however, may be increased to as much as 6 in. at the discretion of the designer, with no ill effects. Although the SPI has suggested a minimum of three gussets, experience has proved that, where possible, a minimum of four gussets should be used. For suggested gusset design and number of gussets, the designer or engineer is referred to Fig. 12-8.

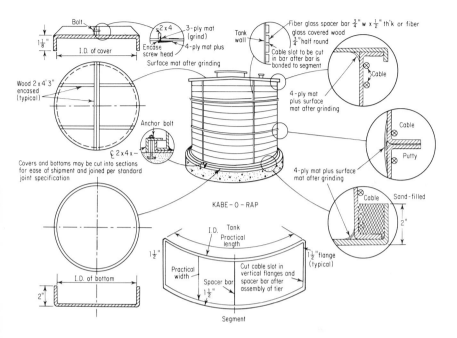

Patents		Registered trademark	United States and Canada — Patents pending in other countries	
United States	3,025,332	Canada	671,320	Italy	662,584
Argentina	133,874	France	1,308,347	Mexico	71,574
Britain	927,664	India	72,541		

FIG. 12-7 Kabe-o-Rap tank-construction details.[5]

Reference is also made to Fig. 12-9, which shows different types of nozzle connections into tank walls. For great strength the nozzle may be gusseted both inside and out. Complete penetration of the tank wall, with internal and external gusseting, provides an excellent arrangement, although it is not necessary for strength purposes in many cases. Probably one of the greatest advantages in providing complete nozzle penetration of the tank wall is the leading of the discharge of the corrosive solution well into the tank to prevent dribbles down the wall, which under some conditions may be harmful. This is the most widely used type of nozzle on reinforced plastic tanks and fabricated equipment.

Bosses   Tank bosses may be used as a point of attachment for flanged fittings being fastened to the tank. There is widely conflicting opinion as to their value, so that the engineer should proceed with caution if he intends to use bosses, and helical inserts in reinforced plastic bosses, as a method of flange attachment to the tank. Normally, the helical inserts are fastened into the helical boss, quite often by some form of adhesive. The bolt is then run through the flange, gasket, and into the boss, so that the flange may be snugged into position. Care must be taken, however, because if the bolt is bottomed, sufficient torque can be developed, literally, to pull the helical insert out of the RP boss. It is extremely important to use torque wrenches when making up into a bossed flange so that overtorquing will not occur. In the event, however, that a helical insert is damaged or pulls out, it may be replaced by drilling out the hole, filling the cavity with a good adhesive, putting in a new insert, letting it all set, and then bolting up once more. While bossed-flange attachments are used by some fabricators, the engineer is on safer ground if he sticks to

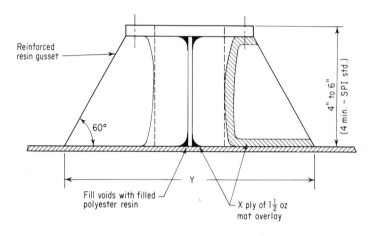

Size (in.)	X	No. of gussets	Gusset thick. (in.)	Y (in.)
2	3	4	$\frac{1}{4}$	12
3	3	4	$\frac{1}{4}$	$13\frac{1}{2}$
4	3	4	$\frac{1}{4}$	15
6	3	4	$\frac{3}{8}$	17
8	4	4	$\frac{3}{8}$	$19\frac{1}{2}$
10	4	8	$\frac{3}{8}$	22
12	4	8	$\frac{3}{8}$	$25\frac{1}{2}$

Notes:
Gussets to be evenly spaced around nozzle. Gussets to be added after complete assembly of nozzle on shell.

FIG. 12-8 Data for nozzle gussets on tanks and structures.

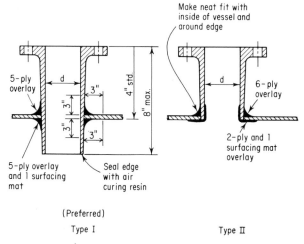

Make neat fit with
inside of vessel and
around edge

5-ply
overlay

d

3"

4" std

8" max.

6-ply
overlay

3"

3"

3"

3"

5-ply overlay
and 1 surfacing
mat

Seal edge
with air
curing resin

2-ply and 1
surfacing mat
overlay

(Preferred)

Type I   Type II

Notes: Each ply 1½ oz mat or equal, all nozzles to be reinforced with gussets

FIG. 12-9 RP nozzle connections into tank walls.

a gusseted flange.   For details of a bossed-flange attachment see Fig. 12-10.

Internal holding rings   It is sometimes necessary to develop flange designs to meet certain special situations.   One of these is shown in Fig. 12-11, a rather complicated yet effective design for mounting Teflon heat-transfer coils between tank top and bottom.   The same design is used for both locations.   In this design the flange is fastened with adhesive to the

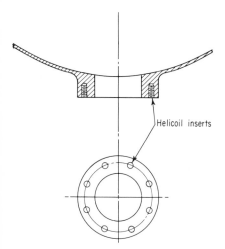

Helicoil inserts

FIG. 12-10 Bossed-flange attachment on fiber-glass-reinforced polyester tanks.

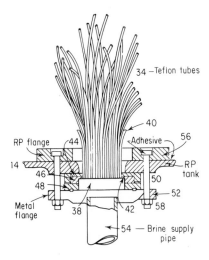

FIG. 12-11   Internal holding rings.

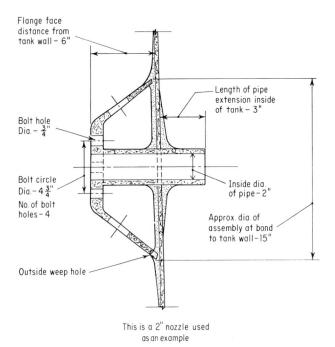

This is a 2" nozzle used
as an example

FIG. 12-12   Conical nozzle design developed by Owens-Corning Fiberglas.

tank surface. This is preferred to using a gasket; it has been found that a gasket presents difficulty in seating. Holding bolts are completely protected from the corrosive solution in the tank by having the heads covered flush with the same adhesive which is used to fasten the flange to the tank bottom. Make sure that the working surface is completely dry and that the temperature is above 60°F. If reasonable precautions are taken, the joint will prove very satisfactory. Note the complete protection of the bolt, which permits the use of a low-grade stainless or black iron in tanks filled with a highly corrosive solution.

Conical nozzle design[6]   The unique gusseted-flange nozzle developed by Owens-Corning Fiberglas provides a continuous 60° conical support from the flange back to the tank wall. Shown in detail in Fig. 12-12, it is reported to withstand a bending torque of 1,500 ft-lb. The pipe nozzle itself goes completely through the tank wall. This type of nozzle attachment is extremely rugged and has much to recommend it. Such a design is reported to be approximately 40 percent stronger than the standard gusseted nozzle. This is developed through full circumferential bonding to the vessel wall. Access to the bolt holes in the flange is obtained by means of several holes cut into the sides of the conical portion. A drain hole in the cone is provided. It is further claimed that such nozzles are more economical to install. It should be pointed out, however, that unless the plant forces are in an advanced state of training, the standard gusseted nozzle will be easier to construct. In the factory fabrication shop in which conical nozzles of this type are built, it can be readily observed to be a highly specialized operation since sizes are fabricated on special molds. This specialization can be purchased only from Owens-Corning at this time, and would be beyond the capability of the average plant shop.

12.8 MANHOLES

Manholes in tanks may be considered from several aspects:

1. In the top of the tank, assuming there are no removable lids, a manhole should be provided to permit observation of the contents by a person outside the tank on an operating platform. Also, this top manhole could serve as a convenient method of access.

2. In tanks over 10 ft deep the engineer should consider providing two manholes, that is, besides the one on the top, one on the side, easily available from the ground or the support upon which the tank rests.

In general, four sizes of manholes are appropriate and may be used at the discretion of the engineer: 18, 20, 22, and 24 in.   The 24-in. manhole will cost about one-third more than the 18-in. manhole, the other sizes lying in between.   Again, at the engineer's discretion, the manhole may extend completely through the tank wall or be simply treated as a large nozzle and strapped into the wall.   For a typical manhole design see Fig. 12-13.   Manhole cover may be flat, complete with handle, or dished if desired.

Two manholes, one on the dome and one just above the base of the tank, provide safety and accessibility arrangements that are hard to surpass.   Accessibility near the base eliminates climbing hazards, permits easy work on agitators or heat-exchange equipment, and possesses

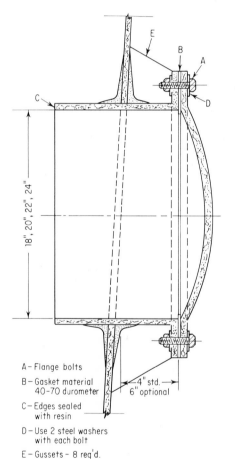

A – Flange bolts

B – Gasket material
    40–70 durometer

C – Edges sealed
    with resin

D – Use 2 steel washers
    with each bolt

E – Gussets – 8 req'd.
    $\frac{3}{8}$" min. thickness

FIG. 12-13   Typical manhole design.   Side-flange manway.

unparalleled ventilation and safety attributes. The two-manhole approach is to be highly recommended as a standard design on all large tanks. No trouble has been experienced effecting a tight seal at the bottom manhole on many installations.

12.9 THE KNUCKLE AREA

Probably the most important item in the fabrication of any tank is the knuckle area, where the bottom and the sides meet. Here with repeated filling and emptying, the tank is subjected to periodic flexure. Probably more study has gone into the knuckle-area design of tanks than into any other single facet of tank construction. The SPI tentative standard indicates that the knuckle radius should be a minimum of $1\frac{1}{2}$ in. Fabrication techniques in bringing together tank bottoms and walls vary from vendor to vendor. Two distinct methods are prevalent:

1. The flat bottom is fabricated integrally with the straight-shell portion of the tank. This practice is used by both filament-wound manufacturers and hand-laid-up fabricators. In addition to this, extra reinforcement is added in the knuckle area and may extend upward for a distance of 4 to 6 in. If the tank is to be used in outdoor service, hold-down lugs are positioned with the extra reinforcement.

2. Other fabricators, as standard practice, make tank bottoms and sides completely separate. Then a heavy-wrapped overlay both inside and out at the point of juncture should be provided. For a typical flat-bottom-tank knuckle-area detail see Fig. 12-14.

12.10 INTERIOR BAFFLES

Often, in order to solve agitation problems, the designer must provide the tank with suitably spaced baffles along the tank wall. If this is not done, the entire contents of the tank, when propelled with an agitator, may simply rotate, little mixing being accomplished. Baffles provide sufficient interference with the rotational flow to ensure proper mixing in many cases. It is not possible to provide a baffle design for each particular installation, but a typical interior baffle is illustrated in Fig. 12-15. The baffle may be built in the shape of a T and suitably fastened to the tank wall by an adhesive. The baffle base should at least be equal to the baffle height. As a rule of thumb the baffle thickness may approximate the tank-wall thickness. The tank wall and baffle should be suitably roughened before applying the adhesive to get the proper bond. Make sure that no raw edges are left exposed.

12.11 ACCESSORY ATTACHMENT LUGS

It is often necessary to fasten attachments to the walls of reinforced plastic vessels. A method of making an attachment through the wall of the vessel is shown in Fig. 12-16.[7] Blocks cut from blind flanges are a suitable source of material. By recessing the bolt head in the flange and then backfilling with adhesive it is possible to provide complete protection. Do not forget to fasten the block to the tank wall with adhesive. As the nut is tightened on the fixture, the entire unit is nicely held in place by compression, and yet is completely protected from the corrosive environ-

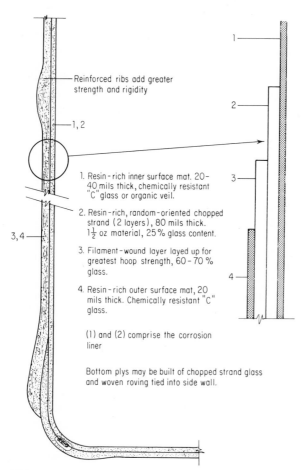

Reinforced ribs add greater strength and rigidity

1, 2

1. Resin-rich inner surface mat. 20–40 mils thick, chemically resistant "C" glass or organic veil.

2. Resin-rich, random-oriented chopped strand (2 layers), 80 mils thick. $1\frac{1}{2}$ oz material, 25% glass content.

3. Filament-wound layer layed up for greatest hoop strength, 60–70% glass.

4. Resin-rich outer surface mat, 20 mils thick. Chemically resistant "C" glass.

(1) and (2) comprise the corrosion liner

Bottom plys may be built of chopped strand glass and woven roving tied into side wall.

3, 4

FIG. 12-14   Typical wall detail, filament-wound tank with composite construction.

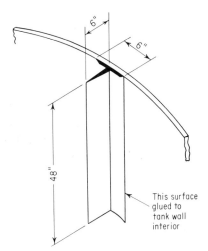

6"

6"

48"

This surface
glued to
tank wall
interior

FIG. 12-15 Interior baffles.

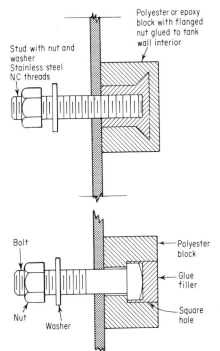

Polyester or epoxy
block with flanged
nut glued to tank
wall interior

Stud with nut and
washer
Stainless steel
NC threads

Bolt

Polyester
block

Glue
filler

Square
hole

Nut

Washer

FIG. 12-16 Accessory attachment lug.

ment inside the tank.   In this manner ladders and steps, and even entire flanges, may be fastened to a tank wall.

12.12 LIFTING LUGS

It is extremely important that tanks and vessels be handled properly. They should be picked up using lifting lugs designed into the tank or by slings.   Do not pick up a tank by its nozzle—it was not designed to serve as a lifting device.   The nozzle would simply be torn from the tank.

12.13 TANK VENTING[8]

Most RP tanks are designed for atmospheric storage conditions.   Actually, the tanks will withstand a nominal $\frac{1}{2}$ oz of vacuum and 2 oz of positive pressure above the liquid at any time without suffering damage. The engineer should take particular care to provide sufficient venting to keep the loading and unloading forces within these limits at all times.   It is of particular importance that this be given full consideration when air-unloading delivery vessels in order to permit proper relief of the "blow-by."   Removal of the top manway cover during unloading where air is used is helpful.   Again, particular watchfulness is necessary where the tank is receiving an emergency discharge from a steam-injection heater. Here uncondensed steam may provide some pressurization within the tank unless suitable venting is provided.

Care must also be taken when emptying a tank to provide for suitable air replacement of the outgoing liquid.   If this is not done, it is a simple matter to pull a good-sized vacuum on a tank.

As a rule of thumb the vent diameter should be one pipe size larger than the tank fill or outlet size, whichever is larger.   Where the tank is installed outdoors, the vent should be carried up a suitable distance above the tank, provided with a gooseneck, and properly braced against wind loads. Many processes require the venting of such a tank into a suitable exhaust system for proper distribution of gases and mists.

12.14 TANK DRAINS

Four different designs of tank drains are shown in Fig. 12-17.   Each of them possesses certain desirable characteristics.

Figure 12-17*A* is a common design used in a hand-laid-up tank. Normally, it is not used in a filament-wound tank unless the height of the drain is above the area in which the bottom is strapped into the sides. With this design the tank will not drain completely.

Figure 12-17*B* is commonly used in both filament-wound and hand-laid-up tanks and will provide complete drainage of the tank. Since it is a bottom drain, special care must be taken on the support design to permit the nozzle and elbow to be properly positioned below the tank.

Figure 12-17*C* permits complete drainage of the vessel and is quite often used in hand-laid-up construction. In this design half the drain is below the tank bottom.

Figure 12-17*D* is quite often used in filament-wound construction to permit almost complete drainage of the tank, by application of the siphon effect, even though the nozzle is elevated above the tank. The tank will drain down to the point where the siphon breaks.

12.15 EXTERNAL DESIGN

Anchorage For tanks to be used in outdoor service, holddown brackets are advised, securely fastened into the reinforcement hoop at the bottom. Use four or eight brackets, depending on tank size, equally spaced.

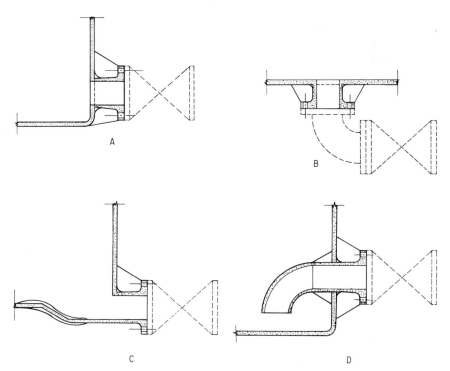

FIG. 12-17 Tank drains, various designs.

WIND–LOAD DESIGN AT 100 MPH[9]

Table 12-2

Height zone, ft above ground	Rectangular tanks, psf	Cylindrical tanks,* psf
Less than 30	25	12–14
30–50	35	17–20
50–100	45	21–25

* Designed wind load is a function of the ratio of height to diameter. For a ratio of 1:1 use the bottom figure shown. For a ratio of 3:1 use the top figure shown. For ratios in between, a simple interpolation will be sufficient.

Anchoring should be capable of holding the tank when empty in a 100-mph wind. This is equivalent to a loading of about 35 psf. To this we must add shape factors which take into account whether we are dealing with a rectangular or cylindrical surface and also the height zone above the ground. The tank designer may find Table 12-2 useful for outdoor tank installations.

Obviously, if the tank is to be installed indoors, no such precautions need be considered. Indoors, it is common practice simply to set the tank on a continuous pad.

Protective coating   The exterior of the tank should be provided with a suitable protective coating to afford maximum protection to the tank exterior against spillage, fumes, weathering, and light degradation. For outdoor service a pigment may be employed in the formulated gel coat, and should be applied to all tank exteriors. Many colors are available, such as white, yellow, blue, green, or gray. The choice of color naturally lies with the individual. White is very pleasing for an outside tank, although if the plant has adopted a scheme of color coding to identify certain types of liquids, such a scheme should be followed. The exterior pigmented gel coat should be at least 15 mils thick and of a special formulation of a thickened flexibilized resin to make sure that all glass is completely buried beneath this protective coating.

On tanks installed on interior service, an exterior protective coating is still important to guard against spillage and fumes and to go along with any desired color-coding scheme.

Tank deck loading   It is common practice in the industry to design the tank roof so that it will withstand a 1,000-lb deck loading without buckling or collapsing.[10]   There are many methods of constructing tank tops.   Any tank installed in a chemical plant, however, should have the capability of supporting a 200-lb man walking on top of the tank with complete safety.   Some of the following more common methods of fabricating tank tops are shown in Fig. 12-18.

1. Hand or sprayed lay-up, dished or flat-top.   Generally cured separately from the tank and then wound into the shell.   Such a top may have diagonal stiffening ribs radiating like the spokes of a wheel from the center.

2. Laminated top consisting of urethane core with fiber glass on both sides, forming a sandwich.   This makes an excellent light and strong lid and is used on flat-top tanks, whether cylindrical or rectangular.   Easily equipped with handles, this type of lid is a natural selection for removal sections.   T bars or angles position the removal sections to prevent lateral displacement.

3. Flat RP sheet equipped with hand grips and suitably stiffened with angles or other RP shapes such as half-rounds, triangles, etc.   Such light lids can be constructed where there is no possibility of a man walking on top of a tank.   They simply serve to contain fumes in a tank, and are not designed for heavy traffic or personnel walking on the top of a tank.   An easily built design of this type is shown in Fig. 12-18E.

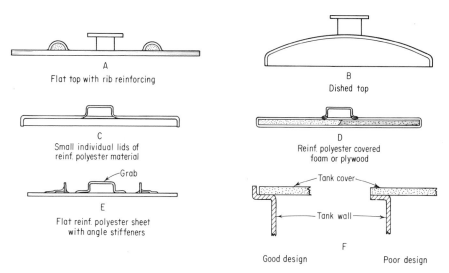

FIG. 12-18   Different types of tank lids or tops.

12.16   SUPPORTING RP TANKS

It is often considered good practice to provide a sand bed approximately 1 in. deep on top of the foundation to ensure continuous support for the bottom of the reinforced plastic tank.   In the construction of any really large plastic tank, the fabricator sometimes has difficulty in obtaining a completely flat bottom due to the expansion and contraction characteristics of the material.   The provision of a sand bed underneath the tank on top of the foundation, providing an even support throughout the tank bottom, is no doubt the ideal situation.   However, once the tank has been put in place on top of the sand, a retaining device must be provided around the tank wall, which often is done with a metal strip to prevent the sand from coming out.   Over a period of time experience has shown that there may be difficulty holding this sand in position, so that after a few tank overruns, the sand may be completely washed out and the tank will settle down onto the original foundation.   Generally, this occurs in an uneven fashion.   As the tank drops, this produces a strain, especially if the tank piping and pump are all tied together.   A strain will occur at the tank wall and in the piping wall at the pump.   Whether additional physical damage ensues of course depends on the particular installation. When a sand bed is used, every effort must be made to eliminate the possibility of the sand washing out from underneath the tank.   See Fig. 12-19*a* for an acceptable design.   Of course, the foundation should provide full and continuous support for the tank bottom, with certain exceptions.

One of the first things that can happen in a tank installation is that the sides of the tank will overhang the foundation walls.   This will produce a severe strain on the knuckle area.   The tank will continually flex on filling and emptying, and failure in the knuckle area will almost always occur (see Fig. 12-19*b*).

Figure 12-19*c* shows a better installation, in which the foundation diameter and the tank diameter are identical.   This provides for substantially complete support for the tank and is satisfactory for indoor installations.   Figure 12-19*d* shows an ideal situation for an outdoor installation, in which the foundation diameter is the tank diameter plus 12 in., constructed in a square shape.   This permits the holddown lugs, of which there may be either four or eight, depending upon the engineer's specifications, to be suitably anchored into the foundation with foundation bolts.   For a typical design of a holddown lug, see Fig. 12-19*e*. In addition, complete support for the entire tank is provided.   Again,[11] care must be taken to provide an absolutely flat bottom for supporting the vessel.

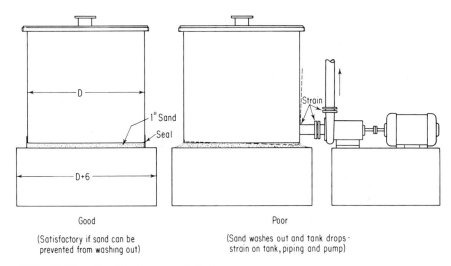

Good

(Satisfactory if sand can be
prevented from washing out)

Poor

(Sand washes out and tank drops-
strain on tank, piping and pump)

FIG. 12-19a   Sand-bed support, indoors.

A third kind of foundation which is sometimes used is a grillage-type support, used either indoors or outdoors, with some minor modifications. Normally, this grillage-type support is used when carrying the tank load between supporting walls or supporting beams (see Fig. 12-19f).   However, in providing a grillage-type support, care must be taken to ensure that the bottom of the tank is of sufficient strength to provide the span over a distance between the I beams.   This type of grillage support is sometimes modified to permit attachment of the holddown lugs furnished with the tank.

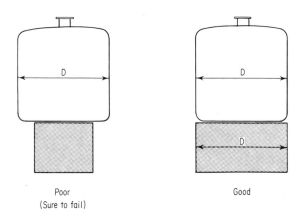

Poor
(Sure to fail)

Good

FIG. 12-19b   Poor tank support, outdoors.

FIG. 12-19c   Good tank support, outdoors.

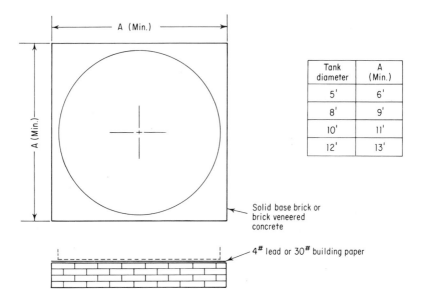

Tank diameter	A (Min.)
5'	6'
8'	9'
10'	11'
12'	13'

FIG. 12-19d Solid-brick or brick-faced foundation, outdoors.

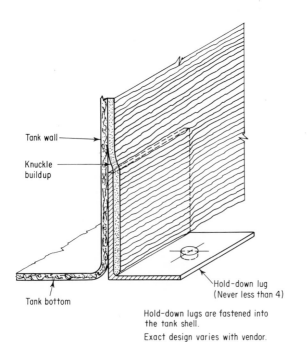

Hold-down lugs are fastened into the tank shell.

Exact design varies with vendor.

FIG. 12-19e Holddown lugs (a typical design).

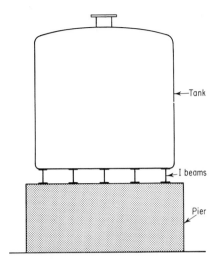

Grillage–type support — indoor (ok with heavy bottom design)
(Quite often minimum cost when carrying tank load between supporting walls)

FIG. *12-19f* Grillage-type support.

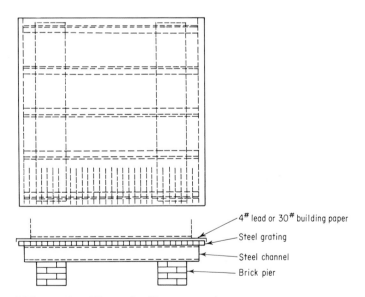

4# lead or 30# building paper
Steel grating
Steel channel
Brick pier

FIG. *12-19g* Pier and grillage support.

Another kind of support which may be used for tank foundations is steel substructure overlaid with a bar-type grating. Ideally, the bar-type grating would then be overlaid with a $\frac{1}{4}$-in. plywood covering or 4-lb lead or 30-lb building paper.[12] This type of supporting plan for elevated-platform vertical tanks is shown in Fig. 12-19g.

Enlarged reinforced polyester flat surfaces   The problem of retaining a true flatness is indeed difficult. For example, the bottom of a 12-ft-dia tank may show some undulation in the tank bottom. If the tank is to be partially filled or filled with liquid at all times, this is no problem, but if not, there may be a certain working of the tank bottom so that repetitive filling of the tank may sound like beating on a hollow drum. This is one of the advantages of having a sand support underneath the bottom of the tank since it will conform to the tank contours. The problem arises with rigid support; and besides, continued multiple flexure of the tank may spell cracking. As little as 2 ft of liquid in the tank, however, will keep the tank bottom lying flat at all times.

12.17   WORKING IN REINFORCED PLASTIC TANKS— PRECAUTIONARY MEASURES OF SAFETY

Extreme care should be taken in working inside tanks. It is good practice for the workmen to wear soft-soled shoes or arctics to prevent scarring and damaging the resin system on the tank bottom. Further care should be taken so that tools are not dropped, starring the surface. Ladder feet should be suitably protected.

In entering reinforced plastic tanks it is the responsibility of supervision to see that the men are completely and adequately protected. Some of the basic simple rules which should be followed are:

1. Make sure that the tank is completely blanked off from all external sources of contamination. Lock all valves and switches associated with the tank. Wash out the tank thoroughly.

2. The tank interior should be tested by competent analysts, immediately before the workmen are to enter, for possible toxic gases and oxygen deficiency (guard against deep tanks). Tank-entry permits should be written, not verbal.

3. A power-driven source of fresh air should be blowing into the tank at all times.

4. All workmen entering the tank should be equipped with a safety harness with a lifeline to the top of the tank. The lifeline should be such

that the man can be hauled from the tank without anyone else having to enter the tank to do so.

5. One man should be detailed to remain at the top of the tank and keep the men in the tank under observance at all times. He must have an adequate emergency plan in mind at all times for the procurement of additional help if required.

6. The responsibility for a man's safety rests directly with the immediate supervisor. The responsibility cannot be delegated or denied.

7. If repair work is going on in the tank and resins and catalyst systems are being used, styrene checks should be made periodically to ensure that the working conditions in the tank remain at safe levels. (For further information, see Chap. 10.)

8. When in doubt use an air-line respirator.

12.18 HEATING IN A REINFORCED PLASTIC TANK

Be cautious in the steam-injection heating of an RP tank. This can be accomplished satisfactorily if the engineer will spend time and effort to minimize vibrations and impingement from the live steam.

Do not impinge the steam on the tank wall.
Do not introduce the live steam through a tank nozzle.

Feed the steam in by means of a separate pipe through the top. Break up the steam flow into many small streams of perhaps ¼-in. diameter rather than have it issue out the end of a 2-in. pipe. Very simple tank heaters can be constructed from RP pipe. Figure 12-20 shows a typical design capable of passing 2,000 lb/hr of 15-psig steam for solution heating. A simple heater of this type can be built for $50. Heavy lead coils have been mounted satisfactorily in RP tanks, but in each case the weight was suitably distributed on scuff pads on the tank bottom.

12.19 INSULATION OF RP TANKS[13]

At the present time some manufacturers are providing preinsulated plastic tanks. One vendor, for example, is supplying as a standard item a line of reinforced plastic tanks with 2 in. of Fiberglas insulation which has a K factor of 0.28. The insulation is completely enclosed and protected by an outside RP laminate. Obviously, some savings in insulation costs could be achieved by this method.

Other companies in the past have attempted to provide double-walled tanks with foamed-in-place urethane insulation. In the designs which have been attempted, shearing forces occurring between the main tank body and the outer protective shell have introduced difficulties in the process. This, however, would not apply to a fiber-glass type of insulation.

Other methods of insulating reinforced plastic tanks in the field which have been successful apply conventional insulations, followed by a vapor barrier or Hypalon outer jacket. Insulations on reinforced plastic tanks may be necessary in process work where the storage or process vessel operates at relatively cold temperatures of 0°C and below and where heat gain might have a measurably adverse effect upon process economies.

For example, the application of 1 in. of foamed-rubber insulation on a 12-ft-dia tank built with ½-in. RP wall decreases the heat gain in the solution by 80 percent, with the agitated solution at 0°C (32°F) and an outdoor temperature of 35°C (95°F). In thinner-walled vessels of approximately ¼-in. thickness, internal solution temperatures of 0°C (32°F) will produce outer-wall temperatures of approximately 10°C (50°F).

This means that reinforced plastic vessels or piping in the 0°C (32°F) to 10°C (50°F) range will sweat profusely on a warm summer day.

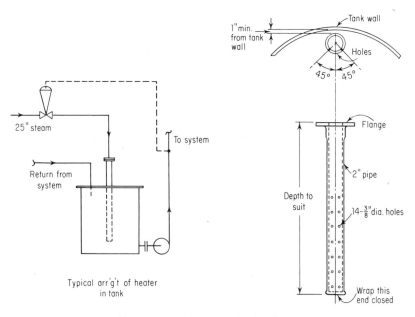

FIG. 12-20 A 2-in. fiber-glass-reinforced injection heater.

Antisweat insulation is necessary to prevent this, that is, if sweating on the exterior surface and subsequent dripping are objectionable. Sweating, of course, is a function of the surface temperature of the vessel or piping and the dewpoint of the surrounding air. Each case must be judged on its own merits. Except for antisweat insulation and insulation in vessels in cold process work or to prevent freezing, there is little need for insulating any other reinforced plastic installation. Solutions at room temperatures and on beyond to that approaching the upper limit of usefulness are not normally insulated except under unusual conditions peculiar to a certain process.

Conversely, it should be recognized that exhaust systems attached to the vessels may sweat on the interior when subjected to cold outside temperatures. Normally, this is no problem when the condensed vapor is either drained off in pockets or carried off with the exhaust gas.

12.20 STORAGE OF FLAMMABLE LIQUIDS

Reinforced plastic tanks may be used for the storage of flammable liquids provided that the conditions are favorable to service life. Flammable liquids are generally associated with the possibilities of explosion. For some of the techniques which have been worked out to minimize the risk from explosion and fire, refer to Chap. 13, which covers not only grounding, but venting procedures. The risk from this hazard can be further minimized by constructing the tank of a resin characterized by a flamespread rating of 25 or less. The chlorinated polyesters to which 5 percent antimony trioxide has been added are a good choice if chemically suitable.

An entire new industry is being created around providing reinforced plastic tanks manufactured from an isophthalic resin for underground gasoline storage.[14] These tanks, built in standard sizes, are offered at a minimum cost for use in areas where underground environment is hostile to a steel tank. Such FRP tanks carry both the Underwriters' and Factory Mutual labels of acceptance. Obviously, no tank-protection system needs to be maintained, nor are tank coatings necessary. The tanks themselves weigh approximately one-fourth that of a comparable steel tank. At this stage of development a 6,000-gal underground tank costs approximately 50 percent more than its steel counterpart. In terms of service life, in seawater areas it is readily conceivable that such tanks would represent the lower cost per year of service.

With the continued advance of the automobile and the service stations needed to look after it, 7,000 to 8,000 new gas stations per year are being built in the United States. Each one of these gas stations will use three 6,000-gal tanks or two 8,000- or two 10,000-gal tanks for gasoline storage. This is a potential market of approximately thirty thousand 6,000-gal tanks per year—small wonder that big business is preparing to penetrate this market at a vigorous rate. Such a development can only result in accrued benefits to the chemical industry in acquiring tank-building know-how on a massive scale. Automation in areas such as this has reduced tank labor costs considerably over the conventional hand-laid-up methods. On a cost-per-gallon basis this is the minimum-cost approach.

12.21 WATER TESTING AND POSTCURE

As part of your specifications request that the tank be water-tested prior to shipment. Many manufacturers are glad to comply at no extra charge. This is good business for them because, in the event of a leak, it will have saved the cost of a field trip.

Most companies, however, charge for a postcure. A postcure is beneficial because it provides maximum chemical resistance and reduces the free radicals by at least 90 percent. Postcuring consists in filling the tank or vessel with hot water at 180°F for 3 hr. It can also be accomplished by blowing low-pressure steam into the vessel for approximately the same length of time, making sure that the vessel is completely vented, so that there is no possibility of pressure buildup.

An oven cure at the same temperatures is satisfactory since this will also cure the exterior of the vessel.

An alternative to postcuring by the vendor is for the plant to perform the postcure after installation or on site. This reduces greatly the problem of resin cracking in transportation. But regardless of where it is done, a postcure is desirable to bring the resin up to its optimum qualities.

12.22 SIGNS AND CAUSES OF TANK FAILURE

1. Guard against overpressuring a tank or pipeline. Typical signs of hoop-stress failure are longitudinal cracks in the internal resin in piping or vertical cracks in vertical tanks.

2. Cracking and leaking laminate is caused by exceeding design conditions such as overpressuring; exceeding vacuum limitations; pulsation, especially from steam-jet heaters; etc.

3. Chemical attack of the laminate may be present in many forms, such as solvent action, dehydration, etc.

4. Nozzle rupture may be due to improper design or inadequate

support. Do not skimp on the nozzles. Even though the tank is designed only for gravity service, specify and use a minimum of 100-psig stub ends, properly gusseted.

5. All tanks and vessels should be inspected on delivery for flexure cracks in shipping. These can occur if the tank or equipment is not adequately supported. There have been cases where large cracks occurred between the body and head of the tank as a result of bad handling.

6. Beware of severe exothermic reactions in plastic tanks. This can damage the tank and cause catastrophic failure. Most systems of this type, in which reactions occur, should be designed with suitable dumping or quenching mechanisms to cope with a violent exotherm.

7. Inadequately grounded systems may cause catastrophic failure with the blowing out of the dished head. Failures of this type are covered in detail in Chap. 13. The dished head is generally the weak point in any tank design. This can be demonstrated in theory by tank design formulas, and has been confirmed repeatedly in practice, whether the tank be reinforced plastic or of metallic construction.

12.23   RP TANKS FOR FOOD SERVICE

In the construction of RP tanks for food service it is absolutely essential that an FDA*-approved resin be used. A number of resins have been approved by the FDA. Begin with this specification.

Next, consideration must be given to the need for ultraviolet sterilization above the liquid level in the tank interior. This poses the problem of RP degradation due to the ultraviolet. Normally, this can be effectively combated with an FDA-approved light stabilizer.

Consider, further, the need for sterilizing the tank. Just before start-up, all tanks must be postcured. This can be done by steam sterilization, which takes 4 hr at 212°F with atmospheric steam. Make sure there is no pressure on the tank.

In the manufacture of tanks for food service, great care must be taken to ensure that the inside surface is entirely smooth and that there are no pockets or pinholes in the interior surface. There have been cases where, as a result of a few pinholes, the tank could not be sterilized. Bacteria continued to remain and grow in these small holes and reinfected each batch of food processed through the vessel. The utmost care must be taken in the finishing up of a tank interior of this type. RP tanks in food service must be capable of complete drainage.

Do not be deterred by the above admonitions from enjoying the

* U.S. Food and Drug Administration.

economies of RP tanks in food service. For example, eight 5,000-gal chilled-juice-storage tanks in 304 stainless steel were quoted at $6,000 each, while the same tank in reinforced polyester could be purchased at $3,100 each, or nearly half. By altering the design slightly and going to a standardized tank, the cost was reduced to $1,800. This is the kind of economy that is to be gained with reinforced plastics.

12.24 LAMINATE LIBRARY

It is a good idea to retain samples of the tank construction. The samples would be furnished by the vendor, cut from nozzles and manways. These can be used for tests and, suitably identified, provide a valuable record for future use. A library of test specimens of this kind should be established and maintained.

12.25 STIFFENING DEVICES—TANKS

There are a great many details to consider in the design of a plastic vessel. The SPI Standard can only be the beginning. Wall strength can be improved by increasing the wall thickness, but it can also be achieved at much less cost by using various types of stiffening devices, some of which are shown in Figs. 12-2 and 12-3. Laminate-covered low-density foam is generally considered a better reinforcement shape material than wood because the wood tends with time to pick up moisture and swell, actually causing damage to the vessel. When encapsulated and subjected to moisture transmittal over a long period of time, wood will split the overlay. Oak and maple, for example, should be avoided as stiffening members. Wood also decays. Balsa wood, however, has generally been found to be an acceptable material of construction as a reinforcing form, due to its low compressive strength. Should it swell and be contained or restricted, it will crush itself rather than damage the FRP covering. In thousands of applications over the past decade balsa wood has proved to be a satisfactory stiffening overlay.

Other methods of stiffening involve covered-steel shapes, angles, pipe, rectangles, covered glass or paper, rope, RP angles, cable-wrapped vessels, and assorted FRP reinforcing shapes such as triangles, half circles, rectangles, etc.[15–17] An overlaid-metal reinforcement has been successfully used by many vendors over the years. A word of caution, however: the overlay or covering must be sound and nonporous; otherwise trouble may ensue. When using a covered-metal reinforcement, make sure that the FRP covering is a minimum of $\frac{1}{4}$ in. thick. Less than $\frac{1}{4}$ in. may lead to ultimate deterioration of the shell unless the lay-up is mechanically sound and free from pinholes and resin starvation.

The honeycomb design as a construction device for reinforced plastic construction in chemical tanks and vessels has not been sufficiently developed at this time to be recommended. Such a design, where it has been attempted, used a sandwiched honeycomb with a laminate on both inside and outside. On trial, failure resulted, due to permeation of the water molecules from inside the vessel into the honeycomb. In one case history, vaporization occurred in the honeycomb, and since the vessel was operating at an elevated temperature, the structure was destroyed. Where a structure is required for strength, the current recommendation is an RP-covered closed-core material such as a foamed urethane.

12.26 BASIC DESIGN CALCULATIONS FOR REINFORCED PLASTIC TANKS[18]

Sample Problem 1

A rectangular tank is 56 in. long × 48 in. wide × 48 in. deep. It will be filled with a dilute acid (specific gravity 1.0). Assuming we use a 2 × 2 × $\frac{3}{8}$ in. steel angle as a stiffener, calculate the maximum deflection which will occur at a stiffener located 16 in. above the bottom of the tank. Since the stiffener is to be continuous around the periphery of the tank, consider the stiffener as a beam with fixed ends.

$$\Delta M = \frac{wl^4}{384EI}$$

$$= \frac{(41.6)(56)^4}{(384)(29)(10)^6(0.48)}$$

$$= 0.076 \text{ in.}$$

where ΔM = deflection, in.

w = load, lb/in. = 41.6 lb

Load on angle = $D \times \dfrac{D}{2} \times \dfrac{62.4}{12} = 41.6$ lb/in.

l = length of long side = 56 in.

E = modulus of elasticity = 29 × 10⁶ psi

I = moment of inertia = 0.48 in.⁴

Now the allowable deflection is $\frac{1}{2}$ percent of the span length, or 56 × 0.005 = 0.28 in. We are thus well within the allowable deflection.

Note that the design of a rectangular tank is treated as a beam formula with fixed ends. It is important that the designer appreciate the significance of this.

Tank corners must be rigidly fixed and of sufficient strength so that they will remain completely stationary. It is paramount in the design of rectangular tanks to ensure that the adjacent sides are tied together rigidly, not only through the wall itself, but through the stiffeners. Steel reinforcing angles should be welded solid at the corners and then ground smooth. Immediately prior to over-laying with RP material, steel angles should be blasted to a white finish, and then covered without delay. Make sure that the only deflection occurring is in the sidewall.

Sample Problem 2

Referring back to Prob. 1, suppose we covered a 3 × 3 balsa-wood section with $\frac{1}{2}$-in. RP as a stiffener. Calculate the anticipated maximum deflection in the tank wall.

$$\Delta M = \frac{wl^4}{384EI}$$

Assume

$$E = (1.0)(10^6)$$
$$I = 5.96$$

Then

$$M = \frac{(41.6)(56^4)}{(384)(1)(10^6)(5.96)}$$
$$= 0.179 \text{ in.}$$

Again, the allowable deflection would be approximately 0.28 in., and this design would be completely satisfactory.

Sample Problem 3

A tank is 10 ft in diameter × 15 ft high and filled with a chemical solution having a specific gravity of 1.2. Assume an allowable stress in the laminate of 1,500 psi. Calculate the wall thickness.

$$t = \frac{PR}{SE - 0.6P}$$
$$= \frac{(7.8)(5)(12)}{(1,500)(1.00) - (0.6)(7.8)} = 0.31 \text{ in., or } \frac{5}{16} \text{ in.}$$

where t = thickness, in.

R = radius, in.

S = stress, psi

E = joint efficiency, 1.0

P = pressure, psi = $\dfrac{15 \times 1.2}{2.3}$ = 7.8 psi

Sample Problem 4

It is desired to build a tank 12 ft in diameter by 24 ft high with a graduated wall. Calculate the wall thickness using a graduated-wall-thickness method; that is, calculate wall thickness at a point (*a*) 6 ft, (*b*) 12 ft, (*c*) 18 ft, and (*d*) 24 ft from the top. Use a specific gravity of 1.2 and an allowable stress in the laminate of 1,200 psi in the top half of the tank and 1,500 psi in the bottom half of the tank. Round off answers to the nearest $\frac{1}{16}$ in.

(*a*) At 6 ft:

$$t = \frac{PR}{SE - 0.6P}$$

$$P = \frac{(6)(1.2)}{2.3} = 3.1 \text{ psi}$$

$$t = \frac{(3.1)(6)(12)}{(1,200)(1) - (0.6)(3.1)}$$

$$= 0.19 \text{ in.}$$

Rounds off to $\frac{1}{4}$ in.

(*c*) At 18 ft:

$$P = 9.3 \text{ psi}$$

$$t = \frac{(9.3)(6)(12)}{(1,500)(1) - (0.6)(9.3)}$$

$$= 0.45 \text{ in.}$$

Rounds off to $\frac{1}{2}$ in.

(*b*) At 12 ft:

$$P = 6.2 \text{ psi}$$

$$t = \frac{(6.2)(6)(12)}{(1200)(1) - (0.6)(6.2)}$$

$$= 0.375 \text{ in.}$$

Rounds off to $\frac{3}{8}$ in.

(*d*) At 24 ft:

$$P = 12.4 \text{ psi}$$

$$t = \frac{(12.4)(6)(12)}{(1,500)(1) - (0.6)(12.4)}$$

$$= 0.60 \text{ in.}$$

Rounds off to $\frac{5}{8}$ in.

NOTE: These calculations agree with the tank table from the Commercial Standard.

Sample Problem 5

To permit good drainage it is desired to install a dished head on the bottom of a tank 12 ft in diameter to carry 24 ft of liquid (specific gravity 1.2). Assume the radius of the dish to be 120 in. Calculate the head thickness required. Assume a laminate strength of 1,500 psi.

$$t = \frac{5PR}{6SE}$$

$$P = \frac{(24)(1.2)}{2.3} = 12.5 \text{ psi}$$

$$t = \frac{(5)(12.5)(10)(12)}{(6)(1,500)(1)}$$

$$= 0.83 \text{ in.}$$

Rounds off to ⅞ in.

Note that domed heads on a tank shall not exceed maximum radius of curvature equal to the tank diameter. In this illustration we have chosen a dish radius somewhat less than this maximum.

Sample Problem 6

It is desired to build a chemical-solution tank 6 ft in diameter in the shape of a sphere by the filament-winding process to withstand a pressure of 50 psig. Assume a hoop tensile of 40,000 psi and a safety factor of 10:1.

$$t = \frac{PR}{SE}$$

$$= \frac{\dfrac{(50)(3)(12)}{(40,000)(1)}}{10}$$

$$= 0.45 \text{ in.}$$

Rounds off to ½ in.

This would be the thickness of the filament winding. Random mat would be added to the inside and outside, plus gel coats for suitable chemical protection. Total wall thickness would approximate $1\frac{1}{16}$ in.

Sample Problem 7[19]

Calculate the maximum stress occurring in the bottom 1 in. of a tank wall 10 ft in diameter by 9-ft straight shell, containing demineralized water with a specific gravity of 1.0. Wall thickness is ½ in.

$$P = \frac{9}{2.3} = 3.9 \text{ psi}$$

$$S = \frac{(144)(3.9)}{(2)(0.5)} = 560 \text{ psi}$$

Allowable laminate stress in a hand-laid-up design would be approximately 1,500 psi. Suppose the above tank had been filament-wound with a design hoop tensile of 40,000 psi and a thickness in the filament winding of 0.25 in. Calculate the stress in the filament winding and the safety factor developed.

$$S = \frac{(144)(3.9)}{(0.25)(2)} = 1,120 \text{ psi in filament winding}$$

$$\text{Safety factor} = \frac{40,000}{1,120} = 35.7$$

12.27 COSTS OF CORROSION-RESISTANT STORAGE TANKS[20]

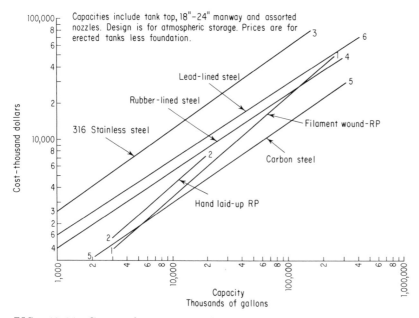

FIG. *12-21* Costs of corrosion-resistant storage tanks—RP versus other construction.

12.28 REVIEW SECTION—RP TANKS—WHAT TO AVOID

AVOID WRONG PRESSURE SPECIFICATIONS. Study the pressure specifications on the RP plastic tank. If you intend it to contain pressure, review the design section to make sure that you have provided sufficient strength in the vessel to meet the proposed specifications. Be particularly watchful of pressure surges. There is nothing more distressing than to find the dome of the RP tank performing in an accordionlike manner, since failure will follow rapid flexure and overpressurization. Normally, the tank dome, or top, is the weakest portion of the RP tank construction. Failure will generally occur in this area if there is overpressurization. Unless specifically designed otherwise, an RP tank will perform best under gravity conditions. Make sure that the tanks are provided with adequate venting and controlled overflow provisions.

AVOID POOR RESIN SPECIFICATIONS. Review the resin specifications to make sure they are what you require. An epoxy tank will generally cost more than a polyester resin tank due to differences in price ranges. Select the resin that will provide the most satisfactory lay-up and one that is properly tailored to meet your service specifications. Do not buy a general-purpose resin and expect it to perform satisfactorily in severe chemical service. The life will be short, and there will be dissatisfaction.

SUPPORT ALL OUTLETS. Make sure that all the outlets, both vents and liquid, are properly supported with either a gusset or a cone. Do not be stingy about gusset thickness. For complete details in this area refer to the section on nozzle supports and their design.

PROVIDE PROPER DRAINS. Make sure that the tank will drain adequately. Again, refer to the design section for a number of suggested arrangements. One of these can probably be utilized to suit your particular conditions. And do not make the drain too small. There is nothing more frustrating than requiring several hours to empty a tank because the drain is too small.

ELIMINATE VIBRATION AND HAMMERING. Review the design and make sure that vibration has been isolated from the tank structure, and hammering from submerged heaters eliminated. Vibration is the enemy of reinforced plastic structures. Under severe conditions it can reduce the useful life of a plastic tank by as much as 80 percent.

SUPPORT PIPING AND VALVES APPENDED TO THE TANK. Make sure that all piping and valves are supported independently of the tank. While reinforced plastics are tremendously strong, continued support of big 6-, 8-, and 10-in. valves on an RP tank is too much to expect. The lever arm on a heavy weight over a considerable distance is sufficient to cause excessive localized stresses, spelling trouble over a long period of time. Good practice has proved this beyond doubt. If valves and heavy piping are supported independently of the tank, there will never be nozzle failure due to overstressing.

FORTIFY THE KNUCKLE AREA. The point at which the bottom of the tank joins the sidewall may develop flexing stresses with repeated filling and emptying. A sufficient overlay will absorb these flexing stresses and literally interlock the sidewall into the bottom. Such an interlock should continue up the sidewall for a considerable distance. In this overlay may also be provided angle clips for fastening the tank securely to a foundation.

AVOID INSUFFICIENT BOTTOM SUPPORT. Consider the various means of adequately supporting the bottom of the tank. A continuous support is preferable, but an intermittent support can be used provided the bottom is made sufficiently strong to bridge the supporting distance.

MAKE TANK LIDS SUFFICIENTLY STRONG. Make sure that your lidding system will support a 200-lb man walking on the top of the tank with complete safety. Assuredly, it will be used to support personnel; so design it to perform this function from the start. It is unsettling to feel a springy lid above a tank of acid. The complete lidding system should be capable of supporting an evenly distributed weight of 1,000 lb.

ESTABLISH GOOD COMMUNICATION. The need for good communication between the vendor and the purchaser is paramount. There have been repeated cases where insufficient communication or understanding has been a contributory factor in vessel failure. Clear specifications are essential. An example can be cited where 200 mm absolute pressure was misconstrued for 200 mm of vacuum. Since the tank was designed for 200 mm vacuum, complete and utter failure resulted when the tank was installed and operation was attempted at 200 mm absolute pressure.

PROVIDE AGITATORS AND PLATFORMS. Good design calls for supporting agitators and platforms completely independent of the RP tank.

DO NOT ALLOW STORED MATERIAL TO FREEZE. The expansion forces of solutions freezing are extremely large. Under certain conditions a storage tank may be badly damaged, if this is permitted to occur.

TAKE CARE OF THE INNER SURFACE. The use of shovels, pickaxes, or other gouging instruments in a tank may craze or damage the inside of the tank. Remember, the tank depends on the inner laminate for optimum chemical resistance.

REFERENCES

[1] Recommended Product Standard for Custom Contact Molded Reinforced Polyester Chemical Resistant Process Equipment, TS-122C, Sept. 18, 1968, pp. 13–16.

[2] "A Guide to Corrosion Control through Reinforced Plastics," p. 16, Monsanto Company, St. Louis, Mo.

[3] Kraus, Norbert J., Justin Enterprises, Inc., Fairfield, Ohio, personal communication, Dec. 21, 1966.

[4] Whiteman, R. L., Carolina Fiberglass, Wilson, N.C., series of personal communications through 1967.

[5] Stone, D. R., Metal Cladding, Inc., North Tonawanda, N.Y., personal communication, May 9, 1967.

⁶ Pearson, L. E., and A. B. Isham: *Materials Protection*, September, 1966.

⁷ Design Brochure, Amercoat Corp., Brea, Calif., Oct. 1, 1965.

⁸ Boggs, H. D.: Tentative Proposed Standards for Filament Wound Fiberglass Reinforced Tanks, Amercoat Corp., Brea, Calif., November, 1966.

⁹ Peter, R. I., and G. A. Stevenson, II: "Design Considerations in Reinforced Plastic Construction," Heil Process Equipment Corp., Cleveland, Ohio.

¹⁰ Pearson, L. E., and A. B. Isham: "Design and Testing of FRP Storage Tanks," Owens-Corning Fiberglas Corporation, Technical Center, Granville, Ohio, April, 1966.

¹¹ "Customer Acceptance Standards," Owens-Corning Fiberglas, Toledo, Ohio, Aug. 9, 1966.

¹² Ref. 7.

¹³ Owens-Corning Fiberglas *Chemical Tank News Letter* 7, Toledo, Ohio, July 6, 1967.

¹⁴ Non-corrosive Tanks, Owens-Corning Fiberglas *Bulletin* 1-PE 3578, Toledo, Ohio, May, 1966.

¹⁵ Milne, William M., Assistant General Manager, Beetle Plastics, Fall River, Mass., personal correspondence, Aug. 31, 1967.

¹⁶ Kauffman, D. K., Amercoat Corp., duVerre Div., Arcade, N.Y., personal communication, Aug. 25, 1967.

¹⁷ Whiteman, R. L., Carolina Fiberglass Corp., Wilson, N.C., personal communication, Aug. 26, 1967.

¹⁸ Roark, Raymond J.: "Formulas for Stress and Strain," pp. 280–281, McGraw-Hill Book Company, New York, 1954.

¹⁹ Engineering Specifications, no. 106, Justin Enterprises, Fairfield, Ohio, Mar. 15, 1966.

²⁰ Boggs, H. D.: Corrosion Resistant Storage Tanks, reprinted with permission from *Chemical Engineering*, Mar. 28, Copyright 1966, McGraw-Hill, Inc.

13

Electrically Grounding
Reinforced Plastic Systems

13.1 HOW STATIC CHARGES BUILD UP

Static electricity is generated when friction occurs between two dissimilar substances. Typical common events generating static electricity are:

1. A belt running over a pulley
2. Nonconductive liquids flowing through a pipe or falling into a tank
3. Walking across a rug on a dry day and touching an area of lower potential

The electrical capacitance of any body is a finite quantity. Ultimately, as the charge builds up, a spark will jump to a source of lower potential. The spark may be so small as to be unseen or unheard, but if it occurs in an area of highly combustible gas or vapor, disaster may ensue.

The electrical resistance of reinforced plastics may be considered as almost infinitely high since they are essentially a nonconductive material. There is no possibility, therefore, of using the reinforced plastic itself as a grounding mechanism for the control and removal of static electricity unless special steps are taken.

If the process is one in which heavy humidification occurs, the inside of the ducts, tanks, or process vessels may be constantly covered with a film of moisture, making a suitable path to ground, provided it is given the opportunity to go to ground. Vapors being carried away from heated solutions will often provide completely wetted tank and duct surfaces, but total reliance on such a possibility as the sole means of protection is inadvisable. There also are many areas in which fire and explosion may be a potential hazard that are not completely wetted all the time.

By far the largest number of applications of reinforced plastic tanks and duct systems lie in the fields of inorganic and organic acids, salts, oxidizing materials, and to a lesser degree, alkalis and organic solvents. In the fields of metallic salts, inorganic acids, and alkalis, electrical conductance in the solution itself is generally good. Even the vapor on duct walls above this material is generally sufficiently conductive since small amounts of the parent solution provide good contamination on the exhaust wall. To repeat, the safest course is to provide complete and adequate grounding regardless of the solution conductivity.

13.2   AN EXPLOSIVE EXAMPLE

The consequence of static charges in nongrounded chemical equipment has been amply demonstrated on many occasions. The following example of the cause and effect of such an explosion is abstracted from an insurance-company report:

> *Occupancy:* Large hydrochloric acid plant
> *Construction:* Tank of glass-fiber-reinforced plastic, in open
> *Wind:* Weather conditions stormy
> *Cause:* Ignition of hydrogen from common vent from process equipment and storage tank, probably by static

> The plant in question produced chlorine and caustic in mercury amalgam cells. Hydrogen from cells and waste chlorine are used to produce hydrochloric acid (HCl). Rated capacity of the HCl plant was 42 tons/day. The plant was highly automated. Static electricity from atmospheric conditions was thought to have ignited the hydrogen exhausting from the final scrubber. The fire flashed back instantaneously through the vent from the scrubber through the storage tank, igniting the hydrogen in the vapor space, and caused the explosion. The reinforced plastic tank in question was a 12,000-gal tank which was thought to be operating at about 9,000 gal working level. The tank was mounted horizontally. [Figure 13-1 shows the tank after the explosion.]

In this explosion note that both ends of the tank blew out, which is what a tank designer would have suspected, since a cylindrical tank barrel, inherently, is of greater strength than the dished ends in normal design.

13.3   REDUCING THE EXPLOSIVE HAZARD—
BASIC PRINCIPLES

What can be done in an explosive atmosphere to minimize the danger of catastrophic failure created by operation in the explosive limit range

FIG. 13-1 A reinforced plastic horizontal cylindrical tank after a hydrogen explosion. Note how the ends have blown out, which is what the designer would expect, since they are the weakest part of the vessel. Static electricity was the probable cause. (Photo courtesy Gagel Foto Service.)

and the destruction which may occur from static charges? The following approaches may be considered basic.

1. In the main storage tank *a system should be provided to keep the explosive gas (such as hydrogen) well below the lower explosive limit.* The basic operation of a simple system based on hydrogen as the explosive potential hazard is shown, conceptually, in Fig. 13-2. In this design an air-sweep duct is brought down into the tank and cut off above the maximum operating level of the tank. Air is continually pulled into the tank by means of negative pressure. Here, with the air spilling into the tank, the relationship of hydrogen and air is similar to that of mercury and water. The hydrogen is pulled out by means of an exhaust duct, and joins additional air from other sections of the process. It is then pumped to the atmosphere through a stack by means of a corrosion-resistant fan. Tests using conventional combustible indicators show that by such means the hydrogen level in the tank can be reduced from a

reading of 100 percent LEL (lower explosive limit) to readings of 2 to 5 percent LEL.

2. *Fresh liquid brought into the storage tank should be introduced below the minimum operating level to eliminate any static electricity induced by a falling stream.*

3. *All the inlet ductwork, and outlet ductwork up to the fan inlet, including the tank and all exposed interior piping in the tank, should be provided with a carbon-loaded reinforced polyester electrically conductive lining.*

4. Blanketing with an inert gas is recommended where it is a feasible procedure, compatible with the process.

5. In the event the engineer is dealing with a heavier-than-air explosive gas, a change in the basic principles is necessary:

    a. Run the exhaust duct down to a point just above the maximum operating level of the tank.

    b. Spill the fresh air into the uppermost part of the tank so as to provide constant displacement.

All the other factors in items 1 to 4 should be used where applicable.

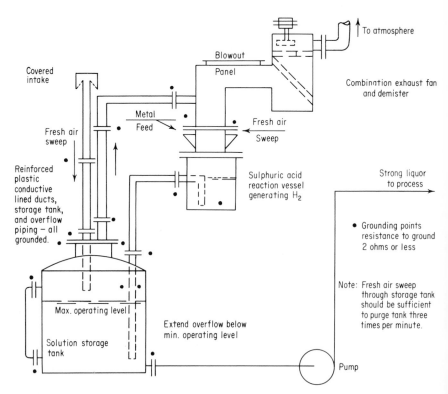

FIG. 13-2   Basic flow diagram for a lighter-than-air, potentially explosive system.

13.4   METHODS AND TECHNIQUES OF GROUNDING

Where the duct is involved and flanges are necessary, jumpers across the flanges may be used. This can be done with tabs of 316 stainless steel, 16 gauge by ½ in. wide, embedded in the flange and covered with the electrically conductive material. In flanges up to 4 in. in size, one tab is considered sufficient; on sizes 6 to 20 in., two strips should be used; on 24 in. and above the designer should go to four stainless-steel strips embedded in the flange. As we have pointed out, the electric resistance of a plastic tank or ductwork to ground with no carbon loading is for all practical purposes infinite. With the proper installation of a conductive lining, the resistance to ground, depending on the point taken in the system, should be less than 2 ohms. If this has been achieved, grounding has been satisfactorily accomplished.

The following general comments may be useful to the engineer in the application of such a grounding system.

1. A carbon loading of about 33 parts of carbon per 100 parts of resin should be used. If the resin mix falls much below 30 parts of carbon to 100 parts of resin, conductivity difficulties will occur. It is disheartening to line a system only to find that the desired results have not been achieved. Insist upon the vendor preparing a specimen in the approved manner and subjecting it to curing and electrical tests before proceeding further. This simple foresight may save trouble later.

2. The carbon to be used should be a fine graphite or a calcined petroleum coke flour, in all instances finer than 100 mesh.

3. The tank should be sandblasted or buffed thoroughly if it is an old tank. If it is a new tank, the lining could, of course, be built right in. On an old tank, in one method, it is suggested that a 10-mil surfacing mat be applied with resin to the ground or sandblasted tank. Immediately following this a 1½-oz mat, resin-saturated, is added. Finally, a 10-mil starched bound surfacing mat, heavily impregnated with a carbon-loaded resin, is used as a surfacing medium. Carbon loading is used on the starched bound mat only; it is not used on the surfacing mat or the 1½-oz mat. The starched bound mat should be applied to the undercoat after the undercoat has set for about an hour and has begun to gel well. The purpose of the undercoat is twofold:

   a. To ensure that a chemical-resistant surface has been restored to the tank

   b. To provide a highly adhesive undercoat system

In a second method carbon-loaded resin is applied to each of the layers of mat, that is, the 10-mil surfacing mat, the 1½-oz mat, and the final 10-mil surfacing mat. Either of the two methods detailed will work satisfactorily.   The second method is shown in detailed sequence in Figs. 13-3 to 13-9.

One other point: Carbon seems to have an inhibiting action on the gel time of the resin.   Briefly, this means that if you treat the resin as you normally would, it will never cure and will remain perpetually sticky.   Obviously, this will not do.   Some of the fabricating companies have developed special resin-catalyst systems to overcome this, which they consider to be proprietary in nature.   Test samples on carbon loadings in which the MEKPO has been increased above normal provided a satisfactory gel.   This, however, illustrates again the necessity of lining a small test piece to make sure the curing and conductivity are satisfactory.

Foam will trap explosive gases and create a most difficult situation. Minimize foam carry-over into a storage tank.   This can be done by taking the overflow from the bottom of the reaction vessel, or antifoam agents can be used where compatible with the process.

Other investigators[1] have suggested alternative methods for grounding nonconductive equipment, and point out that special methods are required to convey electrical charges to ground.   Other methods which

FIG. 13-3   A small-diameter flange being prepared for a grounding strap.   The flange has been completely prepared with carbon-filled resin.   A small recessed area for the grounding strap is clearly visible on the flange.

FIG. 13-4   The grounding strap is embedded in the flange and saturated with a carbon-filled resin.

FIG. 13-5   A finished grounding strap is shown on the 28-in. manway.   This is one of four grounding straps on the manway.

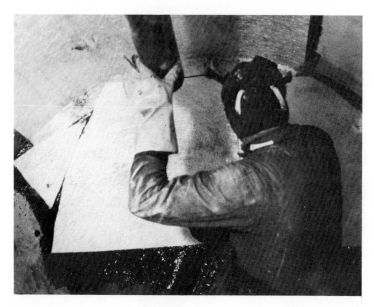

FIG. 13-6 There are several approved methods of lining. In this method a carbon-filled material has been applied to the ground surface, after which a 1½-oz mat is embedded in it.

FIG. 13-7 The 1½-oz mat is saturated with the carbon-filled resin.

have been suggested are:

a. The installation of an internal metal probe or probes extending the full height of the vessel and externally to ground.

b. The use of tantalum inserts through the skin of the vessel, again being led out externally to ground. Where conductive area in contact with the solution or vapor is small, the effectiveness of the inserts may be problematic, depending upon the conductivity of the liquid or vapor.

c. In a highly conductive solution it is suggested that adequate grounding may be obtained by applying a carbon-loaded liner to the wall in three 6-in. strips running vertically from the top to the bottom of the tank; then by passing through the tank wall and grounding these loaded carbon strips by means of a metallic insert properly embedded in the carbon loading.

Now that a well-grounded system has been installed, complete with air sweeps, the good engineer should not turn his back on it and assume

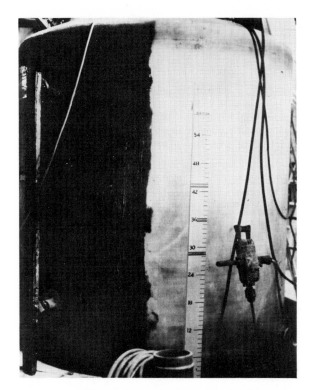

FIG. 13-8 The tank is half-finished. With a strong light inside it, it can be seen where the conductive lining begins and ends.

FIG. 13-9 The job is completed and in service. A light C mat has been added over the E glass, and it, too, has been carbon-resin-saturated. Grounding wires have been added to all flange tabs, and resistance to ground has been checked out at less than 2 ohms.

it will continue in a satisfactory condition indefinitely. It is particularly important that certain checks and tests be carried out routinely in a system of this type, where potentially explosive gaseous concentrations may exist. Some suggested tests and their frequency are shown in Table 13-1.

ROUTINE CHECKS FOR GROUNDING SYSTEM

Table 13-1

Daily: Check area for escaping gas or fumes from the equipment, using an approved gas analyzer or indicator.

Weekly: Check internal equipment, using a combustible-gas indicator to measure explosive concentration as a percent of LEL.

Monthly: Spot-check grounding system.

Thoroughly check area and internal-equipment concentrations, using a different combustible-gas indicator from the one used for weekly checks. Generally, this should be done by a different set of people for maximum reliability.

Check air-sweep system, including airflows and static pressures.

Quarterly: Rigorously check on all grounding.

Inspect wiring, tabs, etc.

Measure process air exhaust and static pressures.

With combustion indicator, analyze air-sweep flows, tanks, reaction vessels, ducts, and stack for gas level.

Use a static-electricity tester of an approved type to determine if static is being generated anywhere in the system. The dial on the meter should give a "safe" reading for a satisfactory check. Points of checking would be at the discretion of the engineer, but should provide a good random coverage.

Yearly: In addition to quarterly checks, inspect exhaust fan, inside of reaction vessel, ductwork where possible, and inside of storage tank.

REFERENCE

[1] Eichel, F. G.: Electrostatics, excerpted with permission from *Chemical Engineering*, Mar. 13, Copyright 1967, McGraw-Hill, Inc.

14

Chemical-Process-Equipment
Design in Reinforced Plastics—
Case Histories

14.1 INTRODUCTION

The foregoing chapters discussed the necessary groundwork for the application of reinforced plastics to the chemical industry. A book dealing with this subject would be incomplete if it did not include a chapter covering working illustrations of the use of this fascinating material. Such a chapter, besides, is a fitting tribute. It would be difficult not to be enthusiastic after having seen RP material looking as good as new at the end of five years, compared with a high nickel-chrome-molybdenum alloy with its service life span of two years. We have seen a good stainless-steel duct system disintegrate in the welds in several years, while its RP counterpart looked as good as new after six years. We have seen $7,000 lead-lined steel tanks being replaced with $2,500 RP tanks which give excellent service. True, tanks and ductwork and piping are a big item in chemical-equipment costs, and the successful solution to the corrosion problem in any one of these facets is a worthwhile effort. But there is so much more to this entire effort than simply tanks and piping and ductwork. This chapter of case histories seeks to demonstrate, through specific applications, successful use of RP material in a widely diversified field. In this field we cover such areas as unique crystallization equipment constructed of RP material, barometric-condenser design and application, massive exhaust-system installations, and a brief preview of the field of pollution-control equipment. Some unique methods that afford substantial reduction in ductwork-installation costs are explored, as is the field of reinforced plastic pumps. But there are also certain pitfalls in the use of RP material. There are whole areas that should be avoided, where the metallic counterpart will give more value for the dollar. The good engineer must be cognizant of these areas.

Reinforced plastic material is also an excellent "do-it-yourself" aid in the field of maintenance. How it has been used successfully makes many an interesting story. It can be used to repair not only itself, but metal structures as well, under many conditions.

The case histories described below are, of course, only a small number of the thousands which have been recorded in the last ten to fifteen years in the chemical industries alone—and this is probably only the beginning. New resins and reinforcing materials will extend the upper limits of usefulness in strength, temperature, and pressure.

The problem of education of the engineering design group in the field of reinforced plastic construction must necessarily be Step 1. If this program is not intelligently planned, the mistakes designers can make while learning may prove expensive. Most engineers and designers are by nature conservative—reluctant to gamble. As a group they tend to stay with the fields in which they have had previous satisfactory experience, even though a premium price may be exacted. Unfortunately, reinforced plastics commonly get lumped into the general group of plastics, and some of the experience with the earlier thermoplastics may have been less than satisfactory. In addition to this, any test program of a conservative nature takes a considerable amount of time to develop. The application of reinforced plastics on a significant scale in any company takes a considerable period of time, extending over a number of years. Too often companies become involved in crash programs immediately following a management decision to proceed with a project. Generally, no responsible person faced with a short deadline will turn to anything other than that which has been tested and tried; his eye is on satisfactory performance ratings, even at a premium price.[1] While the reinforced plastics vendor is generally only too willing to offer recommendations on the application of this material to proper service conditions, his experience, laboratory facilities, and span of test work may be too limited to provide long-term answers. Fortunately, quite a number of the resin suppliers have developed considerable service and case histories which may be useful.

The design ingenuity of the engineering group will undoubtedly expand the range of application. The use of RP material in elevated-pressure work has generally been avoided because it puts it in an unfavorable competitive design with lined metals. There is some hope, however, that the field of elevated pressures in the intermediate range up to several hundred pounds may be penetrated successfully through the use of ultra-high-strength laminates or especially designed honeycomb structures. Meanwhile, present RP applications are mainly confined in the larger process vessels to a range at or near atmospheric. In piping, of course, pressurized systems are easily attained and will operate successfully over extended periods of time. Structures operating under a full vacuum of considerable magnitude can be obtained competitively.

14.2 AIR–POLLUTION–CONTROL EQUIPMENT[2]

It has been said that each year nearly 150 million tons of atmospheric pollution is pumped into the air by mankind. Surprisingly, industry is not the major contributor, being responsible for, perhaps, less than one-third of the total contamination. The major contributor, is, Guess who? . . . You, as an individual! In the form of the most useful product of our Great Society—the automobile. But man is not about to legislate against himself. He will not legislate against his cars, his apartments, his incinerators, and his bad housekeeping. It is much easier to point the finger at industry, because industry is, basically, a small universe which carries little weight at the polls.

As for industry's part, each engineer and plant manager with a responsibility for air pollution must come to the realization, sooner or later, that the control of industry air pollution is a necessary part of his monthly budget. This has long been recognized and accepted in stream pollution. Air-pollution control today is behind stream-pollution control, but the approach, the direction, and the ultimate target are all the same.

The engineer is immediately struck with the inequities of the situation. At the present time laws and codes and enforcement are not uniform. What may be legal in one state is unlawful in another. Being realistic, the plant manager, with his staff, must devote his best efforts to finding solutions which present an economic approach, both with the initial capital investment and a minimum operating picture, and arrive at a solution that will comply with the legal requirements today and, hopefully, for the foreseeable future.

In the past the control of liquid effluent discharges into the receiving streams was, generally, a liability. Such control represented an effort to comply with the legal regulations. The waste-stream area became a part of the cost of doing business, and sometimes an expensive liability in the form of being a good neighbor. But this is not always the case. As research progressed through the years, the waste-treatment processes and their recovered products have been turned into a profit. Almost invariably they have resulted in better process control and in great achievements in process economies in reducing wastes. The social act of being a good neighbor quite often was transformed into a profitable product recovery. This is the case history of many industries. The proof is in the following list of both liquid and gaseous wastes which have been made to yield a net profit for the enterprising corporation.

Type of waste recovery	*Typical example*
The recovery and reuse of metal salts from waste streams	The recovery of zinc by an ion-exchange method and reuse in the process (or any bivalent ion of this type)
The recovery and reuse of acids from waste gas streams	The recovery of chromic acid from an entrainment of chromate mists
The recovery of liquifiable gases and reuse in product manufacture	The recovery of carbon disulfide waste gas streams by an absorption process
The recovery of gases by scrubbing methods	Chlorine recovery to form sodium hypochlorite through scrubbing with sodium hydroxide

The number of applications of scrubbers or air washers to solve chemical air-quality problems is almost limitless, in theory and in actual practice. One recent listing alone contained nearly seventy assorted scrubbers of varied designs to solve air-pollution problems. Some of these were:[2]

HF mist	HCl gas
Kerosene mist	Diammonium phosphate dust
HNO_3 mist	Chrome plating mists
H_3PO_4 mist	Nickel plating mists
H_2SO_4 mist	H_2S gases
Chromic acid mist	CS_2 gases
Na_2CO_3 dust	NH_4Cl dust
Na_2SO_4 dust	HCN gases
SO_2 gases	
NH_3 gas	
Cl_2 gas	

All these were controlled and removed by using reinforced plastic equipment.

There are many different kinds and types of air-pollution-control equipment. The use of reinforced plastics will be found to be most advantageous in the field of water scrubbers,[2] whether they be packed tower, wet cyclone, air washers, spray chambers, or towers, jets, or venturis.[3] A wet-scrubber selection guide should be consulted to identify the type that will produce a good or excellent efficiency (that is, above 85 percent) at a reasonable cost. Here the engineer must identify the kind of problem he faces, that is, gas-absorption mists under 10 microns, entrained liquids over 10 microns, and dust loadings, both size and amount.

Other considerations may be:

Water or liquid rate
Pressure loss
Power consumption
Initial capital investment
Overall operating cost
Reliability
Long-range maintenance picture

In the design of such equipment the engineer must constantly keep in mind the fire-hazard possibilities of the problem.   For further guidance he is referred to additional detailed comments in Chap. 11.   In general, the high-performance chemical-resistance resins, such as Atlac 382, Hetron 72, Hetron 92, Hetron 197, and Laminac 4173, are candidates for this type of construction.*

Some of the various styles of wet-scrubbing devices are shown in Figs. 14-1 to 14-6.

Reinforced plastic fume-handling fans also are well suited to this type of operation, particularly for package units running as high as 50,000 cfm.[4]   Fans of this type have been well developed, and experience in depth over a number of years proves their operation to be satisfactory.

The engineer can also add reinforced plastic pumps to handle the corrosive scrubbing solutions.   These are generally available up to sizes

* Many other good resins are available today which may be equally acceptable. Those named are meant to serve as a reference quality guide only. For a recommendation to suit a specific problem, the fabricator should be consulted.

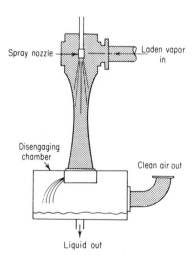

FIG. 14-1   Typical design of reinforced   plastic   venturi scrubbers.

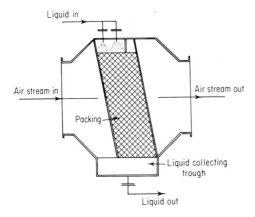

FIG. 14-2 Typical design of reinforced plastic cross-flow packed scrubber.

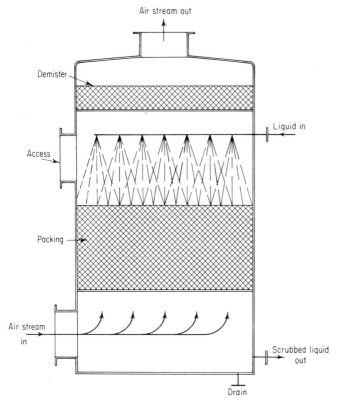

FIG. 14-3 Typical design of reinforced plastic countercurrent-flow packed scrubber.

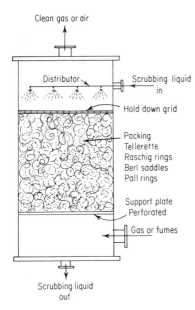

FIG. 14-4 Typical detail of reinforced plastic packed column.

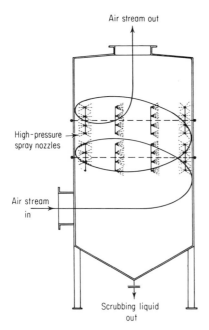

FIG. 14-5 Typical detail of reinforced plastic wet cyclone scrubber.

as large as 600 gpm and with substantial head capabilities. More details on pumps of this type will be found in another case history.

14.3 BAROMETRIC CONDENSER

Barometric, or open, condensers are commonly used with evaporators or crystallizers to provide the open and intimate condensing effect. They are almost a necessity in chemical plants where corrosive vapor and chemical buildup make closed condensers a problem. Many open condensers, especially in corrosive areas, are built of stainless steel or rubber-lined metal. In a recent redo of a chemical plant to increase condenser capacity, bids were received on five 48-in. condensers, each capable of condensing some 10,000 lb/hr of water vapor at approximately 2.6 in. Hg absolute and with a water flow to the condenser of 1,800 gpm at 86°F. Condensers built of reinforced plastic approximately 48 in. in diameter were found to be considerably less expensive than their rubber-lined counterparts. These units, inspected frequently over a 4-year period, have given excellent service.

For some reason one of the condensers was subjected to considerable buffeting, so that a vertical crack approximately 6 in. long developed on one side. This crack was plainly visible to the eye. A reinforced plastic overlay was placed over the cracked area and has operated satisfactorily ever since. Here ease of repair as an advantage of reinforced plastic was remarkably demonstrated. Generally, repair can be done

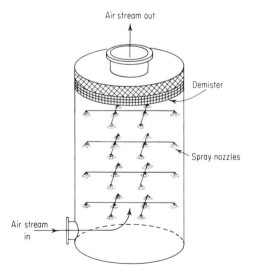

Air stream out

Demister

Spray nozzles

Air stream in

FIG. 14-6 Typical detail of reinforced plastic spray chamber.

the same day, and quite often without taking the piece of equipment out of service. No other material can boast this advantage. Figure 14-7 is a design of a 48-in. barometric condenser.

Calculations covering the design of a structure such as this are developed in the following problem.

A typical barometric condenser design

Problem

Design a barometric condenser to condense 5,800 lb of steam per hr at 25 psig pressure plus entrained vapor (total 10,000 lb/hr). Maximum feed temperature of the condenser feedwater is to be 86°F. Maximum water flow to the condenser is to be 1,800 gpm. Assume a tailpipe rise of 15°F and a maximum absolute pressure in the condenser of 2.6 in. of mercury. Calculate the wall

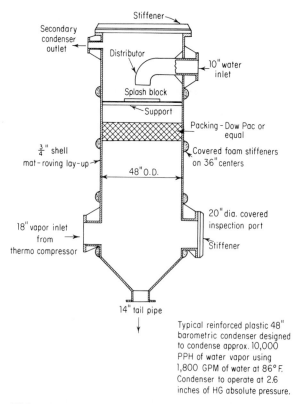

FIG. 14-7   48-in. barometric condenser in RP.

thickness of a 48-in.-dia condenser whose height is approximately 12 ft. Assume stiffener rings are constructed of a 5-in.-dia half-round with $\frac{1}{2}$-in. RP covering, with stiffeners on 3-ft centers. Further, a safety factor of 8:10 is desirable. A wind load of 20 psf is allowable if the condenser is to be installed outdoors.

Assume $E = 1,000,000$

2.6 in. abs $= 0.86$ psia $= 14.7 - 0.86 = 13.84$ psi   collapsing pressure

20 psf wind load $= \underline{0.14}$ psi

       Total collapsing pressure $= 13.98$ psi

$$W_c = KE \left(\frac{t}{D}\right)^3 \text{ psi}$$

$$\frac{1}{r} = \frac{3 \times 12}{24 \text{ (approx)}} = 1.5$$

Assume $t = 0.75$ in. as a trial calculation.

$$\frac{D}{t} = \frac{49.5}{0.75} = 66$$

If $D/t = 66$ and $1/r = 1.5$, then

$$K = 37$$

Based on a K of 37 and a thickness of 0.75 in., we can proceed as follows:

$$W_c = (37 \times 1,000,000) \left(\frac{0.75}{48.0}\right)^3 \text{ psi}$$

$$= 140 \text{ psi}$$

$$\text{Safety factor} = \frac{140}{13.98} = 10.0$$

If the safety factor had been found to be less than the desired 8:10 for this service, the assumed thickness would have been adjusted and the work would have been repeated. We can check the design of the stiffener ring that was assumed in the equation above as follows:

$$EI_c = \frac{W_s D^3 L_s}{24}$$

$$I_c = \frac{(13.98)(49.5)^3(36)}{(1,000,000)(24)}$$

$$= 2.56$$

The amount of inertia of a 5-in.-dia half-round covered with $\frac{1}{2}$-in. RP is 3.15.

Our final design, therefore, would be a condenser constructed of reinforced plastic with stiffener rings on 3-ft centers. Stiffener rings of 5-in.-dia half-round covered with $\frac{1}{2}$-in. RP would be spaced 3 ft apart on the barrel of the condenser. The condenser barrel, 12 ft long, would be $\frac{3}{4}$ in. thick.

Other considerations in the design of the condenser involve flow of water, flow of steam, adequate steam condensing, and proper water flow down the tailpipe. An inspection opening with an abrasion-resistant cover filled with 300-mesh silica flour should be placed opposite the steam inlet to provide a proper inspection port. Only this inspection port need have a silica-flour-filled surfacing agent. The inspection port should be made somewhat bigger than the steam inlet. The final design of this condenser would probably dictate a 10-in. water inlet, a suitable water splash plate for distribution, packing, an 18-in. steam inlet, a 20-in. inspection-port manhole opposite the steam inlet, and a 14-in.-dia tailpipe outlet at the bottom of the conical discharge from the condenser. A small 2-in. secondary ejector attachment nozzle would probably be used at the top of the condenser.

Similarly designed condensers have performed satisfactorily in service for 4 years up to the present time. Both service and inspections indicate the design to be structurally sound.

14.4 COOLING TOWERS

The use of cooling towers in the chemical industry is widespread, from small-unit assemblies to extremely large towers, providing cooling for tens of thousands of gallons of water. The need for the internal use of water is certain to grow as the demands on our current water supply expand with the growing population. Water reuse therefore will have to be practiced more and more intensively. As an example of such reuse, one large chemical plant draws some 13,000 gpm of freshwater from a mountain river, yet the total demands for the plant, including cooling, are nearly 90,000 gpm. The difference between water intake and total demand is made up by the effective reuse of water for cooling purposes on a continuous-recycling basis. For coping with large continuing water requirements, cooling towers are a paramount consideration in the overall solution to the problem.

Over the years cooling towers have generally been built of redwood or cypress. Wood has many advantages, including a relatively low capital cost, ease of availability, and certainly a long history of reasonably successful application. However, anyone who has operated wood towers in a severe environment recognizes that year-to-year maintenance costs

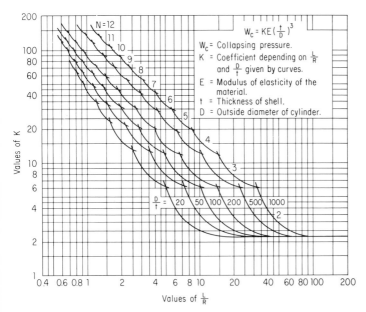

FIG. 14-8 Collapse coefficients: round cylinders with pressure on sides and ends, edges simply supported; $\mu = 0.30$.[5]

may be considerable, especially as the tower begins to age. The wood is subject to aggressive biological invasion, which quite often requires chemical-preservative treatment. Also, wood by its very nature is a relatively combustible material. A major fire-insurance underwriter, reporting on 100 cooling-tower fires since 1930, furnished the following interesting statistics:[6]

50 percent of the fires started while the tower was in operation.

38% of the fires were caused by sparks from an external source, such as stacks and incinerators.

20% occurred from cutting or welding.

13% were caused by violations of smoking prohibitions on the tower.

10% were electrical malfunctions.

19% were from unknown and miscellaneous causes.

A fire-hazard classification test more commonly used is the Underwriters' Laboratory tunnel test (ASTM E-84). In this test the flame-spread rating of various materials has been measured under draft conditions. Some of the reinforced plastic laminates made from the chemical-resistant chlorinated-type polyesters, although not as cheap as wood, are an excellent construction material for consideration in cooling-tower applications. Properly engineered, RP material is resistant to weathering, for all practical purposes is immune to tower corrosion, is not subject to biological deterioration, and by commonly accepted tunnel-test standards is rated as self-extinguishing. A comparison of flame-spread rating of some typical cooling-tower materials is presented in Table 14-1.

14.5   CORRECTING AN ERROR THE EASY WAY

How often the designer, with all the benefit of hindsight, must have wished he had made the tank 50 percent bigger! With RP material such a correction is relatively easy. See Fig. 14-9. Here two regeneration tanks needed to have their capacity enlarged by 50 percent because of an error in some process-regenerate calculations. Difficult to correct? No—not with reinforced plastics. In this particular case two tanks, approximately 5 ft tall, were cut off 6 in. below the rim, and a 30-in. barrel extended the tank with two butt joints. Total cost of the 50 percent increase per tank was less than $500. Note that the lid was

FIRE–HAZARD CLASSIFICATION—MATERIALS OF CONSTRUCTION FOR COOLING TOWERS[7]

Table 14-1

	Thickness, in.	Flame-spread rating, ASTM E-84
Asbestos-cement board	$\frac{1}{4}$	0
Treated plywood	$\frac{1}{4}$	35
Treated white pine lumber	$\frac{1}{4}$	30–65
Red oak	$\frac{1}{4}$	100
White pine	$\frac{1}{4}$	130
Plywood		100–200
Non-fire-retardant polyester sheet, corrugated, general purpose	$\frac{1}{16}$	400
Hetron 93LS, corrugated	$\frac{1}{16}$	35–75
Hetron 93LS plus 5% Sb_2O_3, corrugated	$\frac{1}{16}$	35
Hetron 92 plus 5% Sb_2O_3, flat	$\frac{1}{10}$	21

NOTE: Fire-resistant reinforced polyesters of the good chemical grades (the last three) are currently being used in cooling-tower construction for drift eliminators, spacers, fan stacks, louvers, basins, fans, fan decks, fill, partitions, and water-distribution systems.[7]

reused. The butt joint was made inside and out and finished with the all-important C glass veil.

14.6 CRYSTALLIZERS[8]

In 1966, a new approach to crystallization was conceived and developed, created from materials and techniques which have become available only within the last few years.

The new approach combined the best techniques of the reinforced plastics and the fluorocarbon resins to produce a significant breakthrough in this area of crystallization. Such a crystallizer was constructed using a 10,000-gal filament-wound reinforced polyester tank. Positioned in a "fairy ring" arrangement are 14 Teflon bundles of 650 tubes, each properly relaxed to provide fibrillation and gentle swaying in the agitated vessel. See Fig. 14-10 for a conceptual drawing of this unique crystallizer.

In the manufacture of rayon, the spinning process generates sodium sulfate and water. The water is removed by evaporation. The sodium sulfate is removed as Glauber's salt ($Na_2SO_4 \cdot 10H_2O$). Crystallization

FIG. 14-9 The ease of enlarging the capacity of a plastic tank by approximately 50 percent is demonstrated here. The lid has been cut off, and a 30-in. section inserted and strapped into the tank.

is normally done by a batch process in a vacuum crystallizer in which the batch is cooled from some 50 to 0 + 5°C.   Most rayon plants are equipped with the conventional vacuum crystallizers, which produce the vacuum through thermocompressors.   Normally, a vacuum crystallizer is a reasonably efficient unit at about 50 to 10°C.   Below 10°C its efficiency drops rapidly.   From 5 to 0°C efficiency is very poor and the

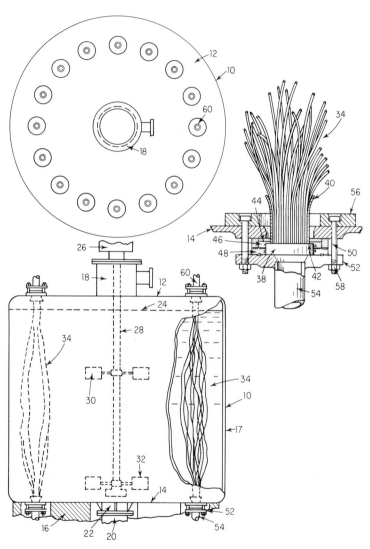

FIG. 14-10   10,000-gal filament-wound RP—Teflon crystallizer.

cooling curve slows up considerably. In the higher temperature ranges 1 lb of motive steam will evaporate nearly 1 lb of water. As the temperature drops, this ratio becomes increasingly poor, so that at temperatures from 5 to 0°C it may take 6 to 7 lb of steam to evaporate a pound of water.

For several of these uniquely designed crystallizers the entire project went from pilot plant to on-stream units in approximately six months. With vacuum crystallizers it could probably not have been done in less than 14 months.

Heat-transfer coefficients in the order of 60 to 65 Btu/(hr)(ft²)(°F) are continuously generated. Calcium chloride brine at −10°C in small-diameter tubes of 0.08 in. ID provides the cooling medium.

This design has scored a breakthrough in the crystallization field in many areas:

1. One of these 10,000-gal reinforced plastic crystallizers can remove two to three times more sodium sulfate than a conventional 10,000-gal vacuum crystallizer, which was our standard of comparison.

2. The total cost of this installed unit was about half the price of a vacuum crystallizer.

3. The capital outlay in terms of pounds of salt produced was approximately one-fourth to one-sixth that of a conventional vacuum crystallizer.

4. The unit performs about as well under winter conditions as it does under summer conditions, so that we no longer have poor performance due, simply, to high condenser water temperatures.

5. In terms of utility requirement, the brine crystallizer removes sodium sulfate at a cost approximately one-fifth that of a steam-driven vacuum crystallizer.

6. The brine crystallizer has been run for long periods of time, not only as a batch unit, but as a continuous-operating vehicle. Continuous crystallization provides improved process control over batch operation. Vacuum crystallizers are commonly operated by the batch method. Operating them under low temperature on a continuous basis would involve operating them in the poorest part of the cycle and be most inefficient.

7. Long-term studies have shown no decrease in heat transfer rates in these crystallization units. Within the limits of measurement no appreciable tube fouling occurred.

8. Operator attention to these units is minimized under constant operation, as a cursory check on the instrument control panel indicates at a glance whether operation is proceeding normally. Coolant flow in, coolant temperature in and out, acid temperature in and out, are con-

tinually monitored along with crystallizer levels and magma flow to the filters. A recording of brine tonnage indicates at a glance the work performed.

9. Operation on a continuous basis provides a constant demand on the refrigeration systems. Operation of the brine crystallizer on a batch basis provides a cyclic demand on the refrigeration systems which needs to be avoided if possible.

10. The use of such units as these provides much greater chemical process efficiencies as entrainment is completely eliminated and the constant output of the unit provides much simpler system control.   All of these result in higher chemical efficiencies of the spin-bath systems and considerable additional savings.

11. The maximum sodium sulphate yield per hour is achieved on a batch basis. Figure 14-10*a* shows a 10,000 gal reinforced plastic crystallizer being swung into position. Figure 14-10*b* shows three of the sixteen Teflon tube bundles in their operating positions.

FIG. 14-10*a*   A 10,000-gal filament-wound RP—Teflon crystallizer is swung into position.

FIG. 14-10b   Teflon tube bundles are shown in the brine crystallizer.

14.7 FRP AS A MAINTENANCE REPAIR MATERIAL

Every engineer is at some time or other faced with the problem of containing a leaking water or process line. Although minor leaks may be banded and the situation held in check for a few days, or perhaps somewhat longer, downtime for complete replacement is quite often expensive. Physical containment for long periods of time is an economic necessity.

The possibility of chemically "spinning a cocoon" around a leaking process line is a concrete reality. Bad pipe sections, and if necessary complete systems, may be renovated by encapsulation. No published standards for pipe encapsulation exist, but experience has established its own standards in areas where success has been achieved.

In the hand-laid-up encapsulation the finished laminate should contain about 70 percent resin and 30 percent glass. In a typical day's time a repair crew of two men may apply up to some 60 ft² in an 8-hr day. As the layers are applied and subsequent curing occurs, they merge into a solid homogeneous laminated structure. One corroded water line, 225 ft long and, principally, 60 in. in diameter, which was leaking badly, was repaired by this method.[9] The procedure is to get the pipe dry, sandblast it if possible, and go on with the laminated buildup. The choice of the resin depends upon the liquid to be contained. Here we have the advantage that the job can be done without restricting the flow of the water or process liquid in any way. And what is more, if directions are followed according to Table 14-2, the engineer need have little concern as to whether corrosion occurs in the metallic understructure. Even if the metal disappears, the plastic shell should still be able to contain the flowing liquid.

In wetting and curing, the laminated polyester will tend to shrink, so that adhesion to a dry pipe is not difficult. The temperature of the pipeline itself may have a profound effect upon the work being done. A constantly sweating line should be avoided. A line with the process at 180°F, on the other hand, will considerably accelerate both the setup and cure time. The higher temperatures also accelerate the emission of styrene fumes, so that additional ventilation should be provided.

It is important to encapsulate flanges, nuts, and bolt heads. Cap the flange edges with a piece of black friction tape, and this will become your mold. It is surprising how easily this can be done. Depending on the line size, the applicator may encapsulate 15 to 30 ft² per man per 8 hr.

USEFUL HINTS

Leaking lead-lined steel pipe conveying acid materials has been given a new lease on life by an overlay of a reinforced polyester. If the

ENCAPSULATING SPECIFICATIONS

Table 14-2

Size, in.	Laminate thickness 25 lb	Code	Laminate thickness 100 lb	Code
2	$\frac{3}{16}$	1–1–1–1	$\frac{3}{16}$	1–1–1–1
3	$\frac{3}{16}$	1–1–1–1	$\frac{3}{16}$	1–1–1–1
4	$\frac{3}{16}$	1–1–1–1	$\frac{1}{4}$	1–1–1–1–2–1
6	$\frac{3}{16}$	1–1–1–1	$\frac{1}{4}$	1–1–1–1–2–1
8	$\frac{3}{16}$	1–1–1–1	$\frac{5}{16}$	1–1–1–1–2–1–1
10	$\frac{3}{16}$	1–1–1–1	$\frac{3}{8}$	1–1–1–1–2–1–2–1
12	$\frac{3}{16}$	1–1–1–1	$\frac{7}{16}$	1–1–1–1–2–1–2–1–1

CODE: 1—1½-oz mat, approximately 0.045 in. thick.
2—24-oz woven roving, approximately 0.085 in. thick.
NOTE: The final thickness will depend on how tight the mat is rolled or pressed out. In actual practice, in heavier laminates the thickness may be 10 to 20 percent less than the thinner laminates, due to compression and roll-out.

internal solution is hot, so much the better, since it provides rapid kickoff and cure.

Leaking lead ductwork can be easily repaired by using a suitable overlay.

Guard against spot internal repairs to rubber-lined steel vacuum equipment. It is extremely difficult to get the corrosive chemicals completely out of the rubber; especially under a vacuum, such patches have a tendency to fall off.

Nevertheless, it is completely possible to repair a rubber-lined steel tank in toto with an RP polyester. Here a small amount of flexibilizing resin will be found to be beneficial. It has been possible for the engineer to build a tank within a tank simply by using the rubber-lined structure as a form. Any technique using the other vessel or pipe for a form has a much greater chance of succeeding.

All sorts of vacuum equipment can be provided with temporary repairs on the outside by the quick addition of an RP polyester patch over an offending hole, provided service conditions are suitable.

The secret of success in any of these repairs is surface preparation. A clean dry surface will ensure a successful repair. Guard against repairs made under wet or high-humidity conditions or cold temperatures, that is, below 60°F.

14.8 THE REPAIR OF RP STRUCTURES

Once in a great while the engineer is confronted with the deterioration of a reinforced plastic structure. Such deterioration can come from:

1. An inherent manufacturing defect, which may have been caused by:
   a. Inadequate roll-out and air occlusion
   b. Improper resin control
   c. Inadequate laminate construction (for example, no C glass on the surface)
2. Damage prior to or after installation caused by:
   a. Severe buffeting and pounding from steam-injection heaters
   b. Improper supporting of the structure or its appendages
   c. Improper allowance for expansion

Assuming that the service specifications of the structure are adequate, the engineer may be greatly mystified to find a number of structures performing quite well, in a solution to the problem, only to have, perhaps, the sixth or tenth structure delaminate or come apart. In the event that repair to the interior of the structure is found necessary, the following guidelines, based on practical experience, may be helpful:

1. Cut away all loose sections of RP material.
2. Dry the understructure as well as possible. (Refer to Sec. 7.1, under butt joints.)
3. Sand or buff the remaining RP material until you have achieved penetration into the completely dry material in all areas. At this stage of the proceedings, the engineer must be particularly careful to see that this preparation has been complete. For example, it is possible to buff into an area which appears to be completely dry and satisfactory for the repair lay-up. Nevertheless, a checkup is necessary to make sure that the corrosive solution has not penetrated farther into the lay-up. This may show as blossoming wet spots, which will appear at random places within the structure, indicating that all the water or destructive solution has not been removed. In such a case, if practical, heat both sides of the laminate with heat-producing lamps or blowers to raise the laminate temperature to 150 to 200°F to drive out the solution. If this is not done, subsequent failure and delamination of the repair is almost certain. In dealing with an acid solution, scrubbing the surface with weak ammonia is very helpful. To repeat, for a successful repair, a completely dry basic area on which to make the repair is essential.
4. In any repair one must guard against attack from the edges. This can be minimized by feathering the mat. It is satisfactory to cut the edge of the mat to the approximate dimension required and then pull out and fragment the ends of the mat. This will provide a tapered feather edge which will help prevent endwise attack.
5. If we assume that surface delamination has occurred, perhaps extending into the structure $\frac{1}{8}$ in., the corrosion resistance of the structure may be restored by:
   a. Sanding down to a good dry workable base.
   b. Washing the workable base with a very dilute ammonia solution and letting it dry completely.
   c. Laying down and saturating with the appropriate corrosion-resistant resin two $1\frac{1}{2}$-oz mats.
   d. Applying in succession two layers of 10-mil C mat.
   e. Finishing up with a hot coat. (NOTE: The terms "hot coat" and "wax coat" are synonymous. Hot coat is the term generally used by the fabricators.) This ensures an adequate cure. In dealing with a vertical surface, CAB-O-SIL should be added to thicken the resin to eliminate drainage from the vertical surface.

If additional structural strength is required, the above plan may be modified to include alternate layers of $1\frac{1}{2}$-oz mat and 24-oz woven roving to provide a laminate of high impact strength. Laminate thickness can be built up to any strength desired. However, laminate buildup greater than $\frac{3}{8}$ in. at any single lay-up should be avoided due to exotherm problems. If the engineer intends to go to a heavier wall thickness than $\frac{3}{8}$ in., the hot coat should be held off until last. Subsequent laminate repairs or additions, where necessary, should never be added until the hot coat is removed through buffing.

In dealing with the repair of a structure from the outside in, consider applying several layers of 10-mil C glass before applying the $1\frac{1}{2}$-oz mat. The C glass is all-important as a corrosion barrier. Make sure that all bad laminate is cut out and the fresh raw edges are sealed with resin.

As described in Chap. 3, repairs to piping or small circular structures are easily achieved by putting a complete wrap around the object of repair. Here the engineer is taking advantage of the inherent shrinking qualities of the polyester resin to obtain a drum-tight fit.

14.9 HIGH–COST RP ARRANGEMENTS

Piping Avoid complicated RP arrangements. Where it is necessary to provide a large amount of RP fabrication in a small area, the cost may be excessively high. For example, one installation projected putting 43 1-in. nipples on a 2-in.-dia header some 60 ft long. Each one of the small nipples was fitted with a $\frac{1}{8}$-in. orifice which could be removed for cleaning. In RP construction this involves too many joints to be profitable. In this particular case the header cost about $600. The same header could have been "cobbled up" out of lead for about $300. In any RP arrangement, the fewer the joints, the greater the profit.

Tanks A second type of high-cost RP construction is the use of rectangular tanks. A 5,000-gal RP tank built in a rectangular shape may very well cost $3,500; the same tank capacity in a cylindrical configuration can be bought for approximately $2,000, and perhaps less.

Elevated-pressure process vessels The greatest economies in process vessels can be achieved in gravity vessels or those structures designable in RP at moderate wall thicknesses, be it pressure or vacuum operation. When we go to larger-diameter vessels under elevated pressure, wall thicknesses become increasingly heavy, so that RP material becomes less attractive and lined-metal structures present a more favorable economic picture. True, some process vessels have been built in RP to operate at substantial process pressures, but these are special cases and not the general rule.

14.10   RP PUMPS

To make an economic comparison of glass-reinforced plastic pumps, either polyester or epoxy, quotations were obtained on a specified pump from nine different vendors, providing 16 alternatives in both plastics and exotic metals.   The quotations requested were for a pump to handle 300 gpm of a sulfuric acid solution, containing also sodium and zinc salts at a 90-ft head.   The solution would have a specific gravity of 1.27 and a temperature of 60°C.   As was to be expected, quotations varied from vendor to vendor, even for the same materials of construction. In general, a 3,500-rpm pump is less expensive than a 1,750-rpm pump because size is reduced.   Quotations were provided with the mechanical seal and motor as an addition so that the costs of the bare pump and base plate could be definitely pinpointed.

Table 14-3 indicates the type of material and the relative cost for the pump described above.

ANALYSIS OF SIXTEEN BIDS.   While 316 stainless steel is listed in Table 14-3, it would not be a satisfactory material of construction for the solution in question.   Worthite would be borderline in this application. Only the glass-reinforced plastics (or silica-filled epoxy), Carpenter 20, carbon, and the extremely expensive high-nickel alloys such as R-55 and Hastelloy C could be considered satisfactory.   Operational experience with glass-reinforced polyester in this solution over a period of 24 months has been excellent.

CHEMICAL–SOLUTION PUMP PLUS
BASE—RELATIVE COSTS

Table 14-3

Material of construction	*Relative cost*
Glass-reinforced polyester	1.0
Glass-reinforced epoxy (or silica-filled epoxy)	1.0–1.5
316 stainless steel	1.0–1.4
Worthite	1.5
Armored polypropylene	1.6
Carpenter 20	1.7–2.3
R-55	1.7
Carbon	1.8
Hastelloy C or Y-17 alloy	1.9–2.2

Entirely on a custom basis, some glass-reinforced polyester pumps of a relatively large size have been constructed handling thousands of gallons per minute. Standard lines of glass-reinforced epoxy pumps are being marketed in sizes up to 700 gpm and heads of 400 ft.

The maximum field experience exists in the glass-reinforced epoxy pumps or composite-plastic materials. The basic advantages of these pumps are as follows:

1. They are compatible with many liquids commonly used in the chemical-processing industries.

2. They possess an excellent resistance to erosion and will quite often outlast metals in this area.

3. They are considerably lighter in weight and much easier to handle than their metallic counterparts.

4. The inside surface of the pump is glassy smooth, which contributes to minimum internal friction.

5. The fact that many resins are available along with various resin/glass ratios permits a wide customizing of the individual pump to meet the purchaser's problems.

6. They are, of course, nonmagnetic and nonsparking.

7. They combine high performance with a relatively low cost.

One of the greatest assets of a plastic pump is its ease of repair. In the event that the pump is inadvertently damaged, repairs are easily made, often the same day, to restore it to an "as new" condition. This has been repeatedly demonstrated in the initial development programs on these pumps. Consider the alternative. It is impossible to think of repairing a high-nickel-alloy pump which has suffered erosion, chemical attack, or mechanical damage. A damaged high-nickel pump can only be sold for scrap. Not so reinforced plastics. With a minimum of training of personnel, these pumps can be restored to their original condition at a low cost, and quite often returned to service the same day, if desired. This is an asset that can be claimed by few pumps in the chemical-resistant field. It is difficult to quarrel with the purchase of high-nickel performance for about the same price as 316 stainless steel. Such is the demonstrated capability of reinforced plastics material in chemical service. Bare pumps which ordinarily cost $1,200 to $1,400 in a high-nickel alloy can be bought for $600 in glass-reinforced plastics. This, of course, is for the particular size shown in Table 14-3. In a large chemical plant, where pumping maintenance costs and capital expense costs can be considerable, this factor alone may offer sizable savings.

Certain additional development programs are still necessary to realize the full capability of this phase of reinforced plastics application. Most probably, these developmental programs will continue to progress.

Pumping efficiencies of a reinforced plastic pump are equal to the highest available in normal commercial practice.

Table 14-4 will acquaint the reader with a complete line of pumps currently marketed in a cast epoxy which shows excellent corrosion resistance and is priced somewhat higher than a glass-reinforced polyester pump. Pumps of the capacity shown in the table, however, are immediately available on the open market in the complete line shown.

14.11 RP ROTARY DRYER[10]

A dryer installed in August, 1959, in a large eastern chemical plant has been operating successfully ever since. At one stage of the chemical process, solids wet with hydrochloric acid are treated in this continuous rotary dryer, where they are subjected to air heated to 225 to 290°F. Vapor composed of free hydrogen chloride and water passes through the exit housing for separation from the material being dried. The Hetron 72 dryer replaced the original metal-alloy dryer.

The dryer was fabricated by duVerre Products, Amercoat Corporation, Arcade, N.Y. Included with the dryer unit were Hetron 72 exit-gas separator housings, hot-air inlet ducts, product-discharge housings, and internal dryer lifting vanes. See Fig. 14-11. This photograph was taken after 6 months service, and it can be seen that there are no signs of corrosion or erosion of the lifting vanes. Another important feature was

RECENTLY MARKETED PUMPS OF CAST EPOXY

Table 14-4

Pump size, in.	Pump capacity at maximum efficiency	Pump cost	Motor size based on 1,750 rpm and water, hp
1½ × 1	55 gpm at 48-ft head	$ 360	1
1½ × 1	110 gpm at 160-ft head	360	7½ (3,500 rpm)
1½ × 1½	130 gpm at 52-ft head	670	3
2 × 2	220 gpm at 98-ft head	820	10
3 × 2	250 gpm at 86-ft head	930	10
4 × 3	500 gpm at 80-ft head	1,020	15
4 × 3	700 gpm at 65-ft head	1,020	15

NOTE: The pumps may also be equipped with 1,150-rpm motors and will run at a reduced capacity. These prices are for the *bare pump only*. Base, coupling, and motor costs are additional.

FIG. 14-11 RP rotary dryer with internal flights being used to dry solids wet with hydrochloric acid at temperatures of 225 to 290°F.

that the product itself did not become contaminated.   Fire retardancy was a specific requirement for this application.

14.12 A SECTIONALIZED TANK

Quite often the engineer is faced with the problem of replacing a tank in a room.   On looking about the installation he finds that, short of knocking a hole in the wall, the only path of entrance is through a door, which may measure 30 × 96 in.   To remove a worn-out lined tank is no particular problem—a blowtorch may be sufficient; or the tank can be dismantled in sections quite expeditiously and lugged out through the doorway.   To put a tank in place, however, is more difficult.   One of the solutions which have been developed is the sectionalized tank.

Figure 14-12 is a good example of the rectangular sectionalized tank. Within limits the same principles can be adapted to circular tanks.   In Fig. 14-12 the tank has been fabricated in three sections to permit field assembly.   The lower section is placed on a suitable foundation fully supported.   Next, section 2 is mounted and bolted in place.   Note the stainless-steel bolts along the section flanges.   The top section is then put in place on the center section.   The previous design was such that the maximum deflection on the tank wall does not exceed ½ percent of the wall length.   Half-rounds have been added for stiffening along the tank wall in a vertical direction.   In some designs the sections are made to nestle in each other to a depth of 3 or 4 in.   Upon assembly of the tank sections, an internal wrap is made at the joint along the tank wall.   Such a wrap is equal in thickness to the tank wall itself, generally 6 to 8 in. wide, and finished off with a layer of C glass and a final good wax coating to which CAB-O-SIL has been added.

Generally, the weakest points in a rectangular tank are the upright vertical corners.   For this reason great care must be taken to ensure that the design is heavy enough in this area.   In the author's experience, failures in the tank corners are much more likely to happen than failures through deflection on the long side.   If, as is done in some cases, steel angles are used for stiffening members, they can be welded at the corners to form a continuous structure.   Following this, overlaying with RP

FIG. 14-12   This rectangular sectionalized tank permits building a big tank in a small space, eliminating the problem of passing through narrow doors and passageways.

material at least ¼ in. thick will provide suitable protection. It should be borne in mind that a reinforced plastic tank has to be corrosion-resistant outside as well as inside, since many tanks fail from the outside in. For this reason everything should be suitably protected from the ravages of external corrosion.

Note in Fig. 14-12 that the nozzles are gusseted to provide adequate strength. Also, lidding for the tank has been provided, and RP hand grabs can be seen. The tank shown has been in service for several years under severe corrosive conditions with zero maintenance requirement. A recent inspection indicated that it was in excellent condition.

14.13 SKY DERRICK CUTS RIGGING COST OF RP DUCT 80 PERCENT

Many chemical plants are faced with unusual material-handling problems. Suppose you had to place on the roof of a large plant some 500 linear ft of ductwork with nothing but the blue sky above you for rigging. How do you get it there? The conventional approach would be to lift it up onto the plant roof by means of a crane; then, by using following rollers, work the duct slowly to the point where it is to be raised into previously erected cradles. Next a stiff-legged derrick or gin pole would be used, or if the length of ductwork was 40 or 60 ft, two stiff-legged derricks would have to be set up. We could further complicate the matter and assume a situation where we had to go over buildings on top of buildings, 400-lb steam lines, main-plant water lines carrying the lifeblood of the mill, and, for good measure, a few 93 percent sulfuric acid lines plus assorted transformers and electric power lines.

One large chemical plant installed a large carbon disulfide recovery system as part of the scheme of overall process improvement for cost reduction. Such a process, involving activated carbon beds, required 800 ft of 40-in.-dia ductwork. Some 500 ft was over the heart of the plant, all above the roof, in distances 12 to 30 ft above the roof line. Placing this ductwork in position required careful planning.

Such a duct system was constructed of a reinforced chlorinated polyester (Hetron 92 with 5 percent Sb_2O_3 added) to withstand the continuous evacuation of some 28,000 cfm of acid-laden exhaust gases containing small quantities of carbon disulfide to be piped into a CS_2 recovery system for reuse in the plant. It was estimated that the physical installation of such a duct system into hangers previously prepared would require the services of six men, plus a crane operator (and a crane) and a foreman, for a period of 30 workdays. At current prices the bill for this operation would run approximately \$10,500. In addition, it is necessary to get together the gin poles, lines, hoists, block and tackle, rollers, and assorted plywood sheets to minimize damage to the plant roofing as the material is rolled across.   It will be seen that the $10,500 is probably conservative. Small wonder that it took a lot of figuring to see how this cost could be reduced.   The method finally decided upon, and used, was a helicopter, serving as a stiff-legged derrick.   Let us look at the economics of that operation:

The rental of a Sikorsky S-58 helicopter complete with a crew
   of three for 1 day.................................... $1,800
One foreman and four men at ground lifting position....... 140
One foreman and five men at roof receiving position........ 165
Rigging and slings..................................... 100
    Total... $2,200

Now let us compare the two in cost.   By the conventional method a cost of $10,500 was anticipated; the use of a helicopter with the ground support personnel cost $2,200.   We have not only saved money, but more important, we have saved time.   Time is of the essence in competitive industry. We have now released personnel to do more important work simply because we have "worked smarter, not harder."   Here we have the opportunity to reduce placement costs by 80 percent by using modern materials and up-to-date techniques, and instead of taking 30 days for placement on the roof, it has actually taken 1 day.

We shall now take a closer look at this operation.

The helicopter and crew   The helicopter used was a Sikorsky S-58 machine with a crew of three.   The pilot, an ex-Marine veteran with 10 years experience with this particular type of aircraft, came with the all-important ground crew.   One of the men was assigned to the ground at the ductwork pickup area.   The second man was assigned to the roof (or placement end) of the operation.   While this particular aircraft has 4,000-lb lift capability, it is wise to limit the lifting requirement to some 3,500 lb, so that the aircraft will be truly responsive under all conditions. Such an aircraft service is available in many of the big cities.   The aircraft in coming from its home base had a normal cruising speed of 90 knots.

The lift   The problem was to use this helicopter with crew and ground support personnel for placing in previously prepared saddles some 29 pieces of 40-in.-dia $\frac{1}{4}$-in.-wall reinforced plastic duct.   Much of this duct was not straight pieces, but consisted of offsets, 90° ells, etc.   The longest piece was a 40-ft section, to which had been added two 90° ells and several straight sections, so that the helicopter pilot was faced with handling unwieldy sections. Many of the other pieces consisted of 40-ft-long straight sections. The simplest piece was a 40-in.-dia 90° ell. Weights ranged from a minimum of, perhaps, 300 lb to a maximum of 1,500 lb. Weight alone is not the final criterion of the difficulty. Many chemical plants have stacks, pipelines, adjacent buildings, electrical transformers, etc., all sticking above the roof, making the project more difficult, so that tag lines plus spreaders can easily run to 75 ft. The aircraft is thus at least this distance above the roof, with the pilot relying on the direction of his ground crew for guidance.

Typical time schedule   Let us study the typical time schedule of airlift operation. All the ductwork was laid out in a grassy area near the plant, in consecutive pieces. See Fig. 14-13. Both ground and roof personnel had the master plan, with a plot of each piece, where it fit into the puzzle, giving its number and the sequence in which it would be picked up.

FIG. 14-13 Sky-derrick rigging.   Helicopter picks up a 40-ft section of ductwork previously laid out in a grassy area.

Before the airlift, the design and field groups had made certain that the units would fit.   The following schedule was typical for such an operation:

Time	Event
8:55 **A.M.**	Helicopter arrived on site.
9:10–10:00	Final site inspection by the pilot and his ground crew, along with plant personnel, going over all the possible problems that could be anticipated in the area.
10:05	Start aircraft.
10:10	First-piece lift-off.
10:10–10:55	Five lifts onto the roof. Nine pieces placed into position. (Some of these had been fastened together before lift.)
10:55–11:10	Aircraft refueling.
11:15–11:50	Four more lifts onto the roof, in which 10 units were placed in position (Fig. 14-14).
Noon to 1:00 **P.M.**	Lunch.
1:00–1:30	Three lifts with five pieces of ductwork placed in position.

FIG. 14-14 Sky-derrick rigging.   The 40-ft section is carried over the plant roof.   Note the long spreaders.   It is laid in previously prepared cradles.

1:30–1:50	Aircraft refueling.
1:50–2:20	Five lifts with five 40-ft sections put in place.
2:20	Completion of job.
2:20–2:50	Aircraft refueling.
2:55	Aircraft leaves plant property, heading back to Philadelphia.

HELPFUL HINTS ON THE USE OF HELICOPTERS IN THIS TYPE OF WORK.

1. An experienced pilot and ground crew are essential. Use only the best.

2. Make sure that planning for the project is adequate. Nothing will better ensure the safety of a project than good planning.

3. Safety

   a. All personnel at both ground and roof locations must wear hard hats with chin straps and goggles.

   b. A walkie-talkie at both locations is helpful, but not indispensable. While the aircraft is over either of the installations the walkie-talkie will be useless (the pilot and his roof ground crewmen will have two-way radios for communication plus earphones for good reception and to drown out the noise of the aircraft).

   c. Thermal currents over a plant can be deceptive. Cooling towers, stacks, both large and small, massive heated areas, windmill refractions from plant buildings, all contribute to aircraft performance and stability. Thermal currents affect the pilot's ability to perform precise work.

   d. All loose boards, etc., from the roof should be removed prior to the operation because the air velocity from the helicopter blades is considerable.

   e. It is not likely that a slag roof will suffer any damage. This is one of the more common questions asked with the use of this type of equipment.

   f. All windows should be closed in the area of operation. If not, the glass may be broken.

   g. Block off any vehicular or pedestrian traffic in the areas of operation of the aircraft. Keep the spectators at an absolute minimum.

   h. Designate a safe area away from the plant, apart from any area of combustibility, for aircraft refueling. It is to the advantage of the aircraft to operate with a minimum of fuel in order to have the maximum liftability.

4. The loss of ground effect with the aircraft may be great. The aircraft operates on a cushion of air when close to the ground. As it gets farther and farther away from the ground, the cushion effect becomes less

pronounced, so that projected lifts and the distance of the aircraft above a flat surface need to be gone over in detail with the pilot before the work is begun.

5. Tag lines should be put on all pieces to be lifted so that both the ground and roof crews have something to take hold of for positioning the lifted objects.   The question of tag lines should be reviewed with the pilot in detail.   Tag lines that are too long force the pilot to stay too far above his reference points.   Tag lines that are too short make it difficult for the roof crew to position the object.   Tag lines of 15 to 30 ft are generally the two extremes.   In large pieces of ductwork which may be 40 ft long, tag lines should be at least 20 ft apart.

6. The uninitiated ground crews rapidly learn to work with the aircraft.   After several lifts are made, the novelty and fear of the aircraft have worn off; the crews settle down to a smooth working routine.   It is quite conceivable that, although only 500 ft of 40-in.-dia duct was put in

FIG. 14-15 Sky-derrick rigging.   Steam or water lines are obstacles that can be successfully surmounted.

FIG. 14-16   Sky-derrick rigging.   The maximum length of ductwork lifted by the aircraft was 60 ft, including several elbows.

position, in the lift job described above, 800 to 900 ft in a day's time could equally well be achieved for very little additional expense, and the saving would be even greater.

7. The fact that a steam or water line is above a projected installation is not necessarily an insurmountable obstacle.   A good pilot can "thread a needle" with relatively big pieces of equipment.   His dexterity is amazing.   See Fig. 14-15.

8. Do not make the load overly long.   The maximum length of this lift was some 60 ft, and that particular unit was well away from any interference (Fig. 14-16).   Normally, a length of 40 ft is ample.   This will make it possible to prepare many joints on the ground, and the minimum number of joints above the roof.   Large loads tend to sway and oscillate like a pendulum, and their maneuverability becomes increasingly difficult. Sections 80 ft long of large diameter might conceivably represent difficulty, not only due to their total weight, but also to susceptibility to pendulum effects in the air.

It takes only a little more imagination to conceive other such cost-reduction ideas, using such a sky derrick to place items of process equipment in inaccessible areas, as long as space is available in the sky above. Evaporators, tanks, piping, transformers, building materials, all have been placed in inaccessible plant positions by means of helicopters. The reverse is also true: it is sometimes desirable to remove equipment from bad locations. The aircraft will work equally well in either direction. Make sure you stay within the weight-lifting capability of the aircraft (Fig. 14-17).

A few last comments on the subject:

The contracting aircraft should deposit with your company a liability certificate covering possible loss of a reasonable amount on liability coverage. Sometimes this is not enough. Review the proposed airlift with your own insurance company to make sure that you have the benefit of their professional advice in planning the job.

One of the obvious things that makes an operation such as this possible is the light weight and high strength/weight ratio of RP duct-

FIG. 14-17 Sky-derrick rigging. A good pilot can work close to buildings and obstacles.

FIG. 14-18   6-in. duct drops are easily attached to the main 16-in. duct by means of Morris Spanner couplings.

work.   Here again, a new-generation material permits the use of new techniques and devices which a few short years ago existed only in the realm of science fiction.

14.14   SPINNING EXHAUST SYSTEM[8]

To provide additional exhaust in a rayon-spinning area an RP duct-and-fan system was installed to cope with the acid fumes.   The capital cost of this installation, completed in early 1965, approached $250,000.   A chlorinated polyester resin, to which 5 percent antimony trioxide (Sb_2O_3) was added to enhance its already excellent fire-retardant qualities, was the material of choice.   This was installed by a contractor, using Morris Spanner couplings and butt joints in the bigger sizes.   The system was further completely equipped with reinforced plastic exhaust fans.   To simplify system design, the total problem was divided into five medium-sized and five large systems, aggregating in the design total 300,000 cfm. The system was equipped with suitable exhaust stacks and weather heads to discharge the fumes a good distance above the plant. This combination of excellent corrosion resistance and ease of construction and erection, coupled with good fire-retardant properties, makes these chlorinated polyesters superb materials of construction for this type of work. Duct sizes ranged from 6 to 42 in. Except for a little bearing trouble with several of the fans, the system has been essentially trouble-free. For a photograph of the inside of the plant, see Fig. 14-18. Figure 14-19 is a photograph of the outside of the plant showing a number of the 30-ft exhaust stacks.

The viscose rayon process is generally characterized by the installation of a very large duct system carrying large quantities of exhaust air which contains minor amounts of acidic entrainment. Normally, these systems were built of Koroseal-, Neoprene-, or Duroprene-lined metal in sizes varying from 10 to 60 in., and in many cases even much larger. Today, nearly all duct systems in the 10- to 60-in. sizes are constructed of a chlorinated polyester to which 5 percent antimony trioxide has been added. Figure 14-20 shows a replacement of a section of Neoprene-lined steel duct with 16- and 42-in.-dia reinforced polyester duct. Many duct

FIG. 14-19 Four duct stacks complete with hoods and fans. They are part of a battery of 10 individual systems.

FIG. 14-20   Sections of Neoprene-lined steel duct system which had failed are replaced with RP ductwork.

systems of this type have been in service for periods up to 6 years with excellent results.

14.15   UNDERGROUND GASOLINE–STORAGE TANKS[11]

One of the dramatic increases in RP composite gasoline-storage tankage has been in underground tanks. In the last several years more than 1,500 of these have been installed by top gasoline suppliers in place of leaking, rusting metal tanks. It is interesting to look at a map of the United States prepared not by state boundaries or by topographic elevations but by underground corrosion potential for metal tanks. This map is the battleground on which RP composite gasoline-storage tanks have staked their claim. For areas showing saline-water infiltration or other corrosive influences, the map is extremely important. A 6,000-gal underground storage tank (steel) for gasoline normally costs about $660. In volume production an RP composite tank has arrived on the marketplace at $950 to $1,000 FOB the factory. This noncorrosive tank—for use where rust was so prevalent in metal-tank installations—*need not be filtered out before delivery!*

14.16 ZINC ION–EXCHANGE PROCESS[8]

In the recovery of zinc by an ion-exchange process it became necessary to filter the zinc-rich liquor prior to the ion-exchange stage. This was accomplished by using a 12-ft-dia gravity filter loaded with a bed of graded Lachine quartz. This filter material has performed exceptionally well in producing an effluent of the highest quality. Normal filtering rate is about 4 gpm/ft². Filter and ion-exchange operation is on a completely automatic basis and requires the minimum of operator attention. Flow through the filter is normally 400 gpm. The backwash liquor is stored in a 15,000-gal RP tank. The filters themselves are a complete RP design. Recovered zinc solution for return to the process is stored in a 3,500-gal RP tank. The regenerate tanks and piping are also of RP material. The zinc ion-exchange units are of rubber-lined steel due to the high operating pressures. A reduction in price on interconnecting piping of over $7,000 was realized in this installation by the use of RP piping in place of rubber-lined steel. Such a process as this is capable of recovering zinc from a dilute acid solution where zinc concentrations are measured in terms of 400 to 600 ppm and raising it to concentrations of 13 percent $ZnSO_4$. Capital-payout time is on the order of two years before taxes.

For a view of this zinc ion-exchange installation and the use of RP material, refer to Fig. 14-21.

14.17 TEN CAPSULE CASE HISTORIES

• 120 plastic air scrubbers,[12] at a cost of 2.8 million dollars, were installed at Intalco Aluminum Corp., Ferndale, Wash., for the removal of HF gas and abrasive alumina dust in an electrolytic reduction facility. Each scrubber, with a rated capacity of 106,000 cfm, was 40 ft long, 12 ft wide, and 25 ft high and weighed 9 tons. Water was the scrubbing reagent. Surface mat was a modacrylic fiber veil. Anticipated life is 15 to 20 years. The remainder of the construction was glass-reinforced polyester.

• Conventional plate-and-frame pressure filters have been constructed from DAP polyester resins with acrylic fibers and pressure-molded to the required shape. These have proved successful in corrosive applications where leakage from corroded metal plates had previously resulted in a considerable chemical loss. One particularly successful application was the filtration of zinc hydroxide from a saturated sodium sulfate solution at a pH of 10 and a temperature of 190°F.

FIG. 14-21 Two 12-ft-dia gravity filters and a 15,000-gal storage tank in an ion-exchange system.

• A fume stack some 200 ft high and 42 in. in diameter was constructed of glass-reinforced polyester.[13] This stack carried hot gases at 200°F, consisting of SO_2, water vapor, and an entrainment of nitric acid and sulfuric acid. It was designed for wind velocities of 95 mph. At the time of reporting, 3 years after installation, it was performing successfully. In the design of this stack, strengths used were reduced to compensate for the high temperature and possible thin sections.

• Expansion joints of reinforced plastic were used successfully in a duct carrying flue gases from a spent-liquor-recovery boiler.[14] These RP joints replaced previous metals made of stainless steel, Inconel, and nickel, which had a service life of about 3 years. Cost of the RP joints was approximately half of the metal joint.

• A 2,000-gal horizontal settling tank[15] subjected to varying concentrations of salts, HCl, esters, and chlorides, which were alternately acidic (2 to 3 pH) and alkaline (8 to 10 pH), with a temperature of 85 to 90°C, was originally constructed of lead-lined steel at a cost of $5,200 and a yearly maintenance cost of approximately $800. A new tank of glass-reinforced polyester (Atlac 382) at $2,100 was maintenance-free over a 5-year period up to the time the report was written.

• A 750-gal chlorine absorber[15] normally containing a 20 percent slurry of calcium hydroxide in which conversion was made to calcium hypochlorite at a temperature of 25 to 75°C, was originally constructed using a PVC-lined steel at a cost of $3,400. This was replaced by a glass-reinforced polyester absorber at a cost of $2,180. Previous maintenance costs had averaged $480 per year. Except for a nozzle repair at $140 in the RP absorber, there had been no further expenses in 7 years.

• A 150-ft stack, 18 to 30 in. in diameter[15] and carrying mists of phosphoric acid, muriatic acid, hydrogen sulfide, sulfur dioxide, and phosphorus pentoxide, plus residual water, had been constructed of rubber-lined steel at a cost of $5,800. To maintain the stack, the fume scrubber, the demister, and duct, which in aggregate had cost $27,500, required an annual expenditure of $11,000. All these units were replaced with a reinforced polyester at a cost of $34,000 because the lined-steel equipment showed excessive blistering and failing. In metal tests Hastelloy C proved to be the only acceptable alloy, at a probable cost exceeding $100,000. Repairs to the RP equipment have averaged $300 per year since installation.

• In the food industry a large pickle packer[16] using 304 stainless steel changed to a reinforced polyester (Atlac 382) because the stainless-steel tank showed pitting after 6 to 8 months. Attack presumably came from the chloride ions in the presence of acid and lactic acids. The small RP tank cost $150 as compared with $600 for the stainless steel. RP tanks are enjoying an increasing acceptance in the food industry by the packers of fruits, vegetables, juices, vinegar, sauces, and even wines.

• The faulty operation of metallic vacuum equipment, such as condensers, piping, and low-pressure thermocompressors, can sometimes be traced to leaks in the equipment due to corrosion. Welded areas are particularly susceptible to corrosive influences, especially where light-gauge metal has been used. Quite often minor leaks in this type of equipment due to the ravages of corrosion reach the epidemic stage. Where the original design may have been in one of the nickel-chrome alloys, the engineer may find that he can encapsulate the thermocompressor or condenser with four to six layers of $1\frac{1}{2}$-oz mat saturated with a high-grade chemical-resistant flexibilized polyester composition, finishing up with a resin coat to which a thixotropic agent has been added. A flexibilized resin mixture such as 80 percent Atlac 382 and 20 percent Atlac 387 may be used to guard against cycling temperatures. All this can be done for one-quarter to one-fifth of the replacement cost of the original thermocompressor or condenser. Field experience with this type of maintenance has been most rewarding in improving equipment reliability.

• Many engineers have watched with dismay deterioration of heat-exchange shells from condensate corrosion or from tube leakage of a corrosive material into the shell area. Other heat-exchange requirements put corrosive media both in the shell and tubes, so that a corrosion-resistant design of heat-exchanger shells becomes a real problem. Previous attempts at solving this problem have generally tried stainless steel, rubber-lined steel, phenolic-based shell coatings, and sometimes lead-lined steel over the bottom 18 in. of the shell. All these are expensive approaches to a troublesome problem. It is possible to have made, on a custom-design basis, reinforced plastic heat-exchanger shells that will prove good in continuous service to 275 to 300°F and immune to attacks from a whole area of corrosive influences to which the polyesters or epoxies are normally resistant. In smaller shell sizes it is quite possible to use standard pipe, with inlets and outlets and flanges added to suit. Where shell sizes are larger or an in-between size, they can be custom-designed and made to order at a nominal cost. One such special shell, approximately 20 in. in diameter and some 9 ft long, built to operate continuously with 25-psig saturated steam and complete with special 8-in. inlets, 4-in. outlets, and 2-in. vents, and equipped with supports, with all flanges drilled, was purchased for less than $1,000. This was approximately half the price of the stainless-steel shell for the same job.

REFERENCES

[1] Wirfel, E. W.: The Project Engineer's Point of View: Plastics in Chemical Plant Construction, *Material Protection*, August, 1966.

[2] Air Pollution Control Equipment, The Ceilcote Co. *Bulletin* 12-1, Berea, Ohio, 1967.

[3] Rigidon Solid Plastic Ventilation Products, Heil Process Equipment Corp. *Bulletin* B-570, Cleveland, Ohio.

[4] A considerable number of manufacturers have well-developed fan lines in RP corrosion equipment. A few of these are Buffalo Forge Co., The Ceilcote Co., Debothezat, Hartzell, and Heil Process Equipment Corp.

[5] Sturm, R. G.: A Study of the Collapsing Pressure of Thin-walled Cylinders, *University of Illinois Bulletin* 329, p. 32, Urbana, Ill.

[6] Wilson, J. A.: Letter to Committee Members, NFPA Committee on Water Cooling Towers, Factory Mutual Insurance Co., June 28, 1965.

[7] Fire Retardant Fiberglass Reinforced Polyester as a Material of Construction for Cooling Towers, Hooker Chemical Corp., Durez Plastics Div., *Bulletin* 1261-G, North Tonawanda, N.Y.

[8] Mallinson, J. H.: Field Histories and Problems with Reinforced Plastics in the Textile Industry, paper presented at National Association of Corrosion Engineers, Niagara Frontier Sec., Niagara Falls, N.Y., Aug. 10, 1967.

[9] Editorial Staff: *Plant Services*, November–December, 1966.

[10] Szymanski, W. A., Hooker Chemical Co., Durez Div., North Tonawanda, N.Y., personal communication, Apr. 10, 1967.

[11] Pistole, Robert, Owens-Corning Fiberglass Co., Huntington, Pa., personal

interview, October, 1966. For additional information refer to Underground Storage Tanks, Owens-Corning Fiberglass *Bulletin* 1-PE-2780A, February, 1966.

[12] 120 Plastic Air Scrubbers, *Plant Services*, March–April, 1967, p. 4.

[13] Clark, James: Reinforced Polyester Plastics, *Materials Protection*, August, 1966, p. 13.

[14] Yovino, J., and E. Costello: Reinforced Plastics for the Pulp and Paper Industry, *Materials Protection*, October, 1966, p. 19.

[15] Fenner, Otto: Plastic for Process Equipment, excerpted with permission from *Chemical Engineering*, Nov. 9, Copyright 1964, McGraw-Hill, Inc.

[16] "Speaking Out," Atlas Chemical Industries, Inc., Wilmington, Del., July, 1966.

15

The Quest for
Minimum Bidding Prices

15.1 THE PHILOSOPHY OF DESIGN

There is a great deal that the engineer can do to achieve the most satisfactory RP installation at the lowest cost. True, it varies with the type of equipment, but there are certain fundamentals which, if followed, will substantially increase job satisfaction. Ultimately, the engineer must become a participant in the design. Too often the engineer simply specifies the size of the duct and its general configuration, or the shape of the tank, with its overall dimensions and nozzles physically located. Rationalizing that the manufacturer is more knowledgeable than he, he then waits for the bids to come in. Too often he assumes, as he examines the bids, that each unit is similarly designed, and that it remains for him only to select the low bidder. *Nothing could be further from the truth.*

After purchasing a number of vessels or systems (they could hardly be said to have been "engineered," even construing the word "engineered" literally), the engineer or designer finally begins to see the light. With experience he comes to realize that true satisfaction, both for him and for his company, can be achieved only by making his designs concrete and absolute. He learns that he must inject himself into RP designs to the point where he can discuss them knowledgeably with vendors. He must feel with assurance that his approval of the vendor's shop drawings will result in satisfactory equipment which will perform the job as intended. The vendor himself is of course an excellent source of educational information. Cross-fertilization of his experience with the vendor's know-how is one of the prime ways in which the neophyte may learn and the experienced engineer improve his knowledge of the trade.

With this general approach in mind, some of the concrete aspects of engineering design are as follows:

CONSIDER THE SHAPE OF YOUR DESIGN. The largest volume obtained with the least surface area is a sphere, and some tanks have been built in this shape. More conventionally, the largest volume for the surface material is a cylinder. The easiest shape constructed on a dollars-per-gallon basis is the cylinder, because it can be put together rapidly on a rotating mandrel and gives maximum strength.

DO NOT OVERDESIGN OR UNDERDESIGN. Cost of overdesign is the premium paid on an insurance policy carried to cover ignorance. The symptoms of underdesign are shortened service life and premature failure.

THE DESIGN SHOULD BE REALISTICALLY CONCEIVED AND SOUNDLY ENGINEERED. You need to know your resins, your structural design, and especially the SPI specifications. Bids should state that they are in accordance with SPI specifications, or if there is any deviation, a complete description of the laminate, safety factors, and other deviations or alternatives should be clearly stated. RP material should be applied only where you are confident it will perform well. Some of the common errors committed in this area are:

1. The use of partial SPI specifications
2. Pulling SPI specifications out of context
3. Using a less desirable safety factor, for example, 5:1 instead of 10:1 on chemical piping or equipment
4. Improper laminate specifications, such as using woven roving all the way through
5. Using a general-purpose or isophthalic resin where a high-chemical-resistant resin should be specified

A sure way to be trapped is to assume that the inferior offering is equal to an offering with complete SPI specifications. If you are going to accept a watered-down version of the equipment requirement, then in your own best interest go back to all the vendors for a quote on the watered-down basis. Frequently, the amount of money to be saved in these "economy" versions is not enough to justify the elimination of the margin of safety furnished by the extra quality. At the outside the savings in going from a high-chemical-resistant resin to an intermediate-type resin may be 10 to 12 percent. The savings in going from a high-chemical-resistant resin to a general-purpose resin may be 25 percent. The difference between one classification and the next may be the difference between 10 to 15 years of uninterrupted service and problems within 12 months. The application should be considered in this light.

CONSIDER STANDARDIZATION. Ultimately, standardization must lead to lowest cost. Can you utilize the standard designs presently manufactured by some fabricators?   If so, it is to your advantage to use them, since this will introduce real economies in your design.   Assembly-line techniques in standard fans, pipe, and tanks are common.

PURCHASING MULTIPLE UNITS AT ONE TIME WILL REDUCE COST.   This is an extension of the cheaper-by-the-dozen philosophy and is almost universally applicable whether you are buying canned goods, potatoes, or reinforced plastic material.   In the design of a vessel, the second one can probably be bought for 10 percent less than the first one.   This is due to common engineering and setup charges.

APPROVE THE VENDOR'S DRAWINGS PRIOR TO FABRICATION.   If the vendor intends to redraw your drawings for shop work, you should insist on approving his drawings prior to fabrication.   Some vendors make a shop drawing of each drawing furnished them by a purchaser.   Now shop redrawing is not cheap.   Unfortunately, it has to be done in a great majority of cases; the vendor will require it in order to produce a completely satisfactory job.   The vendor must do this because so many of his clients present inadequate drawings, cobbled up by designers with little knowledge of RP construction.   *One of the ways to save money is by avoiding this cost of redrawing.   Make sure that your drawings are sufficiently detailed and knowledgeable to permit the vendor to use them directly for shop fabrication.*   Re-engineering costs $6 to $15 an hour, depending on the complexity of the work involved.   By educating yourself and your staff, you can save these re-engineering costs, which have to be added to the cost of the equipment purchased.

EXAMINE VENDOR'S DRAWINGS CLOSELY BEFORE APPROVING THEM. Rome was not built in a day, and knowledge and experience in any field of specialization come hard.   The chances are that you will be approving vendor's drawings even when such drawings are nothing more than a recopy of what you have done, with, possibly, some alterations to suit RP design.   Study alterations closely to make sure that they have not changed the functional operation of the piece of equipment involved. In one such case the vendor changed the slope and depth of the conical outlet of a large barometric condenser.   The change was not detected, the condenser was built, and when the unit went into operation condenser water levels were so high they interfered with the exhaust steam from the thermocompressor.   The condenser bottom had to be redone, which, fortunately, is not too hard with reinforced plastic material.   The moral of the story is, look at the vendor's redrawing with a critical eye and make sure that his alterations, where done, are satisfactory.

MAKE SURE THE ORIGINAL VENDOR QUOTES ON SUBSEQUENT PURCHASES.   Most fabricators save the more complicated molds.   This is important, since, even if you make your second purchase a year or more later, using the same fabricator can save you dollars.   Do not buy your engineering a second time.

15.2   SELECTING THE VENDOR

KNOW THE CAPABILITY OF THE RP VENDORS WITH WHOM YOU PROPOSE TO DO BUSINESS.   Obtaining bids from four or five good vendors is generally considered sufficient to obtain good bidding, but you owe it to yourself to visit and inspect each of the vendors' shops and become acquainted with their organization.   Only by knowing their strengths and weaknesses are you able to interpret their bidding realistically. Further, a reinspection of the facilities every year or two is good practice, to see what progress they are making.   If such visits are timed to coincide with the inspection of equipment being built for you, your visit will be most welcome.   (However, do not leave the impression that you are on a "mining expedition.")

Your mental checklist on such a visit might include the following items:

1. Become familiar with the vendor's ability to construct adequate low-cost molds.   Knowing the vendor's molding capabilities can be most rewarding.   For example, in the bidding for two pieces, constituting a tapered thermocompressor, one bidder quoted $1,500 for the mold and $1,500 for the thermocompressor, while a second bidder quoted $75 for the mold and $1,000 for the thermocompressor.   Guess who got the job? Now, obviously, there was a large difference in conception as to how to build the mold.

2. Look at his plant construction and his plant safety program.

3. See what kind of research effort he is putting forth.

4. Look at his quality-control procedure.   Does quality control exist in name or in fact?

5. What is his labor situation?   Does he have a high labor turnover rate or does he have a lot of old-timers?   The seasoned worker makes a quality product.   Satisfactory hand-laid-up equipment can be produced with a few months training, but there is a great difference in quality on spray lay-up between a few months and a few years.   The use of female help in this industry is becoming more widespread.   On repetitive work women excel.

6. How long has the vendor been in business?   A minimum of 2 years should be required, 5 years being desirable.

7. Look at the equipment that is being built for others.   What new ideas can you pick up?

8. Do you like the vendor's assembly methods?

9. Are the inner surfaces "resin-rich" or "resin starved?"

10. Is he using C glass on all inner surfaces?

11. How good is his engineering shop?

12. Does he have a technical group or use a consultant?

13. Does he have a test program on laminates to ensure the necessary strengths? Are his laminates substantially free from occluded air? (Remember, a little bit of air will lower laminate strengths a great deal.)

14. Consider his capability of handling big tanks and structures. One vendor estimated that in-shop handling of large tanks cost him as much as 20 percent of the total cost of the tank. Obviously, improving the handling situation would permit him to lower his bidding.

15. It is to your advantage to correlate a vendor's bidding with his capabilities. For this you must know his labor rates.

16. Are you paying for a brand-new plant through his depreciation additions to his job?

17. Has he automated his process to the point where labor-hours on a vessel of constant size is but a fraction of competition's? It is surprising how far one can go in this area. For example, filament winding of large cylindrical vessels lends itself to automation and decreased labor costs.

INVESTIGATE THE VENDOR'S FINANCIAL RELIABILITY. Many of the reinforced plastics operations are comparatively small, employing perhaps 80 to 150 people (and sometimes considerably less). Gross business may run from 0.5 to 4 million dollars per year. (Some large companies are now entering the business, and with aggressive merchandising, this volume in their divisions may be considerably exceeded.) Individual jobs placed with RP fabricators may begin at a few hundred dollars. Many are in the $2,000 to $10,000 category. Some, each year, go into the six-figure size, and occasionally an RP job will run into seven figures. There are a dozen or more shops in every industrial state in the nation. Since transportation can become a considerable factor in the cost of a job, regional competitions are felt to a great degree. Most of a fabricator's business probably lies within a 500-mile radius of his plant. Recognizing this, some companies have spread into multiplant locations or distributing centers. As the size of the project increases and the dollar value becomes greater, regional influences become less pronounced. In this frame of reference the purchaser should consider his probable suppliers. It is useless to go beyond perhaps a half dozen vendors in the quest for minimum bidding prices, and since the purchaser pays the freight, these suppliers should be located within 500 miles of the purchaser's point of use.

Visits to other facilities can be most educational to you in broadening your knowledge of the industry and permitting you to adopt selectively the latest techniques. It goes without saying that you are morally bound to keep trade secrets to yourself.

15.3 OBTAINING CONTRACTOR BIDDING

The design is complete, and you are ready for bidding. Through Purchasing, issue your drawing and inquiry requisition to the approved vendors. Invite all of them to come in on a certain day at a prescribed time to review the job in detail with you. (This, of course, applies to large jobs. One would be reluctant to do this unless the job were of considerable magnitude.) Meet with all the vendors at one time. (The worst way is to have the vendors come in one at a time, since this involves endless repetition and wastes your staff's time, with a good chance that all the vendors will not have had the same story.) With the designer, estimator, responsible field engineer, and purchasing agent if possible, in attendance, the project engineer should review in detail the following items:

THE CONCEPTION OF THE JOB

THE WORK TO BE PERFORMED

LABOR AND RELATED EFFECTS. Make certain, prior to execution of any agreement with the vendor, that the legal requirements with respect to contracting have been met (assuming, of course, that a union represents your production and maintenance employees). The vendor should be made to understand that any union or nonunion problems created by his hiring procedures and for the makeup of his work force are entirely his responsibility and that you will not accept any charge directly or indirectly arising from labor problems of any nature.

ACCEPTABLE ALTERNATIVES. As an example, in duct stiffening there is considerable variance as to what constitutes the minimum-cost satisfactory stiffener from vendor to vendor. Vendors should be given the opportunity to quote a guaranteed alternative.

CATALYST RESPONSIBILITY. It is advisable to assume responsibility for the catalyst; otherwise you will lose all catalyst control and your safety program on catalysts will go with it. Remember, all the vendors may not be as safety-conscious as you are, and if you leave it to them, they may come in lugging the catalyst in gallon containers, where your safety program limited the size to 2 oz.

SAFETY REGULATIONS. Cover these in regard to:

Smoking
Goggles
Hard hats

Tank safety requirements
Electrical safety and the use of grounded electrical tools
Solvent regulations and solvent disposal
Location of dispensary and first-aid equipment, including safety
showers and eye fountains
Safety regulations peculiar to your process

IMPORTANT DATES. These are the dates when you expect the proposal to be returned, when you expect to have material on site, and if the vendor is responsible for field erection, the time limit set for completion of the job. Make these dates specific.

MISCELLANEOUS. Give the designer, the estimator, and field engineer an opportunity to add to the conference in their area of expertise. Permit any or all of the bidders to ask questions on any item of the drawings for bidding which is not clear to them.

On RP work to be performed by the contractor on site make sure that he maintains adequate insurance coverage. Request that he have proof of such coverage submitted to you by his insurance carrier.

No claims for extra compensation should be allowed unless properly authorized and approved by the project engineer.

Finally, summarize, in writing, all discussion, so that everyone is bidding on the same basis.

15.4   ADVANCE PLANNING

Plan your requirements sufficiently in advance to eliminate the necessity of the vendor using overtime for completion. Jobs of this type generally are negotiated on a cost-plus basis. Overtime work is not as productive as straight-time work; and do not forget the labor premium built into the job.

15.5   TRAFFIC

Know your shipping routes and shipping maximums from the vendor's plant to your plant. If immediate delivery is of prime importance, schedule a truck to carry the equipment straight through. There have been cases where it took longer to deliver the item than to build it. This has occurred in the case of some fairly complicated RP equipment. Transportation can be fairly expensive on large pieces of RP equipment and, in the case of oversized loads, may require special trucks, and separate vehicles acting as flagmen, running both in front and in back. In addition, permits must be obtained and routes approved by the states in question.   Try to plan your job so that you can ship the least expensive way.   Investigate both truck and rail shipments.   Be particularly careful about the size of your shipment since 12 ft in diameter is near maximum.   Be particularly careful to keep nozzles inboard; otherwise the tank may arrive with the nozzles sheared off.

16
Afterthoughts

16.1 INTRODUCTION

Probably everyone who has authored a technical book has looked at the finished manuscript and thought of items which were omitted. Obviously, the subject cannot be entirely covered in a single volume, nor is it expected to be. To attempt to do so would lead to regions of research of little interest to the harried engineer or designer at the plant level. Yet even a technical book of this modest approach can, from the experience and correlated research which inspired it, unveil many gems of wisdom.

The remaining bits of information offered in this chapter will, perhaps, shed new light on some of the still dark corners of the reinforced plastics field. Hopefully, it will fill in areas in which little published information exists, especially in the practical aspect.

16.2 BUY UP TO A STANDARD, NOT DOWN TO A PRICE

In nearly every form of business endeavor ethics and business practice vary widely. Common experience has taught us that we can expect to run the gamut from the most ethical corporation, whose word is their bond, to garage-shop operators with little visible assets, a nonexistent overhead, and a portfolio of sharp practices sufficient to give the entire industry a bad name. Size alone, of course, is not an absolute criterion, nor is it expected to be. The RP business, however, can be entered with a small amount of capital, so that there may be operators who were lured into it for a "quick killing."

Regretfully, minimum bidding prices on piping (or other equipment) can be obtained in hand-laid-up work by shading the piping thickness. Cases have been known where this has been done even on relatively large orders. In one case a vendor made some forty-five 20-ft sections of 8- and 10-in. pipe, all of which was turned down on the purchaser's inspection because they were ⅛ in. below Commercial Standard specifications for thickness. The Standard specifications permit occasional spots that are 80 percent of the normal thickness, but this does not mean that the fabricator can continually shade the thickness to improve profitability. The purchase order was immediately canceled, solely because the vendor was trying to sell a 50-lb pipe at a 100-lb price. Obviously, practices like this will not be condoned or tolerated. It is, however, to your advantage, as in every other form of purchasing, to make sure that you are getting value received for dollar expended. This applies not only to piping but to ductwork, tanks, and any RP material for which standards or specifications have been established. As a purchaser you must consider the many check points to ensure that you are receiving a quality product. In addition to thickness and adherence to design specifications, look for poor wet-out, presence of dry spots, air in the mat, voids and nonuniformity, prohibitive voids at the joints, thin gel coats, careless workmanship, absence of quality control, poor gusseting and support, thin flanges, and lack of exterior coating. Where tanks or structures are being purchased, have the vendor ship to you all nozzle cutouts for your examination. This is good practice since it serves notice to the vendor that his laminate lay-up will be subjected to close study. The best assurance of a quality product lies in dealing with a vendor of established reputation.

16.3   COBALT NAPHTHENATE

Published literature on cobalt naphthenate is indeed meager. Normally, it is purchased as a 6 percent solution in mineral spirits. Therefore the solute has added to the original compound the possibility of dermatitis and of fire and explosion. One corporation has been producing cobalt naphthenate for a period of 25 years with no difficulty whatsoever. Basically, they have found that there has been no local or systemic toxic manifestations. It is neither an irritant nor a dermatitic producer. However, in the manufacture of the material, the manufacturer insists that its workers, because of the mineral spirits, and not the cobalt naphthenate, wear vinyl gloves and goggles. The normal mixing of a 6 percent cobalt naphthenate solution with a polyester resin should be done at room temperatures, and certainly no heat is to be applied. Cobalt is toxic only when inhaled, and in the ordinary use of this compound there should be no such danger. A good safety rule needs here to be repeated: Cobalt naphthenate must never be mixed with methyl ethyl ketone peroxide, or an explosion may occur. For this reason storage of the two materials is always in separate cabinets. The 6 percent solution in mineral spirits simply indicates that the mineral spirits should be treated with the same safety precautions as one would treat kerosene, gasoline, and other hydrocarbon solvents.

16.4 THE SPRAY–LAY–UP SYSTEM[1]

It is difficult not to become enthusiastic about the potential of spray-up systems in handling reinforced polyesters as coverings or lining materials in plant maintenance work. Engineering of these systems by contract vendors has progressed to the point where they have become highly dependable low-maintenance operations. The possible application of this equipment is limited only by the imagination of the user. Actually, a spray-up system is not only a maintenance tool, but also a manufacturing tool. Of course, the usual safety precautions would have to be taken to ensure the proper use of this tool.

What is it?   Basically, we are dealing with a spray gun that individually sprays a cobalt-promoted polyester resin, a resin into which a catalyst has been inserted, and chopped-fiber-glass roving. At the discretion of the operator, the system automatically and continually mixes and applies a preratioed intermixture of glass, resin, and catalyst. It is done with an "airless" three-head spray-up gun. Systems can be purchased for anywhere from $1,500 for the most simple system to $8,400 for a completely independent system mounted on a trailer and ready to be towed behind a truck. For example, the writer has seen demonstrated a system costing $5,000 which mounted all the equipment on a portable truck and was complete with a 15-ft boom. This system will apply to any surface a mixture of chopped-fiber-glass roving and resin in a wide range of capabilities. For example:

1. The fiber length can be regulated from $\frac{1}{4}$ to 3 in. long for maximum strength. Experimental work has determined that maximum strength of the finished lay-up lies in the $\frac{15}{16}$- to $1\frac{1}{2}$-in.-long fiber area. Start with $\frac{15}{16}$-in.-long fibers.
2. The ratio of fiber can be varied over a wide range of 8% fiber glass/92% resin to 50% fiber glass/50% resin. For maximum-strength corrosion properties, 33% glass/66% resin is recommended. A laminate

of this makeup will test about 16,000 psi, which is above the SPI minimum standard.

3. Although it is claimed that some spray systems are designed to operate on demand for up to 30 days without intermediate attention, it cannot be assumed that all spray systems will so perform. Obviously, to do this there can be no intermixture of the promoted resin and the catalyst until it leaves the head of the gun. Actually, this occurs about 6 in. from the face of the gun, where the two resins literally surround the glass and spray it in a flat spray onto the surface. In a well-designed spray system the maintenance requirement is simply to wash the face of the gun with acetone and a brush after using to remove blowback. The gun may then be hung up, ready to use again with the resins in the tanks. Spray systems, however, vary in design, and resins exhibit different properties, so that a single prescription cannot be followed for all spray or resin systems. For example, some manufacturers point out that resin-system stability catalyzed with MEKPO is a maximum of 2 to 3 days, so that to reduce the likelihood of resin gelation, the catalyst side should be flushed thoroughly with a solvent such as methyl ethyl ketone once a day. Some systems are furnished with acetone flushing pots for purging the gun in a matter of seconds. Such a flushing system virtually eliminates gun maintenance. Most systems are generally equipped with a device to vary the amount of catalyst being used so that it can be regulated at will over a comparatively wide range. Gun operation, generally, is designed to allow the operator a free hand for other work. Also, only sufficient catalyzed resin to be used in the day's operation should be mixed at one time. It is quite apparent that the individual contemplating using spray-up equipment needs to consider carefully the work to be done in relation to the know-how and skills involved. All this knowledge can be purchased with the services of a qualified applicator. Spray application is, basically, a manufacturing or a contracting technique. Conceivably, the skills involved could be applied as a maintenance tool by a chemical-process company well advanced in the polyester-maintenance picture. Even then such a company would probably want to begin on firm footing by using a contractor to do the work. This step, more than anything else, will ensure a successful application, provided the service conditions can be met.

The material having been applied looks as if it is dry. But as it is rolled out, it is quite apparent that this is not the case. Actually, the operation consists of three parts:

1. A priming coat of resin
2. A mixture of resin and catalyst
3. A finished gel coat

What can it do?   As indicated above, the spray gun is not only a maintenance tool but also a manufacturing tool. Let us list each of the purposes it serves:

1. Maintenance.
    a. Acid tank linings.
    b. Encapsulating corroded equipment.
    c. Fans and humidifiers.
    d. Truck bodies.
    e. Building steel.
    f. Faces of buildings.
    g. Corridor walls. (Plants are widely utilizing this system to dress up an area, thus reducing corridor maintenance and wall cleaning by as much as 50 percent.)
    h. Toilets.
    i. Tank exteriors subject to overflowing.
    j. Patching old roofs or building new roofs.
    k. Providing a tough adhering covering on concrete floors which would be chemically resistant.
2. Manufacturing. Using simple molds, some of which do not have to be any more elaborate than a piece of waxed cardboard, make your own:
    a. Machine trays.
    b. Tote boxes.
    c. Pans.
    d. Push trucks.
    e. Sluice troughs.
    f. Eliminator plates.
    g. Small tanks.
    h. Odd piping configurations for low-pressure work.
    i. Ductwork.

Additional helpful hints:

1. Two percent CAB-O-SIL should be added to the resin where the application is to vertical surfaces.

2. An extra 2 percent styrene in the resin will compensate for spray and evaporation losses.

3. The use of a colored tracer in the roving is commonly practiced to maintain proper thickness and distribution of the laminate.

4. Many satisfactory spray systems are operated using the MEKPO–cobalt naphthenate system.

5. The percent Atlac 382/percent Atlac 387 can be varied somewhat so that successful application of resilient lining systems can be achieved

over a wide range, for example, 70 percent to 85 percent Atlac 382, 15 to 20 percent Atlac 387, and the remainder styrene or other additives. The balance of these two resins is an attempt to achieve high corrosion resistance along with good flexibility and impact strength.

6. The permeability of any of these systems may be reduced by filling the resin with 300-mesh silica flour, especially in the final coat. It is common practice to use about 0.5 to 1.0 percent paraffin wax in the final coat. Depending upon the service, a cure time of 24 to 72 hr is generally desired. The usual precaution of having a clean and dry surface, along with air and surface temperatures above 60°F, is essential for a successful application.

7. A cold metallic surface will inhibit the resin cure and spell failure. It is especially important to note this when working on equipment outdoors and in the wintertime. It is not sufficient to heat the vessel interior. Make sure, by means of lamps or other acceptable methods, that the metal surface is warm.

8. With a simple turn of the knob the glass chopper can be turned off and the same gun can be used for applying the primer coat, and ultimately the gel coat. Ideally, a second and more simplified gun and system could be used for gel-coat application since this would eliminate the bother of cleaning the container. This is not absolutely necessary, however; the gel coat can simply be a heavy layer of resin as a final sealer.

9. The ability to control the thickness of the application is remarkable—by actual demonstration, thicknesses of $\frac{1}{32}$ in., through which you could read a newspaper, to $\frac{5}{8}$ in., using a low-exotherm resin. To cover 3 ft² of approximately $\frac{1}{4}$-in. thickness with the gun took 15 sec, followed by a roll-out with a spiral roller, which took, possibly, another minute. Guns of this type have been built with capacities as high as 100 lb/min.

The spray-lay-up system will, of course, handle a high-chemical-resistant resin, such as Atlac 382, Laminac 4173, the Hetrons, or any of the general-purpose resins. Such resins have a thixotropic material added to prevent running on vertical surfaces. To give an idea of the cost, spraying a general-purpose resin takes approximately 1 lb of glass-plus-resin per $\frac{1}{8}$-in. thickness per square foot.

Typical contract work on lining humidifiers[2]   The system used for spray-lay-up was a composite bisphenol system comprised of 80 percent Atlac 382 to provide corrosion resistance, 15 percent Atlac 387 for tough flexibility, and 5 percent epoxy resin for improved adherence. To this may be added, at the engineer's discretion, antimony trioxide, from 1 to 5 percent, to provide the proper flame-spread rating when required.

Quite often the antimony trioxide is used in the finish coat only. The catalyst promoter system in this case was benzoyl peroxide and dimethylaniline. Claims are made that this catalyst system is better than MEKPO for spray linings. Where the lining system is used on concrete, an epoxy primer is first used. The epoxy-primer system, when made resilient, does not shear the concrete due to continual curing, as is the case with polyester-type resins. Further, the epoxy-resin system will provide a greater moisture tolerance than polyesters. Generally, in lining a concrete tank, a carbon-filled coat is used next to the prime coat so that, when the work is finished, a spark test can be run. Experience has indicated that $\frac{1}{2}$-in.-thick floor coatings are not usually necessary. At the present time very satisfactory $\frac{1}{4}$-in.-thick coatings can be achieved by incorporating 24-oz woven-fiber-glass roving as reinforcing in the toppings. This provides greater strength and impact resistance. In addition, it has resulted in a price decrease in this type of application. Approximate application cost for an in-place system such as this, including sandblasting, is

1. $2 per square foot for the bisphenol system (about $\frac{1}{8}$-in. thick).
2. $2.50 to $2.75 per foot for an epoxy-primer system followed by a bisphenol system.
3. $3 to $3.50 per square foot for an unfilled epoxy system based on five coats.
4. Flooring systems approximately $\frac{1}{4}$- to $\frac{1}{2}$-in. thick may run $3 to $4.50 per square foot. An application like this requires 4 to 7 days curing time, depending upon the environmental temperature.

Again, the reader should realize that temperature limitations do exist on coatings and linings of this type. Bisphenol polyester should be limited to 140°F unless a flaked-fiber-glass coating is applied over the fiber-glass mat or cloth-reinforced lining. When flaked fiber glass is applied over the lining, the temperature can be safely raised to 160°F, with occasional service as high as 180°F.

In an epoxy application using the same system of reinforcing and with the flaked fiber glass applied over the regular reinforced material, immersion temperatures may be safely increased to 180°F. In each case, however, the specific service conditions will govern the final choice of the resin.

It must be stressed that in lined vessels there is a tendency for ion migration or vapor penetration to become a problem at temperatures much in excess of 160°F. However, under the same service environments, hand-laid-up molded-fiber-glass vessels may be used quite safely at temperatures of 200 to 220°F. Further, in lining application, it will be found to be beneficial to limit the lining thickness to $\frac{3}{16}$ in. Linings exceeding this thickness may crack, due to rapid fluctuations in temperatures, even when properly applied.

16.5 THE DESIGN PARAMETERS IN DETAIL

In the design of reinforced plastic vessels and parts, the wide variety of conditions to which the projected piece of equipment may be subjected must be taken into consideration. In designing each piece of equipment all the projected service conditions should be listed, in not only normal, but in abnormal, operation. Be particularly watchful of rectangular corners. Do not let them split open. The following examples will illustrate the problems that can be anticipated.

VACUUM CRYSTALLIZER CONE OUTLET. This is a cone, flanged at both ends, 4 ft in diameter at one end, 8 in. at the other, and about 4 ft long. It can be subjected to:

1. A full vacuum, or nearly so, with a deadheaded system
2. A liquid head of some 25 ft of liquid with a specific gravity of 1.3, caused by overfilling the crystallizer
3. An abrasive action from Glauber's salt for 10 min at a time, five times daily, as 20,000 lb streams out through the nozzles at an average velocity of 6 fps (the salt is in 100,000 lb of solution)
4. An 8-in. lined diaphragm valve bolted to the bottom flange

Now look at the conditions: full vacuum, 15 lb pressure, rayon-spin-bath liquor, abrasion from magma crystals. All these items must be taken into consideration in the final design.

ROTARY-DRUM FILTER-DISCHARGE CHUTE. This is a side-enclosed RP chute some 6 ft wide and 5 ft deep, with a 10-in. outlet at the bottom. Normal operation is for the chute to accept the discharge of Glauber's salt from the filter and guide it with a wash slurry solution to the outlet. But suppose the outlet plugs. In seconds the chute fills several feet above the brim like a big ice-cream cone, with Glauber's salt and the slurry solution falling over the side. Make sure that any design can run brimful, and then some, because sometimes it will have to.

16.6 POLYESTER FLANGED FITTINGS—A DIFFERENT VIEW

The Commercial Standard dimensions recommended by the Department of Commerce are an adaptation of the American Standard Steel Butt Weld Fittings, B.16.9, 1958, for 90° elbows, with exceptions for the 2- and 3-in. sizes. This Standard applies to long-radius elbows. Fittings made to this Standard will generally be interchangeable with flanged lined-metal fittings, except in the 2- and 3-in. sizes. But what about the countless number of installations which have been put together using the standard 125-psi cast-iron flanged fittings (generally lined) or the 150-psi steel fitting? Literally thousands of systems have been assembled using the 125-psi cast-iron standard flange fitting with its basic dimensions. The engineer faced with a replacement of the 125-psi *standard* fitting with a reinforced plastic fitting is at present confronted by difficult market conditions:

1. Only a few vendors are offering reinforced epoxy fittings which will meet the 125-psi cast-iron standard.

2. At least one vendor is manufacturing a mitered fitting in reinforced epoxy which meets the 125-psi cast-iron standard.

3. No vendor is yet offering a complete line of hand-laid-up reinforced polyester fittings tailored to the 125-psi cast-iron flanged-fitting dimensions.

It is quite obvious that such a line of fittings is needed covering 90° elbows, 45° elbows, and tees, as a starter, developing from there 45° laterals, concentric reducers, eccentric reducers, and crosses. Working with established reinforced polyester vendors, the author is developing fittings of this type. Figure 16-1 shows a 6-in. 125-psi standard cast-

FIG. 16-1 A 6-in. 90° elbow, hand-laid-up 125-psi ASA face-to-face dimension.

FIG. 16-2   A 6-in. tee, hand-laid-up 125-psi ASA face-to-face dimension.

iron face-to-face-dimension 90° elbow hand-laid-up in a reinforced polyester. Figure 16-2 shows a 6-in. tee fabricated by the hand-laid-up process to the same cast-iron flanged dimensions.

Here we have an opportunity to extend the high reliability and rugged performance associated with a fitting of this type. All flanges meet the 100-psi specification for thickness and the 125- to 150-psi standard for flanged drilling and bolting dimensions. At the customer's option the flange may be either flat-faced, step-faced, or serrated. Flange

HAND–LAID–UP FLANGED–FITTING PRICES—125–PSI ASA F–F

Table 16-1

Size, in.	90° elbows		45° elbows		Tees	
	100 psi	150 psi	100 psi	150 psi	100 psi	150 psi
2	$ 49	$ 54	$ 43	$ 47	$ 68	$ 76
3	57	65	51	55	81	88
4	68	73	59	63	91	100
6	81	94	71	80	109	126
8	98	115	86	98	148	170
10	129	150	112	127	185	220
12	165	197	140	166		

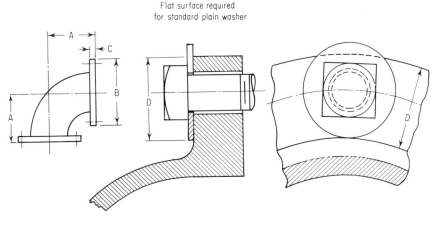

Flat surface required
for standard plain washer

Section through flange
6″ size shown full scale

Size	Inside Dia.	Wall thick.	Dim. A	Dim. B	Dim. C	No. Holes	Dia. Holes	Bolt circle	Size and lg. bolts	Dim. D
2″	2″	$\frac{3}{16}$″	$4\frac{1}{2}$″	6″	$\frac{11}{16}$″	4	$\frac{3}{4}$″	$4\frac{3}{4}$″	$\frac{5}{8}$″ x	$1\frac{1}{2}$″
3″	3″	$\frac{1}{4}$″	$5\frac{1}{2}$″	$7\frac{1}{2}$″	$\frac{13}{16}$″	4	$\frac{3}{4}$″	6″	$\frac{5}{8}$″ x	$1\frac{5}{8}$″
4″	4″	$\frac{1}{4}$″	$6\frac{1}{2}$″	9″	$\frac{15}{16}$″	8	$\frac{3}{4}$″	$7\frac{1}{2}$″	$\frac{5}{8}$″ x	$1\frac{5}{8}$″
6″	6″	$\frac{3}{8}$″	8″	11″	$1\frac{1}{16}$″	8	$\frac{7}{8}$″	$9\frac{1}{12}$″	$\frac{3}{4}$″ x	$1\frac{3}{4}$″
8″	8″	$\frac{7}{16}$″	9″	$13\frac{1}{2}$″	$1\frac{1}{4}$″	8	$\frac{7}{8}$″	$11\frac{3}{4}$″	$\frac{3}{4}$″ x	$1\frac{7}{8}$″
10″	10″	$\frac{1}{2}$″	11″	16″	$1\frac{7}{16}$″	12	1″	$14\frac{1}{4}$″	$\frac{7}{8}$″ x	2″
12″	12″	$\frac{5}{8}$″	12″	19″	$1\frac{3}{4}$″	12	1″	17″	$\frac{7}{8}$″ x	$2\frac{1}{8}$″

FIG. 16-3   90° flanged elbow, fiber-glass-reinforced polyester.

radius is recessed at the bolt holes to provide for backup washers.   Surprisingly enough, such fitting construction can be made and sold at less cost than a reinforced epoxy mitered flanged fitting.   The final design of such a line of fittings involves a combination of 125-psi cast-iron flanged-fitting dimensions plus wall and flange thickness dimensions taken from the current Reinforced Plastic Commercial Standard.   Wall thicknesses can be at the designer's discretion, using either the Commercial Standard 100- or 150-psi wall. Fittings of this type have been constructed and have been found to be tough and durable. Although they are more expensive than a press-molded fitting or a fitting built with press-molded flanges, they represent the current optimum in performance ruggedness. Figure 16-3 illustrates the adaptation of these ideas in a practical, workable design. The illustration is for a 90° flanged elbow.

Table 16-1 gives the approximate current prices for various fittings in either the 100- or 150-psi wall thickness.

16.7   REINFORCING MATERIALS AND PROTECTIVE BARRIERS[3]

Organic-fiber veils are used widely in glass-reinforced plastics, especially with polyesters and epoxies. These may consist of such synthetic fibers as modacrylic, polyester, acrylic, polypropylene, etc. They generally replace C glass in the inner protective barrier and provide a stable reinforcing medium plus improved adhesion and good wet-out when used with polyesters. In addition, some of them are completely transparent in the laminate and virtually defy visual detection.

Organic-fiber surfacing veils have been used successfully in the manufacture of chemical-corrosion-resistant equipment such as pipes, condensers, tanks, ductwork, etc. They will provide increased liquid resistance and, when used on the exterior of a vessel subject to outdoor service, will improve weathering characteristics on cooling towers, boats, and any material subject to outdoor service in a chemical plant. These fiber surfacing veils also improve abrasion resistance and impact strength. Trucks used in the mining and chemical industry, hauling such diverse items as rock salt, coal, titanium ore, and iron sulfate, have been provided with beds utilizing organic-fiber veil-surface RP loading bins.

Gel-coat stabilization in filament-winding processes are improved by the use of this medium in that they permit a uniform thickness of the resin-rich layer. By preventing subsequent winding of filament or glass to cut through to the pipe or mold, they provide a tougher substrate. Then, too, their use permits work to proceed without curing of the gel coat, so that a wet-on-wet coat of the gel may be made, and the probability of air invasion vastly reduced. These fiber veils are normally applied in strips 4 to 6 in. wide, with a small overlap. The addition of a modacrylic veil on the exterior of the pipe, vessel, or lay-up will give an extra-smooth surface and a much better finish. In addition, its exterior abrasion and scratch resistance will have been markedly improved.

Veiling material may be of coarse or fine construction, depending upon the physical characteristics and the usage desired.

Controlled abrasion tests indicate that weight losses due to abrasion are reduced with an organic-fiber overlay.

Veils such as modacrylic, polyester, and polypropylene are excellent for mineral acid applications. Polypropylene fibers are suitable for alkali applications. Some data available indicate organic veils such as polypropylene and the polyesters provide a much higher internal bond strength than C glass veils. Obviously, selection of the veil may vary with the environmental conditions of exposure. If properly selected, the veil reinforcing should be nearly as resistant as the resin itself.

The use of organic veils in place of C glass provides the surfacing system with a reinforcement more in tune with the physical properties of the polyester resin, so that minute crazing of the resin is less likely to occur. This is particularly important in food application. At least one European country insists upon the use of organic-fiber veiling on the interior of all vessels or piping systems made of RP material going into the food industry as an end use. This requirement is intended to eliminate, insofar as possible, all minute crevices for food and bacteria hang-up.

The use of organic-fiber veils embedded in the resin-rich surface should also be considered for service applications subjected to temperature cycling or dehydrolysis. Low-pressure thermocompressors in which 25-psig saturated steam is the prime driving force may be subjected to wide thermal cycling going from temperatures of 215°F to 32°F in periods of nearly 5 hr. Thus the booster diffuser is subjected to repetitive thermal cycles from boiling hot to freezing cold five times a day. Under such conditions and with entraining blasts of steam and evaporated vapor plus crystals, vapor velocities in the diffuser throat may run in the 1,000-fps area. Lead-lined steel boosters last 18 months before they are worn out. High-temperature synthetics or elastomers may go 3 to 4 years. The application of reinforced plastics in an area such as this has had a record of encouraging success with a number of vendors. Prices are generally considerably less when purchased in RP materials than for lead-lined steel, stainless steel, rubber-lined steel, or the high-nickel alloys.

A resilient resin system coupled with an organic surfacing veil may be useful under temperature cycling conditions such as those indicated above. For example, a high-temperature resilient system might consist of 90 percent Hetron 197 and 10 percent Hetron 31 to produce the desired system resiliency. In addition, the polyester surfacing veil will counteract cracks caused by thermal stress and strain and reinforce the structure. In this manner fissures induced by continuous temperature cycling may be prevented.

Organic surfacing veils may run 2 to 3 cents per square foot versus 1 to 2 cents per square foot for C glass. This can hardly be measured in the final cost of a piping system or vessel, but as it is passed on to the consumer, it probably means an increase of 5 cents per square foot.

The designer should also bear in mind that laminate construction can be varied to provide for multiple layers of organic-fiber veils. Such "designs in depth" provide a high degree of chemical and abrasion resistance. A typical polyester veiling used in surfacing work is sold as Type 389-K by the Pellon Corporation, New York. Decorated or printed laminates are also available in this type of material.

Chem-Proof walls[4]   Throughout this book on laminate design we have continually stressed the need for the all-important "resin-rich inner layer." We have dealt in detail with the advantages of secondary barrier systems to provide continued protection in case the resin-rich inner layer was breeched or contained manufacturing flaws. In the case of filament-wound structures, we pointed out that their very survival depended upon the integrity of the inner corrosion barrier. One of the ingenious advances in the technology of glass-reinforced-plastic corrosion-resistant structures has been developed by Poly-Fibre Associates, Bound Brook, N.J. Manufactured and sold under the Chem-Proof wall-structure design, a typical ½-in. laminate would be found to be made up of:

> Inside synthetic layer (modacrylic, polyester, polypropylene, etc.)
> Chopped-strand mat
> Synthetic layer
> Woven roving or cloth
> Synthetic layer
> Woven roving or cloth
> Outside synthetic layer

The laminate construction may use varying resin systems such as Hetron 197, Atlac 382, Laminac 4173, or any of the other good chemical-resistant resins. Strength in the wall is not appreciably sacrificed since the multiple corrosion barriers occupy but a small portion, percentagewise, of the total laminate thickness.

16.8   SUPERPOLYESTER RESINS[5]

The general definition of a polyester is a polymer molecule which contains a number of ester linkages. Scientists in a laboratory, however, have been able to modify the polymer molecule to such an extent that, although it may still be called a polyester, the polymer will exhibit chemical properties which are considerably different from the conventional polyesters. One of the weaknesses of the polyester molecule is that these ester linkages generally occur in pairs. Any work done to reduce the ester linkages in the polymer would provide a beneficial effect.

Unsaturation in polyesters often becomes restricted when a nearby double bond has been tied up in the polymerization operation.   By this means the copolymerization reaction with styrene stops short of complete fulfillment.   Any polymer-molecule redesign that tends to improve the mobility of the reactive group should further improve the overall chemical performance of the polyester resins.   Research along this line has produced resins with superior chemical-resistant properties which may still be cured at room temperatures with the conventional polyester promoters and catalysts.   These resins show particularly enhanced resistance to 50 percent sodium hydroxide, 93 percent sulfuric acid, and various concentrations of nitric acid and chromic acid.   Although still in the development stage, they would appear to be premium-priced resin systems compared with the bisphenol or hydrogenated bisphenol systems. Where short sections of RP piping may be subjected to random surges of 93 percent H_2SO_4 solution due to the malfunction of dilution equipment, such added resistivity could be the difference between success and failure.

Systems such as these would also lend themselves favorably to composite lay-ups in which the inner 100 mils of the piping wall would be constructed of this high-performance resin system.

Flextran—reinforced polyester mortar pipe[6]   One of the larger manufacturers (Johns-Manville) of water and waste water piping has developed a large-diameter flexible conduit for gravity transmission of liquids.   This factory-formed conduit is composed of a polyester resin and siliceous sand, reinforced with continuous roving glass fibers.   As it is constructed of an isophthalic polyester resin the reader may refer to pages 32 to 33 which deal with service conditions to which the isophthalic resins provide a unique resistance.   The sand is a functional low-cost filler which enhances greatly the long-term abrasion resistance of Flextran pipe.   The pipe is normally available in sizes 15 through 48 in. dia and in 10- and 20-ft sections (also see page 199).   Tests conducted with solutions containing large amounts of salt, sand, and gravel and rotated in the pipe for nearly three million revolutions indicated the following wall-thickness loss:

Pipe material	Wall-thickness loss, mils
Bare steel	1.1
Flextran	1.9
Cement-mortar lined steel	18.8
Epoxy coal tar lined steel	20.5
Coal tar enamel lined steel	20.6

As Flextran pipe comes in 20-ft lengths its glass-smooth interior and joint make possible a coefficient of $N = 0.010$.   This can result in flatter grades by the use of smaller-diameter pipe. For example, required for $N = 0.013$ is 12-in. pipe while for $N = 0.010$ only an 8-in. pipe is required.

The reader is referred to page 202 for data on infiltration and exfiltration which should not exceed 100 gal/in./mile/day. The importance of less infiltration—keeping ground water out—is:

1. A smaller-size pipe
2. Less operating and maintenance expenses
3. A smaller treatment plant
4. A smaller capital cost

While capital costs on municipal sewerage treatment plants may be modest they can become severely high where chemical treatment plants are involved. A typical chemical treatment plant using the conventional lime neutralization method may run to a capital cost of \$.70/gal/day. In daily operational costs such a plant may run \$50 to \$150 per million gallons per day. As volume of effluent treated is critically important the significance of a tight sewer pipe specification is important.

Flextran normally weighs about one-seventh that of extra-strength clay pipe and about one-eighth to one-tenth that of reinforced concrete Class 3 pipe.

JOINTING. The reinforced bell-and-spigot single-rubber-ring joint of Flextran is assembled by stabbing plus a suitable puller of a bar and block. Factory-fabricated tees using ABS sewer saddles for larger sizes can be obtained on special order.

MANHOLES AND BRANCHES. Flextran can be easily cut in the field using a portable saw or abrasive wheel. Reference is made to page 201 to show the various methods of treatment at manholes. One of the most popular methods is cutting away the upper portion to provide a smooth well-formed channel.

REFERENCES

[1] The data for this section have been obtained from a wide variety of sources, such as Atlas Chemical Industries, Inc., Wilmington, Del.; Finn and Fram, Inc., Sun Valley, Calif.; Glas-Craft of California, Glendale, Calif.; Plastic Engineering and Chemical Co., Fort Lauderdale, Fla.; and Venus Products, Renton, Wash.

[2] Information on contract work for lining humidifiers, fans, tanks, etc., was obtained from Mathis Fiberglass Coatings, Inc., Anderson, S.C.

[3] Much of the information on organic fiber veiling was obtained from R. Hoehn and W. Heling, Organic Fiber Veils for Glass-Resin Laminates, published in brochure TN 7-9110 by Pellon Corp., New York. Subsequent information was obtained in an interview with Mr. Ralph Hoehn, February, 1968.

[4] Poly-Fibre Associates, Bound Brook, N.J.

[5] Updegraff, I. J., and M. J. Doyle: paper 12-B, 23d Annual Technical Conference, Society of the Plastics Industry, Reinforced Plastics/Composites Div., February, 1968.

[6] Flextran RPM Pipe Brochure, TR-545-A, 10-68, Johns-Manville, 222 West 40th Street, New York, October, 1968.

Appendix A

AUTHOR'S NOTE: The following appendix is a part of TS 122C, dated September 18, 1968:

RECOMMENDED PRODUCT STANDARD FOR
CUSTOM CONTACT-MOLDED REINFORCED POLYESTER
CHEMICAL-RESISTANT PROCESS EQUIPMENT

A1   *Chemical resistance*

A1.1   *Test*—ASTM Designation C581-65T, Tentative Method of Test for Chemical Resistance of Thermosetting Resins Used in Glass Fiber Reinforced Structures* is recommended for the evaluation of the chemical resistance of materials to be used in reinforced polyester chemical-resistant process equipment. The reinforcing materials prescribed in the test laminate are only for the purpose of establishing a uniform basis for comparison. They may not necessarily represent the preferred materials for the particular environment. This procedure may be adapted to test or evaluate components, composition or fabrication variations and production samples. For information on the basis for selection of the standard test laminate, see Appendix A1 of ASTM C581.

A1.1.1   The 10-mil surfacing mat referred to in paragraph 5.1.2.1 of ASTM C581-65T shall be made of chemical resistant glass (Type C or equal).

A1.1.2   The standard test laminate shall be cured at room temperature for 16 hours. Further cure shall be given at room or higher temperature, if necessary, to produce a Barcol hardness equal to the resin manufacturer's minimum specified hardness for the cured resin.

A1.2   *Temperature*—Tests may be conducted at any or all of these

---

* This method is based on a test procedure developed by the Reinforced-Plastics Corrosion-Resistant Structures Subcommittee of the Society of the Plastics Industry, Inc.

temperatures: 23°C, 50°C, 70°C, 100°C (±2°C); reflux temperature; required service temperature.

A1.3   *Reagents*—The following reagents are suggested for use in obtaining general comparative chemical resistance data. The test solutions shall not be agitated, i.e., the exposures shall be under static conditions.

1. 25% sulfuric acid
2. 15% hydrochloric acid
3. 5% nitric acid
4. 25% acetic acid
5. 15% phosphoric acid
6. 5% sodium hydroxide
7. 10% sodium carbonate
8. Saturated sodium chloride
9. 95% ethanol
10. 5¼% sodium hypochlorite*
11. 5% aluminum potassium sulfate
12. Ethyl acetate
13. Methyl ethyl ketone
14. Monochlorobenzene
15. Perchlorethylene
16. *n*-Heptane
17. Kerosene
18. Toluene
19. 5% Hydrogen peroxide*
20. Distilled Water*

A1.4   *Time*—The properties specified in A1.5 shall be determined for specimens immersed in the test solutions for 30 days, 90 days, 180 days, and one year for one set of control specimens immediately following the curing period; and for another set after aging in air at the test temperature for the total test period.

A1.5   *Properties*—Thickness, Barcol hardness, flexural strength and modulus, and appearance shall be determined at each time interval. Appearance observations shall include any surface changes, color changes, obvious softening or hardening, crazing, delamination, exposure of fibers, or other effects indicative of complete degradation or potential failure. Calculation of percentage change in a property shall be based on the property value obtained immediately following the curing period.

A1.6   *Report*—Data shall be reported in tabular form for all parameters tested. The composition, including resin, accelerators, catalysts, and reinforcements, and the fabricating and curing conditions of the laminate tested shall be adequately described.

A2   *Fire retardancy*†—The fire retardancy may be determined in

---

* Replaced every 48 hours with fresh solution.

† Work is in progress to develop test procedures and specifications requirements for application requiring fire resistance.

accordance with ASTM Designation E84-61, Standard Method of Test for Surface Burning Characteristics of Building Materials.*

A3   *Compressive strength* (edgewise)—The compressive strength shall be determined in accordance with ASTM Designation D695-63T, Tentative Method of Test for Compressive Properties of Rigid Plastics.*

* Later issues of the ASTM publication may be used providing the requirements are applicable and consistent with the issue designated. Copies of ASTM publications are obtainable from the American Society for Testing and Materials, 1916 Race Street, Philadelphia, Pa. 19103.

Appendix B

AUTHOR'S NOTE: The following quoted
sections cover Section 4, Test Methods,
and Section 5, Identification, and are
quoted directly from:

RECOMMENDED PRODUCT STANDARD FOR
CUSTOM CONTACT-MOLDED REINFORCED POLYESTER
CHEMICAL-RESISTANT PROCESS EQUIPMENT

TS 122C

September 18, 1968

4. TEST METHODS

4.1   *Specimens*—Tests shall be made on specimens cut from waste
areas when possible; otherwise, the specimens shall be cut from flat laminates
prepared in the same construction and by the same techniques as the process
equipment.   In all cases, the average value of the indicated number of
specimens shall be used to determine conformance with the detailed
requirements.

4.2   *Conditioning*—The test specimens shall be conditioned in accordance with Procedure A of ASTM Designation D618-61, Standard Methods
of Conditioning Plastics and Electrical Insulating Materials for Testing.*

4.3   *Tests*

4.3.1   *Glass content*—The glass content shall be determined in accordance with ASTM Designation D2584-67T, Tentative Method of Test for

* Later issues of the ASTM publication may be used providing the requirements
are applicable and consistent with the issue designated.   Copies of ASTM
publications are obtainable from the American Society for Testing and Materials,
1916 Race Street, Philadelphia, Pa. 19103.

Ignition Loss of Cured Reinforced Resins* except that the specimens tested shall be approximately 1 square inch in area and low temperature pre-ignition prior to placement in muffle furnace is recommended. The average for five specimens shall be considered to be the glass content.

4.3.2 *Tensile strength*—Tensile strength shall be determined in accordance with ASTM Designation D638-64T, Test for Tensile Properties of Plastic (Tentative),* except that the specimens shall be the actual thickness of the fabricated article and the width of the reduced section shall be one inch. Other dimensions of specimens shall be as designated by the ASTM standard for Type I specimens for materials over ½ inch to 1 inch inclusive. Specimens shall not be machined on the surface. Tensile strength shall be the average of five specimens tested at 0.20–0.25 in./min. speed.

4.3.3 *Flexural strength*—Flexural strength shall be determined in accordance with Procedure A and Table 1 of ASTM Designation D790-66, Standard Method of Test for Flexural Properties of Plastics* except that the specimens shall be the actual thickness of the fabricated article and the width shall be one inch. Other dimensions of specimens shall be as designated by the ASTM standard. Specimens shall not be machined on the surface. Tests shall be made with the resin-rich side in compression using five specimens.

4.3.4 *Flexural modulus*—The tangent modulus of elasticity in flexure shall be determined by ASTM Method D790-66 (see 4.3.3).

4.3.5 *Hardness*—The Barcol Impressor (Model GYZJ 934-1) shall be used for determining hardness. Calibration of the Barcol instrument shall be verified by comparing with a blank specimen having a known reading of 85–87. Ten (10) readings on the clean resin-rich surface shall be made. After eliminating the two high and two low readings, the average of the remainder shall be the reported hardness reading.†

4.3.6 *Additional tests*—Recommended test methods for the further testing of reinforced-polyester laminates are given in Appendix A.

5. IDENTIFICATION

5.1 *Labels and literature*—In order that purchasers may identify products complying with all requirements of this voluntary Product Standard, producers choosing to produce such products in conformance with this voluntary Standard may include a statement in conjunction with their name and address on labels, invoices, sales literature, and the like. The following statement is suggested when sufficient space is available:

This ——————————— complies with all of the requirements of Product Standard PS ———, as developed cooperatively with the industry and published by the National Bureau of Standards under the

* Later issues of the ASTM publication may be used providing the requirements are applicable and consistent with the issue designated. Copies of ASTM publications are obtainable from the American Society for Testing and Materials, 1916 Race Street, Philadelphia, Pa. 19103.
† Work is in progress to determine the feasibility of using for these measurements Barcol instruments calibrated with a test disk in the range of 42–46.

Voluntary Product Standards Procedures of the U.S. Department of Commerce.

The following abbreviated statement is suggested when available space on labels is insufficient for the full statement:

Complies with PS ——, published by the National Bureau of Standards.

5.2   *Hallmark*—The hallmark illustrated in Figure 2 has been developed to identify products that comply with all requirements of a particular voluntary Product Standard. Manufacturers or distributors desiring to use this hallmark as a certification mark must request permission in writing to the Office of Engineering Standards Services, National Bureau of Standards, Washington, D.C. 20234.

It is suggested that the hallmark first be used with the language provided in Figure 2 (A) which describes the significance of the hallmark. It is further suggested that the hallmark be used with the language provided in Figure 2 (B) only after such time as the identity of the Product Standard has become recognized within the industry. Finally, it is recommended that the hallmark be used as shown in Figure 2 (C) only after all major distributors, users, and consumers have become completely familiar with the hallmark and the referenced Standard.

The National Bureau of Standards retains the right to withdraw the permission to utilize the hallmark if sufficient evidence is presented which indicates that products identified as complying with a voluntary Product Standard do not comply with all requirements of the voluntary Product Standard.

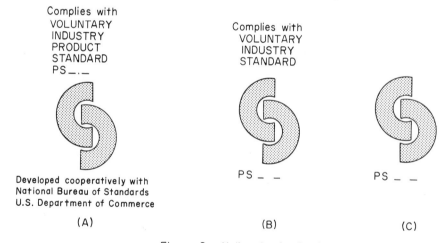

(A)   (B)   (C)

Figure 2 – Hallmarks for Product Standards

USCOMM – NBS – DC

Appendix C

AUTHOR'S NOTE: The following Proposed Standard on Filament Wound FRP Pipe has been prepared by H. D. Boggs, Manager, Engineering Services, Amercoat Corporation, Brea, Calif., and is used with permission.

PROPOSED STANDARD—FILAMENT WOUND FRP PIPE

A/C 1002

December, 1966

INTRODUCTION

Standards for Reinforced Plastic Pipe will come, but it is not possible overnight. As a matter of fact, standards in regard to metal pipe were long after the fact and even then they deliberated many years before agreement.

One of the major problems of standards is this. Standards are a recognized customer need, yet they can be a straightjacket which discourages improvements. Improvements to a comparatively new material and product thereof is a continuing need. Standards, therefore, for standards' sake alone are a danger and a state of mind easy for technical people to slip into.

Suppose we demanded one set of standards covering cast iron, cast steel and steel pipe. It would be impossible because the process and materials are too much in variance. This is the essential problem that has faced the reinforced plastic pipe manufacturers. The processes range from contact molding on temporary or light tools to rolling pipe to centrifugal casting pipe and finally to filament-winding on substantial tools and machinery. The tensile strengths range from 9,000 to 80,000 psi, on the same order as the processes are named above. Structurally, the products differ, as do most of their properties, so you can see we are trying to describe many materials common in name only. Yet each type has certain very definite advantages worthy of consideration.

Standards will be slowly emerging. There are many custom FRP manufacturers who have met separately and together have devised an SPI (soon to be called Commercial) standard for polyester FRP custom goods. Copies of this standard are widespread and used. Such a standard is perhaps crude in terms of our more sophisticated metal standards, but it is a good one and merely a young one which in time will be refined.

Since there is only one producer of rolled pipe, standards at this time will be proprietary at best for rolled FRP pipe. There are a few producers of centrifugal cast pipe, yet these products vary in strength from 9,000 to 60,000 psi, and so they have problems getting together. These standards may be proprietary in nature, too.

Finally, the filament-winders are getting close to being able to point toward standards, although they are engaged in pipe specialization for different markets. At least, now is a time to push for filament standards, particularly as it applies to chemical and industrial piping applications. The standards proposed herewith are for this specific market and they apply to filament-wound pipe.

My purpose is a serious and considered attempt to set up objective specifications for filament-wound goods; and to seek to have it reviewed by those of you most knowledgeable in the field, and later to assist in getting it recognized as a standard, along with other standards being considered.

Amercoat has, in the past, published specifications standardizing the purchase of Bondstrand filament-wound products. This new specification, however, goes beyond this consideration and strives to state certain basic and very meaningful and unalterable limits for filament goods. The specification includes other items that should be stated, but would be stated to suit a given manufacturer's process and goods.

What, then, are the unalterable limits? These I would like to list and discuss. They are as follows:

I. *Non-porosity Edges*

From the onset, the reinforced plastic industry has had to make the best of poor glass finishes. Earlier resins were also a problem, and then finally the very art of filament-winding had to be reduced to properly and thoroughly wetting the glass fibers. All these factors were most serious in connection with polyester filament structures until only about $1\frac{1}{2}$ years ago, when all factors became favorable and it was possible to eliminate all edgewise porosity with continuous fiber structures in polyester resins. Glass finish introduced at that time by PPG was finally and truly compatible.

Earlier, general purpose polyester resins frequently lacked chemical and heat resistance and excessive shrinkage in volume. However, with the bisphenol and chlorinated polyester resins, the resin situation improved several years ago.

Heat cured epoxy filament systems have been suitable since 1954. Suitable glass finishes compatible with epoxies have been available for many years.

Among other things, the epoxy resin experiences low volume of shrinkage while curing. Technics for winding structures without air inclusion, resulting in elimination of edgewise porosity, have been known and used since about 1958.

Once, therefore, it is known how to combine glass finish, resin, and winding technics to make non-porous structures, we have a tool to test for the most paramount value in a filament structure, non-porosity edgewise. Of what value is an 80,000 psi tensile structure if it loses 50 percent of its value due to porosity? A 75 percent strength retention is a positive reachable minimum which can be demanded and obtained at no increase in cost. Knowing, therefore, that such a structure can be so produced with given resins, glass finishes, and winding technics, it follows that new materials can be tested using with them any of the other two given knowns. I mean, by this, that with the proper winding technic and a suitable resin, a new glass finish can be tested. This is the best known test, as a matter of fact, for checking glass finish as received, batch by batch, from the manufacturer. The method is so reliable that glass shipments have been rejected (the glass fiber manufacturer later located the trouble in material sources) where the glass manufacturer could not detect deficiencies by his own quality controls. Likewise, a new filament winder who may be using the proper glass and resin but cannot produce non-porous goods is experiencing such results because of wrong winding technics. A third example is that, when technics and glass finishes are proper, new resins can also be checked. Hence, edgewise porosity testing is a most reliable and meaningful control which should be demanded in specifications.

You may ask, would a little porosity hurt? The answer is yes. The product need not cost more to be non-porous. If porous, the structure will lose strength and have limited life. This cannot be related, in a linear measure, to so much porosity with so much loss, but it will relate to excessive weight gain in water environment and with strength losses. By and large, weight gains of more than 1 percent after 72 hours boiling of samples spells a poor product. Such weight gain is related to that which the resin in the system will absorb, but beyond that, it is a measure of moisture gathering along the glass fiber interface where it destroys adhesion and strength.

I submit, gentlemen, that you want and should demand non-porosity in the specifications.

II. *Liner Thickness and Liner*

A resin-rich liner has been proven to be the best construction for chemical environments. Even non-reactive environments such as water, alums, starch, etc., are best contained in a light liner, although no liner could be used.

In truth, many liner thicknesses could be proposed for different chemicals and temperatures. This would be so complicated that neither the buyer nor the manufacturer could live with the complex selection. Process temperatures frequently are upgraded as well as chemicals; therefore, a pipe system should be made to cover as broad a range as possible. We have chosen to separate the services between light and heavy duty services; thus, liners of .015–.025 mils are light duty ones, and .040–.060 liners are for heavy chemicals. We find that materials in the light class are seldom upgraded to heavy, while the heavy chemicals are upgraded frequently.

Liner thickness in terms of many strong acids and chemicals is the need for a thickness which will not blister in heated service. Blistering is usually a longer term type of failure (frequently a year is required), but it will limit service life and it can be missed and not suspected unless tests have been conducted long enough to pick up the problem. We are sure several producers are stating chemical ratings for thin liners beyond what they should because they have not run tests long enough to develop blistering. Liners well compacted and air-free do not blister if .040–.060 in thickness.

We have checked for this type of failure with many types of pipe and know it is a common problem regardless of the process, technics, etc. Of course, the poorer the product the worse the problem, but no one has yet built a suitable thin liner for heavy chemical services. Like the .100 liner in contact molded FRP goods, the .040–.060 liner in filament goods is a desired feature.

No liner should be made dependent on an adhesive joining with the filament structure. Instead, they should be cross-linked in the resin system throughout the pipe, liner and structural wall together.

Countless thousands of hours and dollars in laboratory fact finding back up these statements. Indeed, the liner system in FRP pipe and fittings is not to be taken lightly and should be standardized.

III. *Chemical Resistance Chart*

The time has come to standardize chemical resistance ratings. To do so, among other things, is to define terms. Rated resistance means what? One year, two years, or what? It is a complicated business since the relationship is a chemical attack on the resin in the system if there is any action worthy of note or liners blistering.

There are many philosophies of how to chart chemical resistance. One may chart everything up to the hilt even if the service life is limited, hoping one gets to tell the customer it is a limited life. If the customer doesn't know and it fails after a year, even when there is an economic service life, meaning lower costs than metal, is the customer happy? Usually not, because he expected more. Thus, we can by standardization of charting bring more clarity to the meaning of the chart. This is proposed in our specifications.

IV. *Other Items*

There are, of course, many other specification matters which can be included. Some are being worked out in Specification Meetings, but I will add this. Data in specifications which covers the technical aspect of the pipe being specified, that is, spans, pressures, burial information, expansion coefficients, unloading instructions, can be altered to suit the product made by each manufacturer. Such things should be stated, however, and they should be factual.

Users and manufacturers are rather tired of viewing catalog information obviously copied from other sources. Each manufacturer ought to deserve the business because he earns it by *caring about these facts*. As ASTM sets up tests, values will change and standardize. Bona fide efforts to test and state facts where methods of tests vary are going to have bona fide differences which can be resolved later.

V. *Materials*

I do not like the use of such statements in specifications as 1062 glass or equal, as is done in this specification, but at this time there is no other way this information can be coded or labeled in a non-proprietary way. We'd rather err in a gusty way than use no code or one that has no meaning.

VI. *Pipe O.D. (Outside Diameter)*

In regard to pipe O.D.'s, it is generally accepted there be I.P. nominal O.D.'s at least through 12″ pipe sizes. However, the outside texture of a pipe wall of various processes, including our own, makes it difficult to set up tolerances with meaning. Because of this, we propose as we do in 6.3, that the O.D. be controlled when the pipe is turned or ground clean of glossy surfaces. This must be done to make a good joint anyway. This approach gets around trying to set wide tolerances for many surface conditions and ties the subject down to meaning, leading to interchangeability in the fitting sockets.

VII. *Quality Control Procedures*

Quality Control Procedures should not be less than we have stated for an industry as new as FRP. Here again, we need uniformity of approach to protect the customer and to raise the quality of goods being made and sold.

Finally, just to clear up the unalterable and proprietary areas of this specification, let me outline those paragraphs which are proprietary wherein the given manufacturer fills in the facts that fit his product. They are as follows:

Paragraph 2.1.1

Manufacturer states the hoop tensile that applies to his product.

Paragraphs 2.3.1, 2.4, 2.5, 2.6, 2.7, 2.8, and 2.9

Manufacturer publishes the data as it applies to his pipe.

Paragraphs 3.2, 3.3, 3.4, and 3.5

Manufacturer conforms as written or changes to the materials he uses, but such materials must meet the edgewise porosity tests and the 3 year minimum requirement of chemical resistance.

Paragraphs 9.2.2 and 9.2.3

Change as per the manufacturer's kit sizes.

LIGHT AND HEAVY CHEMICAL GRADES
PIPE AND PIPE FITTINGS SPECIFICATION
FIBERGLASS REINFORCED PLASTIC

A/C 1002

December, 1966

1. SCOPE

1.1 This specification covers the manufacture of filament-wound pipe and pipe fittings made from thermosetting epoxy and polyester resins and glass-fiber or other reinforcement, together with adhesive necessary for joint assembly.

2. DESIGN CHARACTERISTICS

2.1 Pipe and pipe fittings shall be designed for use according to accepted piping practice.

2.1.1 *Pipe Design Stress*—All pipe shall be designed for a minimum ultimate hoop tensile stress of 80,000 psi and a minimum longitudinal tensile stress of 40,000 psi. Stress calculations shall be based on the fiberglass reinforced wall only. Testing shall be conducted with unrestrained ends per ASTM D1599-62T.

2.1.2 *Verification of Pipe Design Stress*—Minimum design stresses

RATED LONG–TERM WORKING PRESSURE (PSI) (see detailed explanation, pages 415–416)

Nom. dia., I.P. size	Pipe, series 2000, 4000 & 5000	Couplings, filament wound	Elbows		Tees		Crosses		Tapered reducers, filament wound	Plugs & reducer bushing, molded 1 PC.	Special angle elbow, filament assembled
			Filament wound	Pressure molded 1 PC.	Filament wound	Pressure molded 1 PC.	Filament assembled	Pressure molded 1 PC.			
2″	550	550	550³	150	350³	150	...	150	...	350	150
3″	400	400	400³	150	250³	150	...	150	400	250	150
4″	450	300	300	150	300	150	150	150	300	300	150
6″	350	250	250	...	250	...	100	...	250	150	75
8″	300	250	250	...	200	...	100	...	250	150	75
10″	250	200	200	...	100²	...	75	...	200	100	50
12″	150	150	100¹	...	100²	...	75	...	100¹	100	50

[1] Special Q/L fittings available for 150 p.s.i.
[2] Fabricated type.
[3] Available on special order.

shall be verified by manufacturer's test data, based on unrestrained end burst tests in which a minimum of seven diameter length test samples have been hydrostatically tested by a means suitable to cause failure in the fiber structure in less than one minute.

The I.D. of all pipe fittings shall be consistent with the I.D. of the pipe.

2.2.3   *Verification of Working Pressure, Pipe Fittings*—Satisfactory performance as indicated in paragraph 7.2 "Hydrostatic Proof Test" shall serve as manufacturer's verification of compliance with the working pressure requirements for pipe fittings.

2.3   *Temperature Rating*

2.3.1   All pipe and pipe fittings shall be designed to operate continuously at the rated working pressure and temperature as noted on the following charts.

RATED WORKING PRESSURE

The following charts indicate the rated long-term working pressures in pounds per square inch of fluid pressure at temperatures from −60°F through 200°F (the pressure molded fittings are an exception; the rated working pressures shown for these would be applicable up to 300°F). At 250°F use 75% of the listed pressure; at 300°F use 50% of the listed pressure. For temperature limitations in strong chemicals, consult the Chemical Resistance Chart.

ASA 150-PSI FLANGES

Bondstrand flanges are designed for 150 psi working pressure. The use of Buna-N ⅛″ full-faced gaskets is standard. When pressures are greater than recommended for standard gasketing, use series G-6500 B Parker Gask-O-Seals.

Nom. dia., ASA STD	Filament wound flanges			Molded flanges	
	Rated gasket sealing pressure				
	Special gasket	*Standard gasket*	*Bolt torque, ft/lbs*	*Standard gasket*	*Bolt torque, ft/lbs*
2″	. . .	. . .	. . .	150	30
3″	. . .	. . .	. . .	150	30
4″	. . .	. . .	. . .	150	30
6″	150	100	100		
8″	150	75	100		
10″	150	50	100		
12″	150	50	100		

Above ratings are at temperatures from −60°F to 300°F. Seal ratings are based on ⅓ the pressure required to cause a seal leak after short term testing.

In ordering components for a piping system, it should be noted that the fitting with the lowest pressure rating will govern the over-all pressure capability of the entire piping system. When flanged components are being used (such as flanged elbows), the lower rated element (usually the flange) will determine the component's pressure capability.

In testing, a system should be tested at 1½ times the anticipated working pressure. However, under no circumstances should the system be tested at a pressure greater than 1½ times the maximum pressure of the lowest-rated element in the system. (For example, even though a particular pipe may be rated at 550 psi, the lowest-rated element in the system may be a tee having a rating of 150 psi; the maximum testing pressure for the system would therefore be 1½ times 150 or 225 psi.)

2.4   *Span Supports and Hanger Systems*—Required span supports shall not be less than charted below.

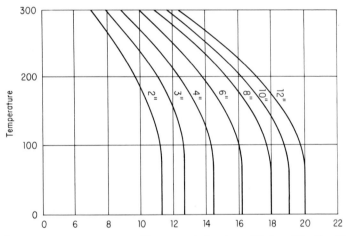

Span vs. temperature. Unsupported pipe filled with liquid (specific gravity 1.5). Maximum deflection 0.5 in.

Recommended hanger data is listed below:

Nominal pipe size	2″	3″	4″	6″	8″	10″	12″
Clamp width	1½″	2″	2″	2½″	3″	3″	3½″

The pipe must not be point loaded. Standard clamps for steel pipe may be used when the width meets the requirement noted above. Clamps with less width than recommended can be used with a light gauge sheet metal shield around the pipe at the clamp. Clamps should be installed loose to allow movement of the line. Should it be necessary to tighten the clamps for blocking the line against surge or controlled expansion a soft (Shore 50–60) rubber pad shall be placed between the clamp and the pipe. Tightening of the clamp shall be such that it does not crush the pipe wall.

2.5 *Coefficient of Thermal Expansion*—The coefficient of thermal expansion shall not exceed 9.5×10^{-6} when measured axially on the pipe with fluids not to exceed 300°F.

2.6 *Thermal Conductivity*—The thermal conductivity shall not be greater than 2.0 BTU/hr/ft²/in. thickness.

2.7 *Vacuum Resistance*—All pipe and fittings shall be capable of operating under total vacuum at 300°F without collapse failure.

2.8 *Ditching Capabilities*—All pipe shall be capable of being laid on a standard grade as prepared by machine ditching equipment.

Where light traffic is encountered (autos, light trucks) the minimum ditch depth for 2″, 3″ and 4″ pipe shall be one foot. For 6″, 8″, 10″ and 12″ pipe the minimum ditch depth shall be two feet. There shall be no maximum limit of depth for 2″ through 8″ pipe. 10″ and 12″ pipe shall not be buried more than 12 feet without special back filling procedure.

Heavy traffic conditions known as H-20 loading conditions require a protective conduit over the pipe.

Back fill shall not include large rocks or sharp rocks of any size in direct contact with the pipe wall.

Compacting shall be done with care to prevent direct blows on the pipe wall. It shall not in any way impact, cut or damage the pipe wall.

2.9 *Flow Resistance Characteristics*—The flow resistance of the pipe with couplings shall not exceed the values given in the chart below:

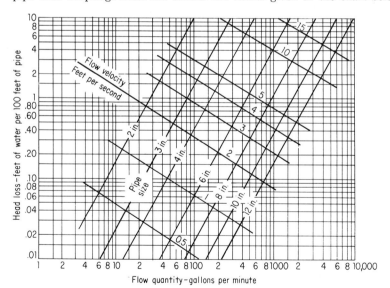

The graph shown above permits the determination of head loss vs. flow rate, based on pipe with water at 68°F (Fanning Formula).

3. MATERIALS

3.1   Selection and approval of resin, hardners, compatible fiberglass and fiberglass finish, and suitable winding techniques in production shall be limited to and be predicated on being able to certify the production of filament-wound products which will pass the edgewise porosity test stated in Section 4 herein. Some "compatible" finishes are not necessarily suitable and only those which are suitable shall be specified and used. Not all winding techniques are suitable and only those suitable shall be used. Not all resin and resin hardener systems are suitable; therefore, only those that are suitable shall be specified and used. Certification, therefore, is as follows:

3.1.1   Sample pipe sections from production lots shall be tested per Section 4 and shall not have edgewise porosity. Since fiber pattern is related to such tests, the winding angle of 50% of the wall shall not exceed 62° to the axis of the pipe. This test is valid for filament-wound polyester or epoxy pipe.

3.1.2   Corollary to the above, the sample product subjected to 72 hours boiling in distilled water shall not show a strength loss greater than 25%. Cut edges are not to be sealed. Use ASTM D1599-62T as a testing procedure.

3.2   *Resin*—The resin shall be 100% solids thermosetting epoxy resin such as Epon 826, as manufactured by the Shell Chemical Company or equal. Polyester pipe shall be made using Atlas Chemical Industries Atlac 382 thermosetting resins or equal. Hardeners for the epoxy resins shall be elevated temperature-cured systems using aromatic amines or a eutectic thereof.

The resin shall not contain pigments or fillers of any type, unless it can be shown that strength and porosity requirements can be complied with.

3.3   *Glass Fiber*—Glass fibers shall be continuous filament, even tension rovings with epoxy compatible size for epoxy resins such as PPG 1062 or equal, and polyester compatible size such as PPG 1064 or equal for polyester resin.

3.4   *Pipe Interior Surface*—Pipe and pipe fittings shall have a continuous, reinforced epoxy (or polyester) resin-rich surface for best flow and corrosion resistance. It shall be a minimum liner .040″ to .060″ in thickness for chemical service. Pipe for light chemical service shall have a minimum liner .015″ in thickness.

The liner resin shall be a 100% solids thermosetting epoxy resin such as Epon 826 as manufactured by the Shell Chemical Company or equal for epoxy pipe or Atlac 382-05 resin or equal as manufactured by Atlas Chemical Industries, Inc., for polyester-lined epoxy pipe. Hardeners for the epoxy resins shall be elevated temperature-cured systems using aromatic amines or a eutectic thereof.

The reinforcement in the resin-rich surface shall be of sufficient strength to prevent cracking, chipping or crazing at temperatures ranging from −40°F to 300°F.

3.5   *Adhesive*—The adhesives shall be appropriately formulated to provide adhesion equal to epoxy cements and chemical resistance compatible to the rating given on the Chemical Resistance Charts for each grade of pipe.

Any additives or fillers must remain uniformly dispersed and suspended in mixed adhesive during its normal pot life.

4. EDGEWISE POROSITY TESTING—APPROVAL OF MATERIAL SELECTION AND QUALITY OF WINDING TECHNIQUE

4.1 *Scope*—This test covers the procedure for determining the presence of edgewise porosity in glass reinforced plastic pipe.

4.2 *Apparatus*

a) Water pressure on edge of sample 500 psi.

b) The test heads shall be arranged as per drawing, Plate 1 attached.

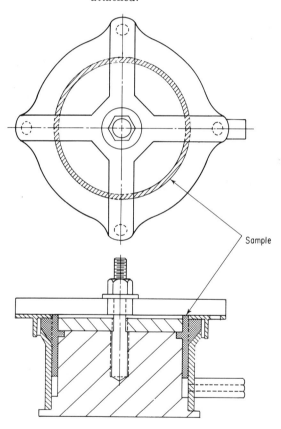

Sample

ATM 1024 Rev 1
End porosity test jig

c) Fluorescein may be added to the water in which an ultraviolet light may be used for leak detection.

4.3   *Procedure*

4.3.1   For all sizes of pipe cut a 2¼″ ± ⅛″ length piece. Do not seal cut ends.

4.3.2   Insert the test piece into the test head.

4.3.3   Pressurize test head so 500 psi water pressure is in contact with the exposed lower pipe end. Test for a minimum of one-half hour.

4.3.4   Check the visible pipe end for water leaks. This inspection shall be visual in a well lighted location with or without fluorescein and ultraviolet light.

4.4   *Results*

4.4.1   Report pipe as passed if no porosity is noted after one-half hour.

4.4.2   Report pipe as failed if porosity is noted at any time. This is cause for rejection of the lot.

5. PRODUCTION

5.1   *Pipe and Pipe Fittings*—The reinforced plastic pipe and pipe fittings shall be manufactured by winding resin impregnated bands of closely spaced continuous glass filament rovings around a polished steel mandrel. Molded fittings shall be compression-molded at molding pressures no less than 1000 psi.

5.2   *Curing*—Pipe and pipe fittings shall be heat-cured at a temperature of between 250°F and 300°F, and after curing shall contain not less than 70% glass fiber by weight. Molded fittings shall contain not less than 40% by weight asbestos fibers.

6. DIMENSIONS

6.1   *Pipe*—Dimensions of pipe shall be in accordance with Table II as measured with a Pi tape. The wall thickness shall not be less than noted as nominal wall thickness.

6.1.1   The nominal outside diameter shall be sufficient so that it is capable of being shaved or ground to fit and lock into the tapered fitting socket, leaving less than 10% of the surface area with unground resin glass surface in low spots; at the same time the first ¼″ back from the end shall

Table II

Nominal pipe size	Nominal outside diameter	Nominal wall thickness
2″	2.375″	.140″
3″	3.500″	.140″
4″	4.500″	.180″
6″	6.625″	.180″
8″	8.625″	.200″
10″	10.750″	.200″
12″	12.750″	.200″

be completely ground or shaved clean, yet be able to lock in place when the pipe is completely inserted in the socket.

6.1.2 The length of the pipe shall be 15' to 20' random lengths, unless otherwise specified.

6.2 *Pipe Fittings*—Fittings shall conform to the dimensions and configurations as indicated in the manufacturer's standard literature.

7. ACCEPTANCE (PRODUCTION INSPECTION)

Acceptance inspection shall be performed as part of the manufacturer's production quality control procedure and shall consist of the tests specified in 7.1 and 7.2. Acceptance inspection shall be performed on each length of pipe and on each pipe fitting to be furnished under a contract. Any pipe or pipe fittings failing to pass any of the tests specified shall be considered as defective units and shall be rejected.

7.1 *Hydrostatic Proof Test*

7.1.1 The pipe shall be filled completely with fresh water and pressurized at 225 psi. These tests shall be made at room temperature and after the internal pressurizing water has had time to assume an equilibrium temperature. The pipe shall remain under pressure for not less than 5 minutes and at the end of that time the pipe or fittings shall be examined for porosity. No porosity to water is acceptable.

7.1.2 An alternate but not concurrent test for pipe is to pressurize the pipe with 40 psi air while the pipe is submerged in a water bath for five minutes. Porosity will be indicated by formation of air bubbles. No porosity to air is acceptable.

7.1.3 Pipe fittings shall be fitted with suitable end plugs and pressurized to 25 psi air pressure. The exterior and cut ends of the fittings shall be coated with a solution of suitable consistency to indicate leakage. These tests shall be made at room temperature with the pipe fittings remaining under pressure for not less than five minutes, during which time the part shall be examined for porosity. No porosity is acceptable.

7.2 Edgewise porosity testing as per Section 4 shall be performed on cut sections taken from production pipe on a statistical but daily basis.

8. PRODUCTION EXPERIENCE

8.1 The manufacturer shall have a minimum of two years production experience in making plastic pipe and pipe fittings using the processes for which acceptance is requested.

9. PREPARATION FOR DELIVERY

9.1 *Pipe and Pipe Fittings*

9.1.1 The pipe and pipe fittings shall be packaged in accordance with the supplier's standard practice.

9.2 *Adhesive*

9.2.1 Adhesive and curing agent shall be packaged in kit form and pre-measured in exact quantities, so as to preclude weighing and measuring at the job site.

9.2.2 Adhesive kits shall be designated as Size I, II and III and shall contain sufficient adhesive and curing agent for joint makeup as shown in 9.2.3.

9.2.3   Size I   —   Four 2″, Three 3″, Two 4″, One 6″ or One 8″ connection.

Size II   —   Two 10″ or One 12″ connection.

Size III   —   Package to suit each joint size for pipe over 14″ in diameter.

9.2.4   Adhesive kits shall be packaged for shipment in accordance with the supplier's standard practice.

10.   ENGINEERING AID

It shall be the manufacturer's responsibility to provide a qualified reinforced plastics Technical Representative to advise the procuring agency, or their duly authorized representative at the job site, on the use of and installation of pipe and pipe fittings procured under this specification.

11.   CHEMICAL RESISTANCE

11.1   The products produced to this specification shall be suitable for use in chemical services as listed in the attached chemical resistance chart which indicates both concentration and temperature limitations. When chemicals are mixed, either in trace or quantities, the chemical resistance may not necessarily be suitable even though each item is shown separately as suitable. Such environments and other conditions may be approved or rejected on a request basis.

12.   STANDARDIZATION OF CHARTING DATA

12.1.1   Chemical resistance, as charted, shall be based on extensive laboratory experience, environmental testing, and other acceptable judgements of a qualified chemist.

12.1.2   The charted data shall include the effect on both the liner and the adhesives in the system.

12.1.3   The charted data shall apply to the products as produced by the manufacturer who is seeking to qualify under this specification, and he shall be able to show evidence that the data has been so developed for his products.

12.1.4   The chart shall show concentrations of chemicals and limitations thereof. If no limitation is shown, it shall mean it has resistance through the concentrated range.

12.1.5   The chart shall show temperature limitations for each chemical.

12.1.6   A solid line shall indicate anticipated resistance for a continuous service life of (3) to (5) years minimum in the environment of a single chemical. Variations including other chemicals, even in trace amounts, must be evaluated separately.

12.1.7   A broken line may be used to indicate areas where a limited service life may be expected, but where an economic pay out may be forecast. This may be a continuous service life of (3) years or less.

12.1.8   Where chemical resistance data has not been established by due process (as per 13.1.1) and there is reason to believe the resistance is suitable, this may be indicated with a dotted line. This shall indicate "Suitable but Subject to Test."

12.1.9   Where chemical resistance is not suitable, it shall be indicated "Not Recommended" in whatever temperature range this may be critical.

Appendix D

AUTHOR'S NOTE: The following proposed Standard on Filament Wound FRP Tanks has been prepared by H. D. Boggs, Manager, Engineering Services, Amercoat Corporation, Brea, Calif., and is used with permission.

LIGHT AND HEAVY CHEMICAL GRADES
FILAMENT WOUND FIBERGLASS REINFORCED TANKS

A/C 1003

November, 1966

1. SCOPE

1.1   This specification covers the manufacture of filament-wound atmospheric storage tanks made from thermosetting epoxy and polyester resins and glass fiber or other reinforcement, together with the adhesive necessary for joint assembly and installing fittings.

2. DESIGN CHARACTERISTICS

2.1   The tanks shall be designed for use according to accepted atmospheric storage practice where the tank is set on a suitable concrete pad.

2.1.1   *Tank Design Stress*—The tank shall be designed for an ultimate hoop tensile stress of 100,000 psi and a minimum longitudinal tensile stress of 20,000 psi.   Stress calculations shall be based on the fiberglass reinforced wall only.

2.1.2   *Verification of Tank Design Stress*—Minimum design stresses shall be verified by manufacturer's test data, based on unrestrained end burst tests in which a minimum of seven diameter length test samples have been hydrostatically tested by suitable means to cause failure in the fiber structure in less than one minute.   Such samples shall be wound in suitable diameter of pipe using the winding pattern and techniques current in the production of tanks.   Reference ASTM D1599-62T.

2.2   *Fittings Design*—All fittings installed through the tank wall shall be of the "interference lock" type of I.P. threaded type.   The fittings shall be flanged on both sides to restore the tank wall to full strength where cut for the opening, and to eliminate the need for fillets around the opening.

2.2.1   Large cutouts for agitator openings and side opening manways shall be suitably reinforced in the tank wall to restore the wall strength in the cutout area and beyond to distribute the stresses of the opening.

2.3   *Tank Bottom*

2.3.1   The tank bottom shall be joined with lapped and cemented joints equally resistant to the chemical environment in the tank.   These joints shall be overwound on the tank O.D. with a heavy filament winding of a low helix pattern to provide a reinforcement hoop and back-up of the joint, having a minimum tensile strength of 120,000 psi.   This hoop shall be no less than $4\frac{1}{2}$ inches wide and $\frac{1}{4}$ inch average thickness.   The lap joint shall be a minimum of 2 inches.

2.3.2   Four right angle hold-down brackets shall be mounted or wound into the reinforcement hoop at the bottom, equally spaced.   Such brackets shall be capable, when tied down to the tank foundation with properly anchored and sized bolts, of holding an empty tank in a 100-mile-per-hour wind.

2.3.3   The tank bottom shall be molded to close tolerances in heated dies.   The bottom shall be suitable to sustain full tank loads when mounted on a suitable foundation level within $\frac{1}{8}''$ between high and low areas. The bottom is not designed to sustain shifting or cracked foundations or foundations not capable of withstanding the load of the contents without changing in shape, nor to be mounted over gravel or rock or other raised projections.

2.4   *Tank Top*

2.4.1   The tank top shall be pressure-molded and cured separately from the tank, but shall be filament-wound into the shell; sandwiched between a minimum of one layer each of helix windings under and over the roof piece.   It shall be joined to the tank shell with the resins used in winding.

2.4.2   The tank top shall have a drain slope of $1''$ in $12''$.

2.4.3   The tank top shall have 360°, $\frac{1}{4}''$ thick filler board between two layers of reinforcement to stiffen the roof so that it will withstand 1,000-pound deck loading without reversing or collapsing.

2.4.4   The seam between the top and shell shall be pressure tight, smooth and neat.

2.4.5   There shall be a $24''$ manway and cover provided in the top at the center.   This shall be sealed and fastened with cadmium plated steel bolts and nuts and a Neoprene gasket which is provided.   Four of these stud bolts shall be $\frac{1}{2}''$ shorter than the remaining bolts, for the attachment of a lift bail supplied with the tank.   The bail shall be used for lifting the tank when moving or installing the tank on the foundations.

2.5   *Other Accessories*

2.5.1   Other accessories shall be available as per catalog or specification at the time of purchase.

2.6   *Painting*

2.6.1   The tank shell and top shall be suitably prepared and coated with a white coating.

2.7   *Tank Walls*

2.7.1   The tank wall shall not be less than .250″ thick the first section (or bottom) one-third in height, .225″ thick the middle one-third, and .200″ thick on the top one-third of height.

2.8   *Preparation of All Surfaces for Cementing and Assembly*

2.8.1   All surfaces of fittings, tank walls, bottoms, tops, or wherever parts are to be joined are to be sandblasted clear, or power sanded or otherwise roughed and made suitable to obtain maximum adhesion of any joint with the adhesives. All cut edges within the tank shall be sealed with adhesive.

2.9   *Preparation for Shipment*—All tanks are to be provided with suitable dunnage acceptable to common carriers to facilitate the handling and shipping of the tanks. All openings are to be closed to prevent wind loading and dirt, water and debris from entering the tank. Unloading and setting instructions shall be provided with each tank.

2.10   *Unloading and Handling*—The tanks shall be unloaded and handled per the attached instructions.

2.11   *Pressures*—These tanks are designed for atmospheric storage conditions. The tanks shall withstand $\frac{1}{2}$ ounces of vacuum and 2 ounces of pressure without damage.

The tank shall be equipped with sufficient venting to keep the loading and unloading forces within these limits at all times.

The standard atmospheric tank design is not suitable for pressures developed when air pressure is allowed to "blow by" during air unload of delivery vessels even with large vents. Proper safety precautions should be exercised when such practices exist. Removal of the top manway cover during loading where air is used is one such precaution. An attendant standing by to close the valve immediately after the liquid empties is the recommended procedure.

3.   MATERIALS

3.1.1   Selection and approval of resin, hardeners, compatible fiberglass and fiberglass finish, and suitable winding techniques in production shall be limited to and be predicated on being able to certify the production of filament-wound products which will pass the edgewise porosity test stated in Section 4 herein. Some "compatible" finishes are not necessarily suitable and only those which are suitable shall be specified and used. Not all winding techniques are suitable and only those suitable shall be used. Not all resin and resin hardener systems are suitable; therefore, only those that are suitable should be specified and used. Certification, therefore, is as follows:

Sample pipe sections for purposes of certification shall be tested per

Section 4 and must not have edgewise porosity. Since fiber pattern is related to such tests, the winding angle of 50% of the wall shall not exceed 62° to the axis of the pipe. This test is valid for filament-wound polyester or epoxy pipe.

3.2   *Resin*—The resin shall be 100% solids thermosetting epoxy resin such as Epon 826, as manufactured by the Shell Chemical Company or equal. Polyester tanks shall be made using Atlas Chemical Industries Atlac 382 thermosetting resins or equal. Hardeners for the epoxy resins shall be elevated temperature-cured systems using aromatic amines or a eutectic thereof.

The resin shall not contain pigments or filler of any type, unless it can be shown that strength and porosity requirements can be complied with.

3.3   *Glass Fiber*—Glass fibers shall be continuous filament even tension rovings with epoxy compatible size for epoxy resin such as PPG 1062 or equal, and polyester compatible size such as PPG 1064 or equal for polyester resin.

3.4   *Tank Interior Surface*—The tank bottom shell and top shall have a continuous, reinforced epoxy (or polyester) resin-rich surface for corrosion resistance. It shall be a minimum liner .045" to .060" in thickness for chemical service. For light chemical service the tanks shall have a minimum liner .025" in thickness. The liners shall be free of voids, smooth and without dry areas.

The liner resin shall be a 100% solids thermosetting epoxy resin such as Epon 826 as manufactured by the Shell Chemical Company or equal or Atlac 382-05 resin or equal as manufactured by Atlas Chemical Industries, Inc., compatible with the resins in the shell, top or bottom. Hardeners for the epoxy resins shall be elevated temperature-cured systems using aromatic amines or a eutectic thereof.

The reinforcement in the resin-rich surface shall be continuous and of sufficient strength to prevent cracking, chipping or crazing at temperatures ranging from −20°F to 300°F.

3.5   *Adhesive*—The adhesive shall be of the filled epoxy compounds appropriately formulated to provide adhesion equal to epoxy cements and chemical resistance compatible to the epoxy resins in the liner. The adhesive for polyester-lined epoxy pipe shall be filled polyester and appropriately formulated to provide adhesion equal to epoxy cements and chemical resistance compatible to polyester resin such as Atlac 382 as manufactured by Atlas Chemical Industries or equal.

The filler must remain uniformly dispersed and suspended in mixed adhesive during its normal pot life.

3.6   In certain rough chemical environments, the bonded bottom joint made with suitable adhesives as noted in 3.5, shall also require a resin-rich 3" wide liner overlay. The joint inner surface shall be prepared by sanding to assure maximum bonding of the liner overlay.

4.   EDGEWISE POROSITY TESTING—APPROVAL OF MATERIAL SELECTION AND QUALITY OF WINDING TECHNIQUE

4.1   *Scope*—This test covers the procedure for determining the presence of edgewise porosity in glass reinforced plastic pipe made with materials and filament winding techniques used to wind tanks. All techniques and materials are to be approved by this procedure prior to incorporating such procedures or materials in producing tanks for the first time.

    4.2   *Apparatus*—(As noted in Plate 1, page 419)

        a) Air compressor set at 40 psig line pressure.

        b) Test heads as per Amercoat drawings (4-0869-1, -2, -3, -4, -5, -6, -7) for 2″ through 12″, respectively.

        c) Dilute soap solution for leak detection.

    4.3   *Procedure*

    4.3.1   For all sizes of pipe cut a $2\frac{1}{4}'' \pm \frac{1}{8}''$ length piece. Do not seal ends.

    4.3.2   Insert the test piece into the test head.

    4.3.3   Connect edgewise test head to the air pressurizing unit and pressurize to 40 psig.

    4.3.4   Coat exposed pipe edge with the soap solution immediately following pressurization. Test is recorded for one-half hour.

    4.4   *Results*

    4.4.1   Report as passed if no porosity is noted after one-half hour.

    4.5   Checks on procedures and/or materials are to be made via this test method when requested.

  5. PRODUCTION

    5.1   *Tank Shell Winding*—The reinforced plastic tank shell shall be manufactured by helically winding resin impregnated bands of closely spaced continuous glass filament rovings around a polished steel mandrel. Molded fittings shall be compression-molded at molding pressures no less than 100 psi.

    5.2   *Curing*—The epoxy tanks shall be heat-cured at a temperature of between 250°F and 300°F, and after curing shall contain not less than 70% glass fiber by weight. Molded fittings shall not contain less than 40% by weight asbestos fibers.

  6. ACCEPTANCE (PRODUCTION INSPECTION)

    6.1.1   Acceptance inspection shall be performed to insure that resin mixes are maintained pure and in proper proportions and thoroughly mixed. Winding techniques shall be of proven quality to produce goods without edgewise porosity. The liner shall be fully wetted and compacted so that it is free of air and bubbles and so the thickness is within specifications in all areas. Fitting, joining, and tank finishing operations shall be checked to assure maintenance of quality practices. Inspection of water testing, painting, packing and loading of the tank shall be complete and recorded for each tank, including a record of any special factors or procedures.

    6.1.2   *Water Testing*—All tanks shall be water tested with a full static head.

  7. PRODUCTION EXPERIENCE

    7.1   The manufacturer shall have a minimum of 2 years production experience in making reinforced plastic tanks by the filament winding techniques and other molding procedures.

8. INSTALLATION

8.1   Foundations shall be designed by the purchaser to meet soil conditions prevalent at the installation site.

8.2   All foundations shall be flat and level. Variations in flatness shall not exceed $\frac{1}{8}''$. Variations in level shall not exceed $\frac{1}{4}''$ in 10 feet. Sloped foundations or foundations for special shaped bottoms shall be checked with the engineering department of the supplier prior to purchase. Such variations must be spelled out in the purchase order.

8.3   All foundations must be inspected at the time of installation to be certain that no rocks, stones, nails, wood or other objects are on the surface where the tank bottom will rest. Layers of standard building paper (30-pound felt) shall be laid over clean foundation prior to tank erection. If tank has been set on the ground prior to setting on foundation, inspection of the bottom must be made to be certain that none of the above-mentioned objects are stuck to the bottom surface.

8.4   In handling fiberglass tanks, no slings or chokes are to be used in lifting the tank. Appropriate lifting devices shall be furnished with the tank and handling instructions shall be furnished by the supplier. It shall be the responsibility of the erector to see to it that these instructions are not violated.

8.5   All fiberglass tanks shall be furnished with appropriate tie-down brackets. The location of the anchor bolts in the foundation shall be furnished by the supplier and these anchor bolts shall be located in the foundation at the time of pour. The orientation of the tie-down brackets shall be specified by the purchaser on the purchase order if his piping orientation is to be maintained. The tie-down brackets shall be made secure to the foundation before any piping to the tank is made.

8.6   All tanks shall be inspected prior to removal from railroad car or truck. Any damage in shipment shall be reported immediately and no tank shall be removed or tie-down brackets changed before outside inspector reviews damage. All claims for damage must be filed with the carrier and reported to the manufacturer.

9. TESTING AND STERILIZATION

9.1   All tanks shall be hydrotested at the point of installation with a full head of water for a period of at least four (4) hours.

9.2   Where sterilization of the tank is required, the customer shall proceed as follows: The tank shall be scrubbed down with fiber scrubbing brushes and detergents. This shall be followed with a water wash. Sterilization shall be accomplished by steam cleaning. Steam temperature at nozzle shall not exceed 300°F and precautions shall be taken to keep steam jet moving at all times to prevent localized erosion of the tank liner.

10. DRAWINGS

10.1   Drawings shall be furnished in accordance with general arrangements and over-all dimensions as spelled out in the purchase order and on customer specification layout sheet.

10.2   Where details are sufficient on customer specification sheet, signature of customer on said sheet shall be construed as approval to proceed with fabrication. Drawings for fabrication shall be forwarded to customer at time of release to fabrication for customer's record.

10.3 Drawings for fabrication requiring customer approval shall be forwarded to customer. Drawings that require such approval shall not be released for fabrication until signed drawings are received back from customer. Delivery schedules shall be predicated on date that approved drawings are received from customer.

10.4 All drawings shall be sufficiently complete and informative as to provide an accurate engineering description of the equipment to be built, to serve as a permanent record and expressed in such terms that will permit checking by the customer.

11. CHEMICAL RESISTANCE

11.1 The products produced to this specification shall be suitable for use in chemical services as listed in the attached chemical resistance chart which indicates both concentration and temperature limitations. When chemicals are mixed, either in trace or quantities, the chemical resistance may not necessarily be suitable even though each item is shown separately as suitable. Such environments and other conditions may be approved or rejected on a request basis with or without time, concentration or temperature limitations as the case may be.

12. STANDARDIZATION OF CHARTING DATA

(See page 422.)

Appendix E

There are presently in preparation by various subcommittees of the Society of the Plastics Industry tentative standards which would further amplify sections of TS-122C or provide better coverage in areas which are not clearly defined at the present time. Some of the standards currently being worked on are:

1. Proposed Recommended Practice to Shipping, Handling and Installation of Custom Contact Molded Reinforced Polyester Chemical Resistant Tanks.

2. Proposed Recommended Practice for Sub Assemblies, Shipment and Installation of Custom Contact Molded Reinforced Polyester Pipe and Ducts.

3. Proposed Product Standard for Filament Wound FRP Chemical Resistant Tanks.

4. Reinforced Plastic Piping Specifications to Cover Filament Wound Pipe.

5. Fire Retardancy, Particularly as it Applies to Duct Systems, Hoods, Fans, Stacks, and Blowers.

6. Development of Quality Standards for Corrosion Resistant Laminates.

7. Proposed Product Standard for Compression Molded Flanges.

Index